Synthetic Detergents

SYNTHETIC DETERGENTS

Seventh Edition

A. S. DAVIDSOHN DIPL. CHEM.
and
B. MILWIDSKY B.SC. (SA)

Copublished in the United States with
John Wiley & Sons, Inc., New York

Longman Scientific & Technical
Longman Group UK Limited,
Longman House, Burnt Mill, Harlow,
Essex CM20 2JE, England
and Associated Companies throughout the world

*Copublished in the United States with
John Wiley & Sons, Inc., 605 Third Avenue, New York, NY 10158*

© A. S. Davidsohn and B. Milwidsky 1987

All rights reserved; no part of this publication
may be reproduced, stored in a retrieval system,
or transmitted in any form or by any means, electronic,
mechanical, photocopying, recording, or otherwise,
without the prior written permission of the Publishers.

First published by Leonard Hill, 1938
    under the title *Polishes* by J. Davidsohn and A. Davidsohn
2nd edition, 1949
3rd edition published under the title
    Polishes and Cleaning Materials by A. Davidsohn, 1956
4th edition published in two volumes as *Polishes* and
    Synthetic Detergents by A. Davidsohn and B. M. Milwidsky, 1967
5th edition of *Synthetic Detergents*, 1972
6th edition published in Great Britain by George Godwin Limited, 1978
7th edition first published in Great Britain by Longman Scientific & Technical 1987

British Library Cataloguing in Publication Data

Davidsohn, A.
    Synthetic detergents. — 7th ed.
    1. Detergents, Synthetic
    I. Title   II. Milwidsky, B. M.
    668'.14   TP992.5
ISBN 0-582-46238-X

Library of Congress Cataloging in Publication Data

Davidsohn, A.
    Synthetic detergents.

    Bibliography: p.
    Includes index.
    1. Detergents, Synthetic.   I. Milwidsky, B. M.
(Benjamin Max), 1923–   . II. Title.
TP992.5.D3   1987   668'.14   86-20035
ISBN 0-470-20722-1 (Wiley, USA only)

Set in 10/11 pt Monophoto Times New Roman

Printed and bound in Great Britain
at the Bath Press, Avon

Contents

Preface to the seventh edition		vii
Preface to the sixth edition		ix
List of figures and tables		xi
1	Development of the detergent industry	1
2	Principal groups of synthetic detergents	10
3	Inorganic components of detergents, builders and other additives	47
4	Sundry organic builders	82
5	Synthesis of detergents	112
6	Manufacture of finished detergents	200
7	Application and formulation of detergents	254
Selected bibliography		309
Index		311

Preface to the Seventh Edition

In presenting this seventh edition we have maintained the guidelines of the sixth edition, but have shifted the emphasis somewhat.

From reports reaching us it has become apparent that the book is being used, in addition to a reference book for the industry, as a training manual for workers new to detergents and as a textbook for the study of a typical chemical process industry.

With this in mind, we have described in somewhat greater detail basic synthetic processes to give the reader a working understanding of the raw or intermediate products he will encounter, including the new materials coming into use. Sight has not been lost of our original aim to cater for the practical man.

The world is also preoccupied with the diminishing reserves of fossil fuels. We have also indicated possible sources of organic intermediates not based on petrochemical sources.

Formulations have been updated to cover the latest trends in the industry, including the newer raw materials becoming available and the latest state of the art plant being manufactured.

The chapter on analytical procedures has been omitted as there are separate works on this subject and it cannot be covered, other than very sketchily, in a single chapter.

A. S. Davidsohn
B. Milwidsky
January 1986

Preface to the Sixth Edition

This is a work by practical men for practical men. It is not intended to be simply a collection of formulae and recipes, nor does it set out to list all the materials which could conceivably be used as basic ingredients. Excellent books of that kind do exist but, in our view, they are often so comprehensive that it becomes difficult to distinguish between references to literature and practical work. We have purposely confined our text to those details which the practical manufacturing chemist will require and, we hope, will be able to use.

Our intention has been to cover all the operations that a manufacturer is likely to encounter in practice. All of the formulations have been critically selected so as to provide a useful starting-point; each manufacturing chemist can then use his own initiative to develop these formulations to suit his own local, or special, conditions. The reader may notice that we frequently quote from technical brochures issued by raw-material manufacturers. We make no apology for this, for the extensive research and development on which these publications are based often make them unique as sources of information for workers in this field.

During the past decade the detergent industry has seen a series of upheavals. It has seen a change from hard to soft detergents in the developed countries, and the large-scale introduction of enzymatic powders (and the subsequent problems). It has also had to face the attack on phosphates. In addition, the rate of growth of non-ionic detergents is increasing at the expense of anionic.

The alterations and many additions to this sixth edition reflect the new developments that have taken place as a consequence of these changes. However, the basic structure of the book remains the same, although, since this edition has been completely re-set, the opportunity has been taken to improve layout, to use metric units wherever practical, and to remove some inconsistencies that have crept in over the years.

Although this edition was prepared during a time of economic stagnation, the detergent industry has continued to develop from year to year. We have endeavoured to cover the technical aspects of this healthy trend and we hope that this book will continue to meet the needs of those in or concerned with the detergent industry throughout the world.

A. S. Davidsohn
B. Milwidsky
November, 1977

List of Figures and Tables

Figures

2.1	Hypothetical illustration of how dodecane can be solubilized and converted into a detergent with varying qualities	13
2.2	EOD of conventional versus narrow range ethoxylate	32
3.1	Synergism of sodium tripolyphosphate with detergents	51
3.2	Reaction of TAGU	66
5.1	Flow chart for high-pressure hydrogenolysis to obtain fatty alcohols from natural fats	114
5.2	Production of linear alkyl benzene schematic flow	128
5.3	Alkylation process using aluminium metal as catalyst, forming aluminium chloride *in situ*—first introduced by Rheinpreussen	129
5.4	Ester interchange process—batch process	133
5.5	Methyl ester production—continuous system	134
5.6	Freezing points of sulphuric acid and oleum	140
5.7	Viscosity of molten sulphur	146
5.8	Flow diagram of Ballestra Sulphurex Process	158
5.9	Ballestra multitube film reactor	164
5.10	Ballestra's film sulphonation	167
5.11	Process for making linear-paraffin-sulphonate detergent	177
5.12	Laboratory ethoxylation plant	184
5.13	Ethylene oxide batch reactor	186
5.14	Pressindustria plant for the manufacture of ethylene oxide adducts	189
5.15	Plant for the production of normal and super alkylolamides	195
6.1	Ballestra continuous slurry preparation system	214
6.2	Ballestra spray-drying unit and finishing system	218
6.3	Ballestra wet scrubber system	226
6.4	Patterson–Kelley V-shaped blender	231
6.5	Patterson–Kelley zig-zag continuous blender	232
6.6	Patterson–Kelley system for increasing spray-drier capacity	233
6.7	Anhydro fluid-mix process	234
6.8	Mixer for liquid and paste-form detergents	242
7.1	pH concentration curves comparing sulphuric acid with other acids	299

Tables

1.1	US soap and detergent sales	3
1.2	Comparative production figures for synthetic detergents in the USA	3
1.3	World production of cleaning materials	4
2.1	Effect of the addition of alkylolamides to dodecyl benzene sulphonate solution	38
2.2	Effect of the addition of ethanolamide on the foaming power of various detergents	39
2.3	Comparison of foam-boosting ability between amine oxides and lauric diethanolamide	40
3.1	Amounts of phosphates necessary to prevent precipitation of unit weight of metallic soap	50
3.2	Influence of hard water on the pH of phosphate solutions	52
3.3	Phosphates used in detergents	53
3.4	Requirements for phosphate substitutes	54
3.5	Properties of common silicates used in detergents (on an anhydrous basis)	57
3.6	Concentration of alkaline and neutral silicate solutions	59
3.7	Baumé concentration table for caustic soda solutions at 15·6°C (60°F)	71
3.8	Potassium hydroxide—density of aqueous solutions at 15°C (59°F)	72
3.9	Influence of builders on surface and interfacial tensions of dodecyl benzene sulphonate	74
3.10	Characteristics of builders used in detergent mixtures	75
3.11	Residual pH of solution after addition of 0·5 mol/litre acid to 100 ml of 0·4 per cent solution	76
3.12	Residual hardness after addition of 0·2 per cent alkaline builder	77
3.13	Effect of builders on clay and grease removal action of non-ionic detergents	77
3.14	Effect of alkaline builders on foaming power of non-ionic detergents	79
3.15	Surface absorptive capacity of builders	80
4.1	Chelating ability of various chelates for calcium in alkaline solution (pH above 10)	95
4.2	Properties of alkylolamines and alkylamines	106
4.3	Properties of the most important chlorinated solvents	109
4.4	Properties of solvents which are also miscible with water	110
5.1	Comparison of alcohol distribution from two types of Ziegler production units	117
5.2	Properties of typical commercial fatty alcohols derived from oils and fats	118

5.3	Properties of pure fatty alcohols	119
5.4	Typical composition and properties of selected Alfol* alcohol blends	120
5.5	Typical composition and properties of selected Epal† blends	121
5.6	Typical composition and properties of selected Ethyl alpha olefins	122
5.7	Typical properties of linear alkyl benzene, produced by the Pacol process	128
5.8	Properties of sulphuric acid	136
5.9	Properties of oleum	138
5.10	Storage and handling equipment for various strengths of sulphuric acid and oleum	142
5.11	Acid sulphonation conditions	150
5.12	Air requirements for sulphur burning	156
5.13	Product characteristics of materials sulphonated in a Ballestra Sulphurex F film reactor	169
5.14	Composition of polyphosphoric acids	178
5.15	Preparation of non-ionic surfactants from ethylene oxide	179
5.16	Physical properties of pure ethylene oxide	180
5.17	Physical properties of pure propylene oxide	181
5.18	Properties of fatty acids	182
5.19	Properties of three important alkyl phenols for the production of surface active ethylene oxide condensates	183
5.20	Setting-point and main uses of various monoethanolamides and monoisopropanolamides	196
6.1	Percentage of water absorbed by anhydrous salts to form crystals	222
6.2	Effect of variation of operating conditions on power characteristics	225
6.3	Density of mixed powders made by the Combex system	230
6.4	Physical properties of ABS solutions neutralized by various bases	237
7.1	Comparison of formulation of heavy-duty detergents	256
7.2	Typical formulations for powders produced by dry neutralization	280
7.3	Typical formulations for powders produced by spray-drying	281
7.4	Alkalis for dairy cleaners	296

* The name Alfol is a registered trade mark of the Continental Oil Company, Oklahoma, USA.
† The name Epal is a registered trade mark of the Ethyl Corporation, Baton Rouge, LA, USA.

1. Development of the Detergent Industry

Although the start of the synthetic detergent industry is not shrouded in the veils of history as were the beginnings of the soap industry, it is nevertheless not easy to pinpoint exactly when the detergent industry, as such, came into being. The primary problem is to decide exactly what is being referred to as a synthetic detergent. The term itself leads to confusion. In the USA the words surfactant or syndet are being used, whilst in Europe the term 'tenside' (for tensio-active material) has come into fashion.

Many definitions of synthetic detergent have been proposed, all of which are very wide. *The Comité International de Dérivés Tensio Actifs* has after several years of deliberation agreed on the following definitions:

Detergent: Product the formulation of which is specially devised to promote the development of detergency. *Note:* A detergent is a formulation comprising essential constituents (surface active agents) and subsidiary constituents (builders, boosters, fillers and auxiliaries).

Surface Active Agent: Chemical compound which, when dissolved or dispersed in a liquid is preferentially absorbed at an interface, giving rise to a number of physico-chemical or chemical properties of practical interest. The molecule of the compound includes at least one group with an affinity for markedly polar surfaces, ensuring in most cases solubilization in water, and a group which has little affinity for water. *Note:* Compositions in general are usually mixtures of such compounds.

Amphiphilic Product: Product comprising in its molecule, at the same time one or more hydrophilic groups and one or more hydrophobic groups. *Note:* surface active agents are amphiphilic products.

We shall, however, in this book continue to use the term synthetic detergent for a material which cleans (or is used for cleaning), but in this definition we do not include soap.

Even so, this is still a wide definition, because, of course, it can refer to the active ingredient, or the solid, liquid, paste or powder compounded from this active matter. However, this should not lead to confusion, as the industry itself as yet makes no distinction in terminology between the basic material and the ready-for-use product.

The first synthetic detergents which fall into our definition of the term seem to have been developed by the Germans in the First World War period to allow fats to be utilized for other purposes. These detergents were of the short-chain alkyl naphthalene sulphonate type, made by coupling propyl or butyl alcohols with naphthalene and subsequent sulphonation, and appeared under the general name of Nekal. These products proved to be only fair to moderately good detergents, but good wetting agents and are still being produced in large quantities for use as textile auxiliaries.

In the late 1920s and early 1930s long-chain alcohols were sulphonated and sold as the neutralized sodium salts without any further additions except for sodium sulphate as an extender.

In the early 1930s long-chain alkyl aryl sulphonates with benzene as the aromatic nucleus, and the alkyl portion made from a kerosene fraction, appeared on the market in the USA. Again, these were available as the sodium salts extended with sodium sulphate. Both the alcohol sulphates and the alkyl aryl sulphonates were sold as such as cleaning materials, but did not make any appreciable impression on the total market. At the end of the Second World War alkyl aryl sulphonates had almost completely swamped the sales of alcohol sulphates for the limited uses to which they were applied as general cleaning materials, but the alcohol sulphates were making big inroads into the shampoo field. An exception was Teepol, a secondary alcohol sulphate which remained popular for some years.

In common, however, with other chemical developments during this century, progress was not in one direction only. The limiting factor is always the availability of raw materials in a particular country. Concurrently with the above developments, there were developed, both in Germany and the USA, the Igepon type of compounds of which Igepon-T, the sodium salt of oleyl tauride is an example, and in Germany the Mersolates, which are alkane sulphates.

Each of these basic materials has its advantages and disadvantages, but in considering the feasibility of production the following factors must be taken into account:

> availability of raw materials;
> ease of manufacture;
> cost of raw materials;
> cost of manufacture;
> suitability of finished product.

We have purposely placed suitability last, as it is only too true that not always is the best material made available.

As a result of its ease of manufacture and versatility, the alkyl benzene sulphonate very quickly gained a foothold in the market, and after the last war the existing keryl benzene was very quickly replaced by an alkyl benzene made from propylene tetramer coupled to benzene (PT benzene).

This PT benzene very quickly displaced all other basic detergents and for the period 1950–65 considerably more than half the detergents used throughout the world were based on this.

To give an idea of the enormous rise in synthetic detergent production, Table 1.1 compiled from figures submitted by the American Soap and

Detergent Association and the German firm of Henkel KGaA shows both soap and detergent sales in the USA for various years.

TABLE 1.1

US Soap and Detergent Sales

Year	Soap sales 1000 tons	Synthetic sales 1000 tons
1940	1410	4·5
1950	1340	655
1960	583	1645
1972	587	4448
1982	545	5090

These figures reveal that immediately after the Second World War synthetics started making inroads into the production of soap, which now seems to have settled down to a constant whereas synthetics have increased enormously.

By 1959 although the US per capita consumption had somewhat levelled out, total production was still rising as shown in Table 1.2 which has been compiled from the Henkel figures.

TABLE 1.2

Comparative Production Figures for Synthetic Detergents in the USA

	1958 1000 tons	1963 1000 tons	1972 1000 tons	1982 1000 tons
Domestic detergents (solid)	1200	1425	2672	2763
Domestic detergents (liquid)	354	640	1773	2327

The broad picture that appears from Table 1.2 is that while solid detergents (among which of course powders are included) are making great strides forward, the liquid detergents are increasing at a much faster rate.

Looking at world production figures for the last few years (Table 1.3), it appears that although soap is growing, synthetics are increasing tremendously. This is because Third World countries are increasing the use of both synthetics and soaps. A random look at per capita consumption shows that in Western Europe in the years 1960 to 1980 per capita consumption rose from 9·7 to 18·9 kg per annum, in North America from 12·8 to 30·1 and in Asia from 1·0 to 2·1 (2·5 for China).

It can be seen that despite a tremendous growth in the past decades, the production of synthetics is still continuing to rise.

No review of the history of the detergent industry can omit mention of

TABLE 1.3

World Production of Cleaning Materials				
	1968 1000 tons	1970 1000 tons	1980 1000 tons	1982 1000 tons
Soaps	6493	6059	8218	8705
Synthetics	7369	8352	16,453	17,889
Other*	1015	1310	2380	2562

* Includes fabric softeners.

the builders, without which the remarkable success outlined above could never have been achieved.

After the Second World War, when detergents started appearing in appreciable quantities on the retail market, it was noted that white cotton goods were not being washed as white as they should be. This was explained by the fact that although the active material was able to lift the dirt from the cloth it could not keep it in suspension. Hence small spots of dirt were being redeposited uniformly over the whole surface area of the cloth while in the wash-tub or machine, thus giving the cloth a grey appearance.

The sodium salt of carboxymethylcellulose (CMC) had been known to industry for many years and, in fact, a French patent had been applied for in 1936,[1] using CMC as an additive to washing materials. However, this patent was not developed extensively until the Second World War, when CMC was used in Germany on a moderately large scale, initially as an extender for soap which was in short supply, and then as an additive to the synthetic detergents being produced as a wartime substitute for soaps. When intelligence reports on the German industry were published, the use of CMC as an additive to synthetic detergent powders was noted and investigated and it was found that this addition eliminated the redeposition problem.

Despite the considerable advances made in the production of the active detergent matter, by the end of the Second World War progress in the use of detergents for heavy-duty (cotton) washing was still relatively slow, although they had already displaced soaps to a considerable extent in the field of fine laundering and dish-washing. To improve the heavy-duty washing properties, manufacturers turned for analogies to the soap industry. Soap for cotton washing had for many years been 'built' with alkaline materials such as carbonates, silicates, borax, and orthophosphates. All of these singly and in combination were tried with moderate success. Condensed phosphates had started appearing on the market in increasing quantities and from 1947 onwards heavy-duty detergent formulations were introduced, initially with tetra sodium pyrophosphate and then with sodium tripolyphosphate with startling success.

With the advent of CMC and tripolyphosphate builders the detergent industry established itself and has never looked back.

Production in the use of sodium tripolyphosphate rose from 100,000 tons in 1947 to over 100,000,000 tons in 1970. In that year environmental problems arose and production fell somewhat, concurrently a search for replacements was undertaken. To date no 'plug-in' replacement has been found.

Propylene tetramer benzene sulphonate held almost undisputed sway as the major ingredient used in washing operations till the early 1960s.

As early as 1952 in the UK, somewhat later in other countries, it was noted that sewage treatment problems were arising. The amount of foam on rivers was increasing and where water was being drawn from wells located close to household discharge points, the water tended to foam when coming out of the tap. This was attributed to the fact that propylene-based alkyl benzene sulphonates are not completely degraded by the bacteria naturally present in effluents, and was further narrowed down to the fact that it is the branched-chain formation of the alkyl benzene which hinders the attack by the bacteria. However, fatty acid sulphates were found to degrade very easily, and since all naturally occurring fatty acids from which fatty alcohols are produced are of the straight-chain variety (as also are the Ziegler alcohols which started appearing in commercial quantities at about this time), it seemed possible that a straight-chain alkyl benzene might be degradable.

Methods of test were developed and it was, in fact, proved that linear alkyl benzene is biodegradable. Germany introduced legislation prohibiting the discharge of non-biologically degradable material into sewer systems. In the USA detergent manufacturers agreed voluntarily to switch over from PT benzene to linear alkyl benzene by June 1965. In the United Kingdom a similar type of 'gentleman's agreement' was entered into.

The change to linear alkyl benzene (which can be considered as a return to a purified form of the keryl benzene in use twenty years previously) gave some rather surprising results. It was found that the detergency in a heavy-duty formulation using linear alkyl benzene sulphonate was approximately 10 per cent better than when using PT benzene sulphonate, solutions of the neutralized sulphonic acid had a lower cloud point, and pastes and slurries had a lower viscosity. The first two results were obviously advantageous and a lower viscosity in slurries had an advantage when the product was spray-dried to a powder, but when the LAS was sold as a liquid or paste detergent, this lower viscosity had to be overcome as sales appeal was lost. The manufacture of powders based on LAS posed some problems, however. Powders became sticky and lost their free-flowing characteristics, whether made by spray-drying or one of the other methods. Mausner and Rainer[2] have indicated that the actual isomer distribution of the linear alkylate has an effect on the stickiness of the powder, with the 2-phenyl isomer giving the greatest tendency to stickiness and the 5- or 6-phenyl isomer the least. Additives to overcome this tendency have therefore been developed (p 78).

The switch to linear alkyl benzene is not, however, complete. In many parts of the world where the problem of sewage treatment is not serious, the PT benzene is still being used in ever-growing quantities. Also the Ziegler alcohols are now competitively priced with the linear alkyl benzenes, and alkane sulphonates are reappearing.

Despite the 'hardness' of propylene tetramer benzene, it is still used to a certain extent. For agricultural emulsifiers, due to administrative difficulties of reformulation and consequent registration, the PT version is still used, but in countries where sewage problems have not yet arisen, there is a switch to the linear product for economic reasons, in that with the fall in production of the one and the increase of the other, the linear product has become cheaper and problems of reformulation are not as complicated for detergent products as they are for emulsifiers.

Modern awareness of the danger of depletion of petroleum reserves has started a tendency to reconsider the use of materials based on oils and fats as the source of the hydrophobe, that is a reversion to the original type of detergents, as the sources are self-replenishing.

Having successfully coped with the problem of biodegradation the industry faced a new attack. It appeared that in certain lakes and ponds algae started reproducing at an unprecedented rate. This was blamed on the extensive use of phosphates which are a food for these organisms, and again the detergent industry became the whipping boy, because tremendous amounts of sodium tripolyphosphate are used and then discharged down the sewer. (The term eutrophication, meaning nutrition by chemical means, has been applied to this phenomenon.) It is not clear whether the blame should be taken solely by the detergent industry, as concurrently with the increase in the use of detergent phosphates there was an increase in the use of phosphate fertilizers, which also find their way into natural water systems. However, with the big international preoccupation with ecology the detergent industry is searching for an efficient substitute for sodium tripolyphosphate. To date a complete replacement has not been found but in the Scandinavian countries particularly, formulations of household powders are beginning to appear with appreciable portions of the phosphate replaced by NTA (nitrilo triacetic acid) which is a better sequestering agent than tripolyphosphate but has none of the other properties exhibited by the phosphate. We are inclined to fear that in time the extended use of NTA might bring new problems of this sort, as it contains nitrogen which is again a good fertilizer and nutrient for algae. We quote from an editorial comment in *Chemical Week* on this subject:[8]

'We caution the legislations and administrations against shooting from the hip. They may blast phosphate out of detergents, but they may not be very proud of their victory when the facts are finally in.'

The search is still going on for a phosphate substitute. NTA on its own will only partially replace phosphates. A mixture of NTA and borax has been suggested as a complete replacement but here again the borax might produce more problems than the phosphate is alleged to produce. Some of the hydroxy-polycarboxylic acids (see p 92) containing only carbon, hydrogen and oxygen are also being considered.

In 1977 the German firm Henkel patented the use of synthetic zeolites as a partial replacement for phosphates. These zeolites, sodium aluminium silicate of a particular lattice structure have the facility, although insoluble in water, of absorbing by ion exchange heavy metal cations. The reaction with calcium is rapid, with magnesium somewhat slower but this increases with rising temperature.

No claims for total replacement of phosphate are made by the proponents of the zeolites, it is emphasized that they be used in admixture with STP[4] as they only soften water and perform none of the other functions attributed to the condensed phosphates.

In actual fact, their use has not reached the actual totals predicted but in Europe a fair proportion of detergent powders contain zeolites, in the USA a somewhat lesser amount; this is despite the fact that there are no ecological or toxicological objections to their use.

The other candidate for a (partial) phosphate replacement is nitrilo triacetic acid. NTA has had a chequered career. In 1970 it was used in large quantities in the 'no' or 'low' phosphate powders as a chelating agent (to complex the hardness ions) but towards the end of that year the US Surgeon-General issued a report stating that it has been found that NTA caused birth defects in rats when their drinking water was contaminated with NTA and cadmium or mercury salts. The report had specifically indicated that the quantities under the test conditions were several tens of times the possible concentration that could be encountered in practice and that there was no health hazard. Use of NTA was immediately phased out in the USA, but Canada and Europe did not react to this report. Further work indicated the possibility of urinary tract and cancer problems, again in high concentrations. In 1980 the Environmental Protection Agency (EPA) in the USA cleared the material for use in detergents, but at the time of going to press, in New York State doubt is again being cast on the safety of the material, but nowhere else is the product considered suspect.

With the pressure on detergent manufacturers to reduce phosphate concentrations, work is being continued on a substitute. NTA is a probable contender but no experience is available of large-scale discharge of this nitrogen compound into large masses of water, it might well be that if the volume comes close to that of the phosphates, adverse effects will be noted. Note that it took some ten years of continual use of the hard alkylate till biodegradation effects were noted, and well over twenty years for the phosphates. Zeolites might only be a partial answer, but being insoluble in water they have no effect on detergency. On the other hand sodium tripolyphosphate not only softens water, but also has a positive effect on detergency (p 50).

Two important reviews have appeared on this subject. In *European Chemical News* (19 November 1976), under the title 'No case for a European ban on detergent phosphates', it is pointed out that, in parts of the USA, although the phosphate ban is in effect, deterioration in washing quality is being experienced with no appreciable improvement in quality of water in lakes and slow-flowing rivers. The Swedish policy of treating sewage water with iron salts to precipitate and remove phosphates is recommended.

Phosphorus and Potassium (No. 90, July/August 1977) concludes that there are no reasonable grounds for banning NTA.

Despite all this, it appears that the pressure on detergent manufacturers will ensure a gradual diminution of the P content of detergent products.

The biggest single revolutionary trend in the detergent industry in recent years has been the use of enzyme additives. Enzymes as aids to washing are not new to the industry. Proteolytic enzymes had been tried as additives to washing powders in Germany in the 1920s with only moderate success and again in Switzerland in the 1930s. Enzymes, which can be called organic catalysts, tend to hasten reactions and the proteolytic enzymes convert or 'break down' proteins wholly or partially into amino acids. The action is rather slow and the production costs high, but with improved methods of production and purification, strains of enzymes, usually in admixture with a proportion of amylase which breaks down starches, were developed which were relatively fast acting. These were added initially to 'pre-soak' detergents and found immediate acceptance in the European countries where washing habits were such that washing was normally soaked for a period prior to the wash proper. Better and better strains of enzymes were developed, with stability to a wider pH spectrum, stability against perborate and quicker action. In the United States detergent manufacturers resisted the incorporation of enzymes into their powders for some years after this type of powder had almost completely swept the board in Europe but in 1968 enzymatic powders started appearing there as well. The position at present is that enzymatic powders are now holding a large proportion of the household detergent market in Europe and some 15 per cent in the USA and formulations have appeared made for machine washing. Washing-machine manufacturers are now producing automatic washing machines with a 'Bio' programme which allows the washing to remain in contact with the detergent solution for an extended period of time at a relatively low temperature before beginning the washing and heating cycle. The production of enzymatic powders has raised its own problems. Certain dermatological problems did occur, when enzymatic powders were used for manual washing. Otherwise no problems have as yet been encountered at the consumer level but it is wise to be cautious. The basic enzyme itself is a mixture of organic material, some 6 per cent true protease, the balance being unidentified organic matter. It must be mentioned that producers of proteolytic enzymes have brought on to the market highly refined, practically dust-free, encapsulated enzymes, which do not contain so much undefined organic material deriving from the fermentation process.

These, then, are the main trends in the development of detergents, but one must not lose sight of the fact that many other types of detergent were produced in large or small quantities concurrently with the few mentioned above. Each has a definite place and use of its own, but the vast majority are modifications of the few types mentioned above.

References

1. *French Patent* 805,718, Kalle, F. P. (29 Apr 1936).
2. Mausner, M. and Rainer, E., *Soap*, **44,** 8, 34 (1969).
3. *Chemical Week*, **106,** 5, 4 (1970).
4. Berth, P., *JAOCS*, **55,** 1, 52 (1978).

2. Principal Groups of Synthetic Detergents

Synthetic detergents, like soap, are materials which dissolve or tend to dissolve in water and in non-aqueous materials under certain conditions. To achieve this double tendency, these materials include two distinct groupings in their molecular structure. One, which is easily soluble in water, is called the hydrophilic group, and the other, which on its own would be insoluble in water, is known as the hydrophobic group.

The hydrophilic group is usually, although not always, added synthetically to a hydrophobic material in order to produce a compound which is soluble in water. However, this solubilization does not necessarily always produce a detergent, since detergency depends on the balance (ratio) of the molecular weight of the hydrophobic portion to that of the hydrophilic portion.

As an illustration, consider the material dodecane, $C_{12}H_{26}$. This is completely insoluble in water. If an OH group is substituted for one of the terminal hydrogens, the new material $C_{11}H_{25}CH_2OH$ lauryl alcohol, is still practically insoluble, but a tendency to solubility has arisen. If now this lauryl alcohol is sulphated to

$$C_{11}H_{23}\overset{\overset{\displaystyle H}{|}}{\underset{\underset{\displaystyle H}{|}}{C}}-O-\overset{\overset{\displaystyle O}{\|}}{\underset{\underset{\displaystyle O}{\|}}{S}}-OH$$

we have a material which is completely miscible in water in all proportions, ignoring, for the sake of development of this theme, hydrolysis which will occur when dissolving the acid product. If this sulphuric ester is neutralized with a caustic alkali, ammonia, or organic amines, the material becomes completely soluble in water and in this instance is a very good detergent.

If, instead of sulphating the lauryl alcohol, it is treated with, say, ten molecules of ethylene oxide:

$$H_2-C\underset{\underset{\displaystyle O}{}}{\diagdown\diagup}C-H_2$$

we obtain the material

$$C_{11}H_{23}CH_2-O-CH_2-CH_2(OCH_2CH_2)_8-OCH_2CH_2OH$$

PRINCIPAL GROUPS OF SYNTHETIC DETERGENTS

This material is again completely water-soluble and a good detergent. If only five or as much as twenty molecules of ethylene oxide had been used, the detergency would fall off, although the materials would still be water-soluble. If less than five molecules of ethylene oxide had been used, the product would be insoluble.

To return to the lauryl alcohol, if it were to be oxidized to lauric acid, that is the original dodecane, with a carboxyl group replacing the terminal CH_3 group, the material is still practically insoluble in water, but if this acid is neutralized with caustic alkalis or certain selected organic amines, the product is again water-soluble, being, in fact, a soap. In this instance it is a moderately good detergent, but if the original chain length had been say 18 carbons, the detergency would be greatly improved. This demonstrates what was pointed out above, ie, that the hydrophile/hydrophobe balance is important for optimum detergency.

It should be noted that in the previous paragraph no mention was made of neutralization with alkaline earths. This illustrates one of the failings of soaps. With the magnesium and calcium salts which are normally present in hard water, soaps form an insoluble scum which lowers the efficiency of the soap. This is a basic defect of the carboxyl group and detergents are so designed that this carboxyl grouping is replaced by something else which does not behave adversely in the presence of salts of calcium and magnesium.

To continue with our analogy of dodecane as the starting material, this in itself can be sulphonated by sulphoxidation (see p 24) to dodecane sulphonic acid

$$CH_3-(CH_2)_n-CH-(CH_2)_{9-n}-CH_3$$
$$|$$
$$SO_3H$$

which is miscible with water. When this is neutralized with caustic alkalis, selected alkaline earths or organic amines, solutions in water are obtained which are detergents.

Now if the dodecane were to be chlorinated and then reacted with trimethylamine the compound obtained is lauryl trimethylammonium chloride.

$$C_{12}H_{25}\overset{+}{N}(CH_3)_3Cl^-$$

again soluble in water and a detergent of sorts.

Finally, if by a rather complicated series of condensations, dodecyl beta alanine

$$C_{12}H_{25}NHC_2H_4COOH$$

is produced, this is soluble in water, even without neutralization of the carboxyl group, and also has detergent properties.

Thus, in every example we have discussed, a solubilizing group has been added to the dodecane. This group or grouping is hydrophilic, while the original dodecane is hydrophobic, and in every case the resultant product is a detergent which is available on the world market. In practice dodecane itself is not often used as a starting material, very few, if any, of the reactions discussed are as simple as shown. This discussion was purely illustrative of the theme we wish to develop, which is to illustrate how an insoluble hydrophobe can be solubilized and converted into a detergent with varying qualities.

The above reactions are shown schematically in Fig. 2.1. In addition, this diagram includes another very important reaction which we have not mentioned previously, as it was not necessary for the illustration we were developing. The reaction consists of chlorinating the dodecane, coupling this product with benzene by means of the well-known Friedel-Crafts reaction, and then sulphonating it to dodecyl benzene sulphonic acid. This process not only provides the most important detergent on the market but is also one of the few cases when dodecane can be used as the starting material.

From the figure it will be noticed that end products 1, 3, 4 and 7 are all acidic and need to be neutralized with a base before being used. No. 5 is a basic material which in the process of manufacture is produced as the chloride (ie, it is in effect neutralized with hydrochloric acid) and No. 2 needs no neutralization at all. If the molecular structure of compound No 6, for which we have taken as an example dodecyl beta alanine, is examined, it will be noted that it contains both a carboxyl (acid) and an amine (basic) group. The carboxyl group can be neutralized with an alkali, or by the amine group from an adjacent molecule. Similarly, the amine group can be neutralized by a strong acid, such as hydrochloric acid, or by the carboxyl group from an adjacent molecule.

Classification

There are four main classes of detergent: anionic, cationic, non-ionic, and amphoteric.

Anionic detergents are compounds in which the detergency is vested in the anion, which has to be neutralized with an alkaline or basic material before the full detergency is developed. In *cationic detergents* the detergency is in the cation, and although in the manufacturing process no neutralization takes place, the material is in effect neutralized by a strong acid. *Non-ionic detergents* contain no ionic constituents, as their name implies, and *amphoteric detergents* include both acidic and basic groups in the same molecule.

Let us now discuss each of these groupings separately.

Anionic Detergents

This is by far the largest class of detergents. Sisley,[1] in fact, lists sixty groups and sub-groups of anionic detergents, and of these only seven are neither sulphated nor sulphonated products. In practice, and for the

ANIONIC DETERGENTS

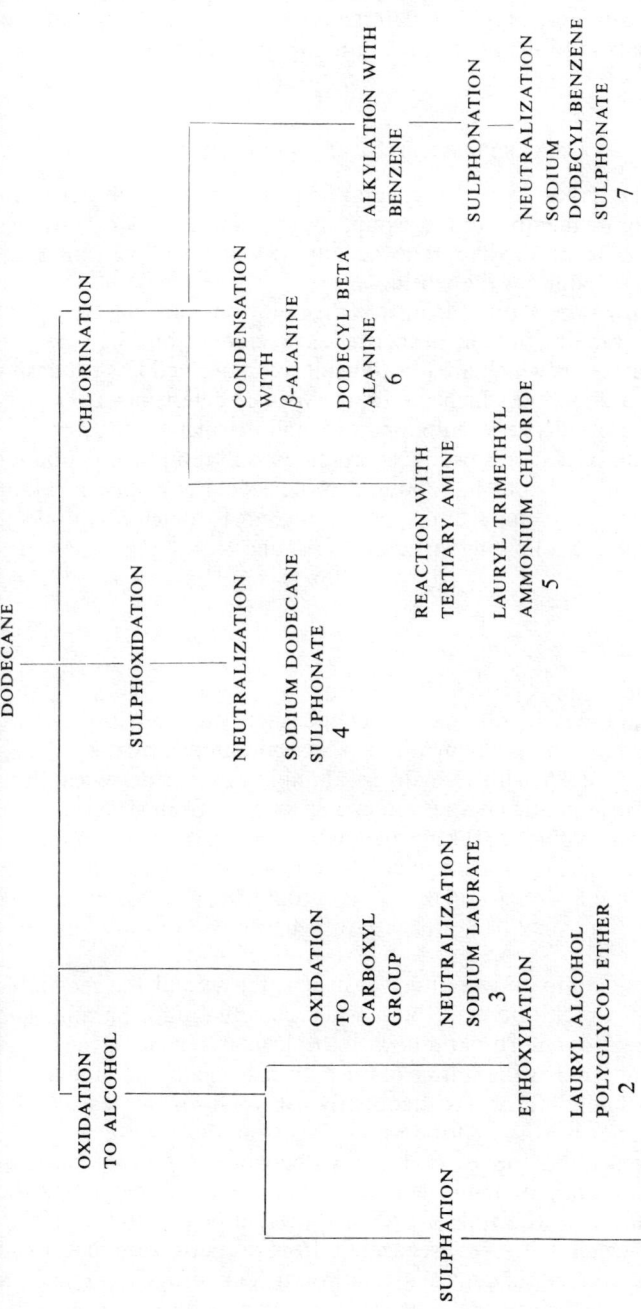

Fig 2.1 Hypothetical illustration of how dodecane can be solubilized and converted into a detergent with varying qualities

purposes of this volume, which does not claim to deal with the very specialized or proprietary types of detergents, we can safely say that the anionic detergents fall into the following main types of grouping:

1. Alkyl Aryl Sulphonates

This is by far the largest group in general use, since dodecyl benzene sulphonate, the leading member of the group, either as the propylene tetramer benzene or as the linear alkyl benzene, accounts for some 40 per cent of all detergents used throughout the world.

There are many important reasons for this. Not only are the alkyl aryl sulphonates outstanding in their properties as detergents, but they are also based on raw materials which are less difficult to obtain and less expensive than the raw materials on which the other types of detergents are based. The main source of alkyl aryl sulphonates is the petroleum industry.

As the name implies, these products are based on aromatic compounds combined with an aliphatic chain bound to the aromatic nucleus. Taking as an example the most important type of alkyl aryl condensate, dodecyl benzene, this has the following structural formula:

$$\bigcirc\!\!-\!\!C_{12}H_{25}$$

whether the side chain is branched or linear.

This compound may be obtained by condensing benzene with a monochlorinated aliphatic chain having about 12 carbon atoms, by the Friedel-Crafts reaction, usually with anhydrous aluminium chloride as catalyst. The source of the aliphatic chain is either a special cut from straight-chain petroleum, for which purpose the petroleum from Pennsylvania is especially suitable, or a synthetic hydrocarbon chain obtained by condensing a lower molecular aliphatic hydrocarbon (generally propylene, which is condensed to its tetramer). This may be coupled with benzene without any previous chlorination.

Because of the comparatively high cost of chlorine and the inevitable production of hydrochloric acid, both when chlorinating a paraffin and when condensing this chloro-paraffin with the aromatic nucleus, the use of an olefin was preferred as the source of the alkyl side chain and in this case the catalyst is usually HF or, less frequently, BF_3. This is the reason why propylene tetramer held sway for so many years as the alkyl group.

For the reasons given above, when the switchover to LAS took place in certain countries straight-chain olefins were frequently employed as the alkyl source. However, the bulk of LAS is at present being produced from linear alkane which has been separated from a petroleum cut by a molecular sieve and subsequent chlorination, or dehydrogenation.

The aromatic nucleus is usually benzene, but occasionally it is naphthalene, toluene, xylene, or even phenol. When naphthalene is used, the side chain can be as low as propyl. However, the sulphonated product is not suitable as a detergent, because of the low-molecular-weight side chains on the naphthalene nucleus. Nevertheless, by increasing the length of the side chain the detergent properties are increased.

To achieve the full hydrophile/hydrophobe balance when the lower homologues are used as the side chain, two members are necessary. Thus propyl naphthalene sulphonate is unsuitable either as a wetting agent or detergent. Dipropyl naphthalene sulphonate begins to show some wetting properties. Dibutyl naphthalene sulphonate is considerably better than butyl naphthalene sulphonate and, in fact, has fairly good detergent properties. Mono capryl naphthalene sulphonate is a very good detergent. With a side chain over C_{10}, however, the solubility of the alkyl naphthalene sulphonates starts falling off.[2]

Thus, as the molecular weight of the side chain increases, it is necessary to limit the final product to one side chain. Conversely, if the size of the aromatic nucleus is decreased, it is necessary to increase the length of the side chain. As a rough guide, nonyl naphthalene sulphonate is equivalent to dodecyl benzene sulphonate. Thus it can be generalized that, as the side chain on the aromatic nucleus increases, the solubility decreases, and the detergency reaches a maximum and then starts decreasing. In practice, dodecyl benzene sulphonate was found to have the best balance of characteristics.

As stated above, the dodecyl benzene sulphonate seems to give the best all-round performance. However, factors vary in different parts of the world, and also conditions and requirements vary. It should be stressed that when we discuss dodecyl benzene it is meant to be a material whose average molecular structure suggests something close to dodecyl benzene: that is a product with an average molecular weight of the order of 246. This does not, in fact, indicate that the product is exactly that. In fact, if the base material is 'propylene tetramer', the propylene itself could have had small amounts of ethylene and butylene present as adulterants, and when this is polymerized to the tetramer appreciable amounts of the trimer and pentamer of each of these ingredients can also be produced as well as possible co-polymers of two or three of each of the ingredients. Many combinations are now possible and each has several possible stereo-isomers, and when all of these are coupled to benzene, coupling is then possible at many different points of the side chain; so one can appreciate that the total possible amounts of isomers and homologues are enormous. With the switch to linear alkylates, the alkyl portion is less varied, and can be in the range C_{11} to C_{15}.

It is therefore becoming standard practice to refine the raw alkylate to one product with an average molecular weight in the range of 233–45 and to another with the average molecular weight in the range 257–65, the exact figure depending on the manufacturer and his possible sales of the alternate product. The first is called commercially dodecyl benzene and the second sometimes tridecyl benzene. Tridecyl benzene sulphonate, in general, has better detergency and foams better in soft water, whereas dodecyl benzene has a lower cloud point and viscosity in liquid formulations. At approximately 180–200 ppm water hardness, the difference in detergency falls away.

Although the branched and the linear alkylates are both usually dodecyl benzene, it has been common to call the sulphonated branched product ABS and the linear one LAS.

With the changeover to linear alkyl benzene certain changes were observed. In this case as the alkyl portion is straight-chained it has no substantial amount of isomers but again contains homologues. When coupled with benzene, isomers arise in that the coupling can be at any of the positions between 2-phenyl and 6-phenyl, and it has been found that detergency is dependent on the position of the phenyl group. Rubinfeld and Cross[3] have suggested that for optimum light-duty detergency, linear alkyl benzene should have a mean chain distribution of lower than 12, the molecular weight should be in the range 231–41 and it should contain 15–25 per cent 2-phenyl isomers. For heavy-duty formulations they suggest a mean chain distribution of higher than 13, a molecular weight range of 258–66, the 2-phenyl content to be less than 20 per cent and the 5- and 6-phenyl content to be at least 40 per cent.

In the refining of the alkylate a certain amount of polyalkyl benzene is left as 'bottoms' in the still. This is a highly complex mixture of materials with an average molecular weight of 300 or more and on sulphonation it in turn yields a product which is a very good emulsifier, but poor detergent.[4]

Three principal ways to use LAS and ABS are open to the manufacturer of cleaning materials: (1) He may produce his end product from basic detergent raw materials, such as dodecyl benzene, by sulphonation processes and finishing processes requiring rather high initial investment, but giving him the widest scope for developing a complete range of cleaning materials. (2) He may use a detergent material that has already been sulphonated, the sulphonic acid. This method still provides a fairly wide scope for development and avoids the necessity of erecting an expensive sulphonation plant which can only be operated by skilled labour and under expert chemical supervision. (3) He may purchase ready sulphonated and neutralized detergents in a highly concentrated form, either as a paste-slurry or in the form of spray-dried, or drum-dried, powder or flakes. This material has then only to be diluted to obtain liquid products, etc., or be mixed with suitable builders to provide products ready to be marketed. The last-mentioned way of simply mixing detergent concentrates is the least expensive, from capital costs but also the least interesting, and does not, in fact, constitute a genuine manufacturing process. However, even this method affords certain possibilities to the small-scale producer. By careful selection of those materials with which he compounds the detergent concentrates, quite interesting and valuable cleaning materials may be obtained. This will become clearer during the description of the composition of cleaning materials later in this volume.

2. Long-chain (Fatty) Alcohol Sulphates

Before the last war fatty alcohols became available from the catalytic hydrogenation of fatty acids, and the first of the alcohol sulphate type of detergent, Gardinol, came on to the market in Germany in the early 1930s, and soon afterwards in the USA. More recently, alcohols have been produced by the Ziegler process, a growth reaction using ethylene and trialkyl aluminium. This process yields even-numbered alcohols (those produced from natural fatty acids are invariably even-numbered); any desired length of chain can be produced, and the alcohols can also be blended in proportions not found in nature. Thus myristic alcohol (C_{14}) is not in great supply from natural sources, but can be manufactured in unlimited quantities by this process. Despite the fact that the Ziegler alcohols have to a large extent replaced the lower (C_{12}) natural alcohols, the higher ones (C_{18}) made from tallow are not necessarily more expensive. Several household and industrial detergents have been based either wholly or partially on sulphated tallow alcohols for many years and at the moment it is not clear whether synthetic or tallow alcohols are being used. Also included in this group are the OXO alcohols.

The OXO alcohols, being made by a hydroformylation reaction using an olefine and CO, are mixtures of both even and odd numbered alcohols with some side branching. Catalysts are now being used which limit the branching to 25 per cent 2-methyl to give a structure

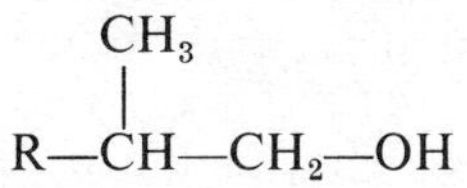

$$\begin{array}{c} CH_3 \\ | \\ R-CH-CH_2-OH \end{array} \qquad *$$

There was a misapprehension that this side-branched alcohol is a secondary alcohol but as can be seen from the above structure it is an iso-alcohol.

OXO-type alcohols are both ethoxylated and sulphated. It must be pointed out that some manufacturers use a type of process which does give some secondary alcohols

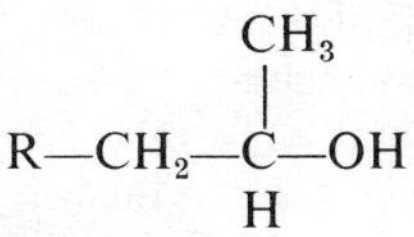

$$\begin{array}{c} CH_3 \\ | \\ R-CH_2-C-OH \\ H \end{array}$$

and these are not readily sulphated.

Although exact figures are not available, many of the large 'soapers' are now using[5] in their heavy-duty washing powders both LABS and a high molecular (C_{12}–C_{18}) alcohol sulphate. This is achieved relatively easily by a firm which does its own sulphonation, not so easily by a firm that has to buy its detergent raw materials.

* Typical structure of Neodol alcohols manufactured by Shell Chemical Company.

Despite the fact that the sulphate bond —C—O—SO_3 is not as stable as the sulphonate bond —C—S—O_3, the stability of the alcohol sulphates is sufficient for all normal purposes, and there is not much to choose between the detergency of, say, cetyl alcohol sulphate and dodecyl benzene sulphonate. Like all sulphuric acid esters, however, the alcohol sulphates suffer from one serious drawback (apart from price) in that they are not stable as the acid material. Therefore the manufacturer who wishes to 'sulphonate' must immediately neutralize the product, and the manufacturer who buys a ready sulphonated detergent for further compounding can only buy a ready neutralized paste or powder. Neither material is a highly concentrated product and the best the manufacturer can hope for is a 60 per cent material, the balance being either water or an inert filler such as sodium sulphate. Drum-dried or spray-dried powders are available with a concentration of upwards of 90 per cent active matter, but the high cost of production allows the use of this type of powder in only very special products. The use of the products for normal detergents does not therefore give the manufacturer the latitude that the use of 100 per cent LAS can give him.

3. Sulphonated Olefins

Early attempts at producing commercial sulphonated olefins were based on internal olefins reacted with sodium bisulphite. Sulphonated internal olefins do not give good detergents, rather wetting agents and the bisulphite reaction is rather difficult to carry out. With the advent of alpha olefins, initially from wax cracking and then from the Ziegler process (see p 120) attention was drawn to the possibility of sulphonating olefins with SO_3.

The first commercial sulphonation of olefins with SO_3 involved the use of an SO_3-organo compound complex, either produced separately or *in situ*. Nowadays best results are obtained by sulphonating with uncomplexed diluted SO_3 in a film reactor (see p 162).

The reaction between alpha olefins and SO_3 is not straightforward, mixtures of alkene sulphonic acids, sultones, alkene disulphonic acids and sultone sulphonic acids might be formed (see p 166). The sultones need to be hydrolysed (saponified) to produce hydroxysulphonic acids. The materials sold under the name AOS are all therefore neutralized, usually with soda, and consist of mixtures of

$$R-CH=CHCH_2SO_3Na \quad \text{and} \quad R-CHOH(CH_2)_nSO_3Na$$

where n can be both 2 and 3.

Despite what was said above about internal olefins, a patent assigned to Ethyl Corporation[6] claims that the presence of internal olefins together with alpha olefins gives an AOS with syngergistically improved detergency over pure alpha olefin sulphonates.

AOS has not yet made great strides in the heavy duty laundry field but is being used successfully for light duty detergents, hand dishwashing, shampoos, bubble baths and synthetic soap bars.

At the time of going to press there is controversy about skin sensitizing by materials containing AOS. One party to the dispute maintains that after twenty years of commercial use no ill effects have been noted, whereas the other party states that if hypochlorite bleaching is used together with the alkaline hydrolysis, traces of sultones in ppb quantities can remain and cause this skin sensitization.

The newer film reactors (p 162) which produce light-coloured olefin sulphonates without bleaching are therefore the answer.

4. Sulphated Monoglycerides

In the late 1940s heavy-duty household powders were sold, in which sulphated monoglycerides were the active material. These were a moderate success, but, being based on fats, which are expensive and in short supply, they were quickly superseded by the cheaper alkyl benzene sulphonates. Quite apart from the economic aspect, the sulphated monoglycerides do not lend themselves easily to the manufacture of suitable powder products, although they are still being used occasionally in light-duty liquid formulations. Because of their sensitivity to alkaline hydrolysis, no real advantage is seen for them, especially as the manufacture of the monoglyceride sulphate is rather a tricky process and the number of detergent manufacturers producing this material is limited.

5. Sulphated Ethers

Non-ionic detergents of the ethylene oxide condensate type, which will be discussed later, are in general excellent detergents. They have two disadvantages, however, in that they produce weak or unstable foams and although completely soluble in water, they exhibit 'invert solubility' in that they are more soluble in cold than in hot water. Thus when solutions of non-ionics are heated, at certain temperatures the solution becomes cloudy and in the limiting instance can separate into two phases. Fatty alcohol sulphates on the other hand are high foamers but their sodium salts do not produce clear solutions unless in very low concentrations. To produce clear solutions of fatty alcohol sulphates, recourse must be made to cations other than sodium with concomitant difficulties.

If a fatty alcohol is ethoxylated (see p 29) the ether produced still has an —OH group at the end able to be sulphated. This class of ether sulphates has become the fastest-growing group of anionic detergents. In practice the alcohol is not ethoxylated to the degree where it becomes in itself a detergent; only two to four molecules of ethylene oxide are added and the unsulphated material is still water-insoluble. The material is then sulphated by chlorosulphonic acid, or SO_3 and neutralized normally by caustic soda.

The sodium salts of these ether sulphates have very low cloud points even in relatively high concentrations. In addition, their wetting properties are lower than other anionic detergents, while their foaming properties are considerably higher, making them excellent raw materials for hair shampoo, and in combination with other anionic or non-ionic detergents, they are being used more and more for household dishwashing. They are used to increase the detergency of liquid detergents based on alkyl benzene sulphonate or n-paraffin sulphonates. Generally 20 per cent of the total active matter is replaced by sulphated ethers.

Lightly ethoxylated alkyl phenols are also being sulphated. In this case, however, the molecule contains two possible points of attack. The terminal —OH group can be sulphated and the benzene ring can be sulphonated. As the hydrophilic group in this class of product is the 'ethoxy sulphate' it is desirable that sulphation and no ring sulphonation takes place. A complete exclusion of ring formation can only be achieved by the use of sulphamic acid,[7] which, however, produces simultaneously with the sulphation the ammonium salt. The use of SO_3 can under the best of conditions give a mixture of 20 per cent ring sulphonate but if great care is not taken di-sulphation, ie sulphation of the —OH group and also sulphonation of the ring of the same molecule, can occur. Alkyl phenol ether sulphates have been recommended as bases for toilet preparations but their main use is as components of light duty household detergent liquids, and for emulsion polymerization.

These lightly ethoxylated alcohols are more complex mixtures than one would consider at first glance. The ethylene oxide content is only a mean and for these low values appreciable portions of the alcohol will remain unethoxylated (25 per cent when 2 moles EO are added, 18 per cent for 3 moles), thus the sulphated product will have alcohol sulphate as well. Figures can vary for different manufacturers so nominal ethylene oxide from two manufacturers will not necessarily have the same properties, not taking into account that the basic alcohols might also vary in their microcomposition. Ethoxylation of alkyl phenols follows a different path, all the phenol is ethoxylated before multiple ethoxylation starts.

6. Sulphosuccinates

Materials of excellent detergency can be obtained by condensing a hydrophobe containing an —OH group with maleic anhydride

$$R-OH + \begin{array}{c} HC-C \overset{\nwarrow O}{\underset{\searrow}{}} \\ \| \quad \quad O \\ HC-C \underset{\searrow O}{\nearrow} \end{array}$$

to produce the half-ester

$$\begin{array}{c} O \\ \parallel \\ C\text{---}OR \\ | \\ HC \\ \parallel \\ HC \\ | \\ C\text{---}OH \\ \parallel \\ O \end{array}$$

and reacting this with sodium sulphite or bisulphite to produce the sulphosuccinate

$$\begin{array}{c} O \\ \parallel \\ C\text{---}OR \\ | \\ H\text{---}C\text{---}H \\ | \\ H\text{---}C\text{---}SO_3Na \\ | \\ C\text{---}ONa \\ \parallel \\ O \end{array}$$

Sulphosuccinates, depending of course on the choice of the R-group, are excellent high-foaming detergents and ideally suited for toilet preparations as they are non-irritating to the skin.

This is a process which does not require large capital outlays and allows the small or medium sized manufacturer to do his own synthesis. The choice of the hydrophobe is wide, it can be an alcohol, a lightly ethoxylated alcohol or an alkanolamide (which also has a terminal —OH) or a mixture of any of the possible variations of these. The maleic anhydride (the common choice but other unsaturated anhydrides can be used) can be reacted with 1 mol of the (mixture of) hydrophobe(s) to produce the half (or mono-) ester or with two molecules to produce the di-ester. In the case of the di-ester the properties of the finished sulphosuccinate are more wetting than detergent and they are used for industrial purposes. The half-esters are excellent high-foaming detergents, mild to the skin and ideally suited for toilet preparations.

If the di-ester is produced, the sulphonation is performed with sodium bisulphite; for the half-ester sodium sulphite is used.

The sulphosuccinates suffer fom one disadvantage, their solubility is not good (depending of course on the choice of hydrophobe). This does not allow of the preparation of clear liquids if the concentration is over 10 per cent active matter.

Solubility can be increased by using monoethanolamine sulphite in place of sodium sulphite. This monoethanolamine sulphite is not an article of commerce but can readily be manufactured by passing SO_2 into a solution of monoethanolamine in water to form either the bisulphite or sulphite as required:

$$HOC_2H_5NH_2 + SO_2 + H_2O \rightarrow HOC_2H_5NH_3SO_3H$$

$$2HOC_2H_5NH_2 + SO_2 + H_2O \rightarrow (HOC_2H_5NH_3)_2SO_3$$

Allied to the sulphosuccinates are the sulphosuccinamates, where the maleic anhydride is reacted with a primary or secondary amine:

$$\begin{array}{c}HC-C=O\\ \diagdown\\ O + R'R''HN \rightarrow\\ \diagup\\ HC-C=O\end{array} \qquad \begin{array}{c}O\\ \|\\ HC-C-NR'R''\\ \|\\ HC-C-OH\\ \diagdown\\ O\end{array}$$

Maleic acid amide
or maleamic acid

The second carboxyl group can also be amidized to form the di-amide or it can be esterified with an alcohol, ethoxylated alcohol, alkanolamide, etc. to produce the mixed ester-amide.

These are again sulphonated in the usual way, using either sulphite or bisulphite.

The possibilities of all the possible variations of the sulphosuccinamates have not yet been fully explored but they are said to be even kinder to the skin than the normal sulphosuccinates (see also p 173).

7. SULPHONATED METHYL ESTERS

In considering the aforementioned and following detergent raw materials, it will be noted that the vast majority are from petrochemical sources. The world has now come to realize that this is a diminishing resource and even the switch to coal, both as a source of energy and as a basic starting point for chemical syntheses will only put off the evil hour for a century or two.

Fats and oils are self-replenishing and attention is being paid to their use as detergent intermediates. Alcohol sulphates are one such product and now attention is being drawn to sulphonated methyl esters.

They are already being used in powders and a surprising spin-off was noted in that the material has the ability to sequester calcium and magnesium ions, even in the presence of soap, without the use of phosphates.[8]

The methyl esters need to be of high quality with a minimum of double bonds, indicating the use of refined coconut or hydrogenated tallow as starting materials. Production of methyl esters will be discussed later.

The sulphonation reaction is considered to be:

$$\text{R-CH}_2\overset{\overset{\displaystyle O}{\|}}{\text{C}}\text{-OCH}_3 + SO_3 \rightarrow \text{RCH}_2\overset{\overset{\displaystyle O}{\|}}{\text{C}}\text{-OCH}_3 \rightarrow \text{R-CH-}\overset{\overset{\displaystyle O}{\|}}{\text{C}}\text{-OCH}_3$$

$$\downarrow \qquad \qquad \qquad \text{SO}_3\text{H}$$

$$O=\overset{\overset{\displaystyle \|}{\text{S}}}{\underset{\underset{\displaystyle O}{\|}}{}}=O$$

that is, an intermediate complex is formed and this rearranges to the sulphonate.

There are two competing reactions. If double bonds are present, an olefin type of sulphonation will occur, particularly as that rate of reaction is considerably faster than the methyl ester sulphonation. It is, therefore, necessary to produce the methyl esters from material with very low iodine values. The second reaction is a double sulphonation of the ester which on neutralization produces a disodium sulphocarboxylate, a sodium carboxylate and sodium methyl sulphate.

Fortunately the second reaction can be controlled by sulphonating conditions. However, on neutralizing the sulphonated methyl ester neutralization conditions must be such that no ester hydrolysis (saponification) occurs. This will be dealt with in the section on sulphonation.

8. Alkane Sulphonates

The Mersolates, one of the very early types of detergents marketed, fall into this class. They were made according to the Reed reaction by chlorosulphonation of petroleum fractions, but were quickly superseded by sulphoxidation and by alkyl benzenes when these appeared in commercial quantities. When the problem of biodegradation of propylene tetramer dodecyl benzene arose, Esso developed a new process for sulphonating straight-chain alkane fractions, which uses gamma rays from a cobalt-60 source. The method of synthesis is not straightforward but appears to be the initial formation of a peroxy-sulphonate as an intermediate, and this acts as an initiator for the sulphonation reaction (see pp 175-177).

Gamma irradiation has been replaced by the less dangerous ultra-violet irradiation system. It is very important that the n-paraffins used are practically free from aromatics or alkyl aromatics, whose presence retards the reaction. The removal of aromatics and alkyl aromatics is best accomplished when the n-paraffins obtained by the molecular sieve process (see p 123) are treated with SO_3 converter gas. A large plant of this type has been put up by Ballestra, Milan, in the south of Italy, treating 60 tonnes of n-paraffins per hour.

In Europe three large production units for the sulphoxidation process have been set up by Hoechst of Germany. Another sulphoxidation plant by the Société Nationale des Pétroles d'Aquitaine (SNPA) is now at planning stage (see Boy et al.[9]).

The SNPA process consists of the following basic steps: n-paraffins are reacted with SO_2 and O_2 under ultra-violet radiation with a wavelength essentially between 3300–3600 μ with a power level of several kW. Unfortunately, at the same time as paraffin sulphonate is formed, SO_2 reacts with O_2 and H_2O to form H_2SO_4. The overall reaction can be written as follows

$$RH + 2SO_2 + O_2 + H_2O \rightarrow RSO_3H + H_2SO_4$$

where RH represents n-paraffin and RSO_3H the sulphonation product on a secondary carbon. Operating temperature and pressure are close to ambient.

The next step is to remove the sulphuric acid and some dissolved SO_2 from the mixture of paraffin sulphonate and unreacted paraffin (the latter to be recycled). SNPA has developed its own separation process which uses a solvent or solvent mixture of a 'slightly polar nature' with alcohol of at least five carbon atoms. No exact details of the SNPA process have yet been revealed. A flowsheet attached to the article,[9] however, shows the basic principle. The paraffin sulphonate is netralized with NaOH and the product marketed in the form of a 30 or 60 per cent past ('Hostapur' Hoechst, Germany). Ballestra, Milan, in collaboration with ATO, France, has developed a complete photochemical system for the production of SAS (sodium alkane sulphonates).

The detergency of alkane sulphonates is very similar to that of LABS. Stability against hardness and biodegradability are even better. Liquid detergents based on alkane sulphonates are being manufactured in combination with ether sulphates (see p 19), where the ether sulphate constitutes 20 per cent of the active matter. Powders have not been produced so easily by spray-drying detergents solely based on a high percentage of alkane sulphonates because the material is soft and somewhat hygroscopic. This, however, can be overcome by the use of sodium toluene sulphonate and by the addition of colloidal silica.

Another method of producing alkane sulphonates is by the bisulphite addition to alpha olefins. This is a well-known reaction but until recent years complete reaction of the olefin was not attained, necessitating extraction of the unreacted olefin.

Present work on this is indicated by two patents[10,11] by which the olefin, the bisulphite solution and an initiator are all reacted together in the presence of a water-miscible organic solvent, which is eventually stripped off. Yields are close to 100 per cent based on the olefin.

It is of interest to note that if alpha olefins are the starting material, terminal sulphonates will be formed as bisulphite addition to olefins occurs in a manner opposite to that predicted by Markownikoff's rule[12] and the Marathon Oil patent claims that a high quality shampoo material is formed when alpha olefins are used as a starting material.

9. Phosphate Esters

A series of interesting and highly specialized detergents are the phosphate esters.

Two phosphating reagents can be used, P_2O_5 and polyphosphoric acid.

If one takes as an analogy sulphonation with oleum, where oleum is considered to be SO_3 dissolved (or loosely combined) with H_2SO_4, we have the same phenomenon.

P_2O_5 when dissolved in phosphoric acid (or in a dearth of water) gives the complex phosphoric acid

$$4H_3PO_4 + P_2O_5 \rightarrow 3H_4P_2O_7$$

In this reaction we have shown the production of pyrophosphoric acid, but of course in practice higher condensed acids are also produced and commercial polyphosphoric acid is sold as 115 per cent (calculated as orthophosphoric acid).

The reactions between an alcohol and polyphosphoric acid are

$$R-OH + HO-\underset{\underset{OH}{|}}{\overset{\overset{O}{\|}}{P}}-O-\underset{\underset{OH}{|}}{\overset{\overset{O}{\|}}{P}}-OH \rightarrow R-O-\underset{\underset{OH}{|}}{\overset{\overset{O}{\|}}{P}}-OH + H_3PO_4$$

or

$$2R-OH + H_5P_3O_{10} \rightarrow R-O-\underset{\underset{OH}{|}}{\overset{\overset{O}{\|}}{P}}-OR + 2H_3PO_4$$

Thus both mono- and di-esters are formed and also free phosphoric acid.

Conditions can be varied but as a general rule when using 115 per cent polyphosphoric acid two-thirds mono- and one-third di-ester are formed. The free phosphoric acid is not normally removed from the ester.

When using P_2O_5 as the phosphating agent the reaction is different

$$3R-OH + P_2O_5 \rightarrow R-O-\underset{\underset{OH}{|}}{\overset{\overset{O}{\|}}{P}}-OR + RO-\underset{\underset{OH}{|}}{\overset{\overset{O}{\|}}{P}}-OH$$

and yields equimolecular amounts of di- and mono-esters and no phosphoric acid (in practice a small amount of orthophosphoric acid is always formed).

Depending on the choice of the hydrophobic portion and the phosphating reagent, the possibility of varying the properties of phosphate esters is enormous.

Theoretically any compound containing reactive —OH can be phosphated. More often than not ethoxylates rather than pure alcohols are used. These ethoxylates need not necessarily be the lightly ethoxylated ones as used for ether sulphates, the true detergent grades, both alcohol and alkyl phenol based, are often used. Also if low molecular weight materials, not necessarily water-insoluble, are phosphated, wetting agents and good hydrotropes (see p 97) are obtained.

These materials are sold as the acid, or partially or wholly neutralized salt and are offered in the USA by General Aniline under the name GAFAC and by Witco under the name Emphos and in Europe by Hoechst.

The phosphate esters possess good detergency, especially on hard surfaces, are low foamers with good acid and alkali stability, and are biodegradable. They find application in metal cleaning and plating, and since they are very soluble in organic solvents they are used in combination with solvents as dry-cleaning detergents, for hard surface cleaning and as agricultural emulsifiers.

10. Alkyl Isethionates

These are made by reacting ethylene oxide with sodium bisulphite

$$CH_2\underset{O}{-}CH_2 + NaHSO_3 \rightarrow HO-CH_2-CH_2SO_3Na$$

This sodium isethionate is then dried and the —OH group is esterified with a fatty acid.

Alkyl isethionates, which are gentle to the human skin, have been used successfully in synthetic toilet soap bars and for hair shampoos.

11. Acyl Sarcosides

Early attempts to improve soap were based on the principle of adding hydrophilic groups to the carboxylic acid. One of the first products of this type, and a rather successful one, was acyl sarcoside, in which an amido linkage is interspersed between the carboxyl group and the hydrophilic radical.

A method of proparation is to react a fatty acid chloride with N-methyl glycine (sarcosine), an amino acid:

$$R-\overset{O}{\overset{\|}{C}}-Cl + HN-CH_2COOH \rightarrow R-\overset{O}{\overset{\|}{C}}-N-CH_2COOH + HCl$$
$$\phantom{R-\overset{O}{\overset{\|}{C}}-Cl + H}\underset{CH_3}{|} \phantom{COOH \rightarrow R-\overset{O}{\overset{\|}{C}}-N}\underset{CH_3}{|}$$
Sarcosine

The reaction is normally carried out in alkaline solution so the sodium salt is naturally produced together with salt as a by-product.

The sarcosides are gentle to the skin and are affected considerably less than carboxylates by hard water ions. They are used for personal care products, as dyeing aids and for textile finishing.

12. Alkyl Taurides

Allied to, but somewhat different from, sarcosides, the taurides are manufactured by a somewhat similar process. Taurine (amino ethane sulphonic acid) or methyl taurine are reacted with a fatty acid chloride:

$$R-\overset{O}{\overset{\|}{C}}-Cl + R'HN-CH_2-CH_2-SO_3H + 2NaOH \rightarrow$$

$$R-\overset{O}{\overset{\|}{C}}-\underset{\underset{R'}{|}}{N}-CH_2-CH_2-SO_3Na + NaCl + H_2O$$

where R' can be hydrogen or either methyl or ethyl groups.

The taurates have all the properties of soaps: foaming power, lathering, kindness to the skin and emulsifiability, without the disadvantages associated with carboxylates. They are used in synthetic toilet 'soaps', in personal care items, in textile finishing and in agricultural wettable powders.

It will be noted that the last two items are manufactured from fatty acid chlorides, with the production of salt, whereas the alkyl isethionates are shown to be made from fatty acids without any salt as a by-product.

Alkyl isethionates can also be made by reacting a fatty acid chloride with sodium isethionate in alkaline solution with the production of salt. Conversely, the taurides and sarcosides can also be made by reacting a fatty acid with the anhydrous, molten sodium salt of taurine or sarcosine.

The difference between the two routes is that using the fatty acid chloride the reaction is done in a water solution at a temperature below the boiling point and goes to completion quite easily. Sodium chloride is formed. Using the fatty acid route, no sodium chloride is formed but temperatures over 200°C are required. It is extremely difficult to reach completion and unless special precautions are taken, the finished product will be of a darkish colour.

13. Fluorosurfactants

A new development in the surface active field is fluoro chemicals, which are at the moment very expensive but have unique properties:

(a) They lower surface tension of aqueous systems to $2 \cdot 0$ N m² (most detergents give a surface tension in the range of $2 \cdot 8$—$3 \cdot 1$ N m^{-2}).
(b) They are effective in concentrations of the order of $0 \cdot 01$ per cent.
(c) They show surface activities in organic systems.
(d) They are chemically stable in hostile environments.

Most modern methods of manufacture[13] involve reacting tetrafluorethylene with a fluoride ion (caesium, potassium or tetra alkyl ammonium fluoride) in polar aprotic solvents (for example dimethyl formamide). Polymerization takes place, and the pentamer (the most abundant)

$$(C_2F_5)_2-\underset{\underset{CF_3}{|}}{\overset{\overset{CF_3}{|}}{C}}-C=C\overset{F}{\underset{CF_3}{\diagdown}}$$

is used for further synthesis.

This pentamer is coupled with phenol

$(C_2F_5)_2CF_3CC(CF_3){:}C(CF_3)F + C_6H_5OH \rightarrow (C_2F_5)_2CF_3CC(CF_3){:}$
$C(CF_3)OC_6H_5 + HF$

This phenol group can be sulphonated to give an anionic detergent; also it may be chlorosulphonated to give the sulphonyl chloride, then treated with $N'N'$-dimethylpropanediamine to give a tertiary amine and then quarternarized with methyl iodide to produce a cationic detergent. The tertiary amine can also be reacted with propiolactone to produce an amphoteric surfactant.

Non-ionic fluorosurfactants can also be made from these phenyl compounds or else by reacting the pentamer with ethylene oxide condensates and rearrangement of the molecule (with the removal of one molecule of HF).

Cationic Detergents

The cationic detergents are of relatively limited interest to the cleaning materials manufacturer. Compared with the anionic and non-ionic detergents their detergency is relatively poor.

The term cationic detergents is somewhat of a misnomer. These materials do exhibit surface active properties but are seldom used in cleaning formulations. Their detergency is relatively poor and the cost high.

The principal uses for cationic detergents are as germicides, fabric softeners and specialized emulsifiers. Anionic and cationic detergents are mutually incompatible, as the anion and cation precipitate each other. However the Witco Chemical Corporation produces a series of cationic detergents (Emcol CC-36) which can be formulated in conjunction with anionic detergents and are powerful antistatic agents.

These cationic detergents are almost invariably amino compounds and the most effective of the group are the quaternary ammonium salts, with one long chain attached to the nitrogen nucleus, or quaternary pyridine based salts. Here the possible permutations are again enormous. For example, cetyl trimethylammonium chloride, which is a typical normal quaternary ammonium salt, can also appear as the bromide and then, instead of the cetyl group, any of the long chains available both from fatty or synthetic sources from lauryl to stearyl can be used. One methyl group can be replaced by another cetyl or benzyl (or even a naphthyl) radical and then one or all of the methyl groups can be replaced by an ethyl group and so forth.

For the reasons mentioned above, the cationics have not found much favour as detergents and so very little work has been done on the investigation of structure against optimum detergency.

Non-ionic Detergents

In discussing anionics we mentioned that most of the anionic detergents are sulphonates or sulphates of one sort or another. Similarly, the vast majority of all non-ionic detergents are condensation products of ethylene oxide with a hydrophobe.

This hydrophobe is invariably a high-molecular-weight material with an active hydrogen atom, and the non-ionic material can be one of four main reaction products:

1. FATTY ALCOHOL AND ALKYLPHENOL CONDENSATES

The reaction proceeds thus:

$$R-OH + CH_2\underset{O}{-}CH_2 \rightarrow R-O-CH_2-CH_2OH$$

Reaction of one molecule of ethylene oxide is not sufficient to produce a water-soluble detergent product and the reaction is continued:

$$R-O-CH_2-CH_2OH + n(CH_2\underset{O}{-}CH_2) \rightarrow$$

$$R-O-CH_2-CH_2-(OCH_2CH_2)_{\overline{n-1}}OCH_2CH_2OH$$

It will be noted that an ether linkage is obtained. Since this linkage is basically a strong one, the material is not subject to hydrolysis and, as there is no possibility of ionization, it is not affected by metallic ions. In general for an average hydrophobe when n is approximately 6, the product becomes water-soluble and starts assuming detergent qualities.

The above reaction can be continued indefinitely but, in practice, optimum detergency is found in the range of 10–15 molecules of ethylene oxide per molecule of fatty alcohol. The source of the alcohol varies greatly, but natural, ZIEGLER and OXO alcohols of various molecular weights are being used and up to 50 molecules of ethylene oxide are being added. Another source of alcohols is the one used by Union Carbide. The process has not bee disclosed but it is considered that their feedstock is n-paraffins (from kerosene) which are then oxidized to secondary alcohols, both even and odd numbered. These alcohols do not lend themselves easily to ethoxylation and it is considered that the ethoxylation is done in two stages,[14] firstly, with an acid catalyst (possibly boron trifluoride), to produce a product containing 1 mol of ethylene oxide. After this stage the catalyst and the unreacted alcohol are removed and ethoxylation is proceeded with in the usual manner. (It is worth pointing out that these ethoxylates are also rather difficult to sulphate with SO_3 and chlorosulphonic acid.)

Although chemically very different from fatty alcohols, the alkyl phenols behave in exactly the same manner as the fatty alcohols. Here again the choice of one particular alkyl phenol (or naphthol) is quite wide, but the preference usually falls on nonyl (sometimes octyl) phenol. With 4 molecules of ethylene oxide, nonyl phenol is still completely insoluble in water; at 6–7 molecules it is completely soluble in water at room temperature; at 8–12 molecules it exhibits excellent detergency.

Two further variations are now also being used, dodecyl phenol ethoxylate for certain agricultural emulsifiers and dinonyl phenol ethoxylate (where problems of non-ionic biodegradation have not yet arisen) as low or non-foaming ingredients of household washing machine powders.

Up to now, the alkyl phenol ethylene oxide condensates have been the most widely produced non-ionic detergent. However, with the recent international preoccupation with biodegradation they have fallen into slight disfavour and fatty alcohol ethylene oxides condensates are gaining. The situation is not completely clear because of conflicting claims as to the biodegradability of the alkyl phenols (which by virtue of their method of manufacture are branched chain), largely because there is not at present an internationally accepted method for chemical determination of non-ionics in micro quantities.

These ethylene oxide adducts are solubilized by the ethylene oxide units forming hydrates with water. Compared with all other materials which dissolve in water, these products show an apparent anomaly: their solutions are completely clear in the cold, but when heated they become turbid and, if the temperature is raised sufficiently, separation into two phases can even take place. This is explained by the fact that at an elevated temperature the hydrates are destroyed; so much so that this temperature is a function of the amount of ethylene oxide molecules present in the molecule.

It should again be stressed that, although the number of molecules of ethylene oxide present is considered a basic property of the material, this number is only an average, and if, say, there are 10 molecules stated to be present, the actual molecular configuration can vary from 4 to 15.

With the advent of polyester fibres for use in clothing and bedding, the use of a non-ionic constituent in laundering increased tremendously as it was found non-ionics will remove soil from these fibres better than the hitherto used anionics, but most laundry powders or liquids use a combination of the two.

With the latest trend to 'cold water' washing (see also p 279) a further discovery has been made. As stated above the ethylene oxide distribution follows a Poisson curve around the mean. It has now been found that if 'peaked' ethoxylates are used the performance in cold water is improved.[15]

Normal ethoxylation is done with an alkaline catalyst, usually KOH, sometimes K_2CO_3 or sodium methylate. These give the normal distribution of condensations. Kravetz[16] has indicated that these peaked narrow range ethoxylates are produced commercially in two ways. The first way is to distil off the more volatile elements of the condensate, in which case if the alcohols are the base material the unethoxylated alcohol and the 1 or 2 mol products are 'topped'. The topped portion being returned for further processing. The distribution of a normal and topped ethoxylate is indicated in Fig 2.2.

The other method is to use catalysts specifically designed to produce a narrow distribution, in this case less of both the lower and higher adducts. It is considered that the catalyst used is of the alkaline earth type.

Both of these types of narrow range ethoxylates (NRE) are available commercially.

An added advantage is that the 'pluming' encountered in spray-drying powders containing non-ionics is lessened if the NREs are used because the lower fractions which cause pluming have been either removed or are not present. One of the manufacturers of dinonyl phenol ethoxylates makes the same claim for its ethoxylates.

Ethylene oxide reacts with water to form ethylene glycol according to the reaction

$$CH_2\!-\!CH_2 + H_2O \rightarrow CH_2\!-\!CH_2$$
$$\underset{O}{\diagdown\diagup} \qquad\qquad \underset{OH}{|} \;\; \underset{OH}{|}$$

This ethylene glycol has two —OH groupings, each of which is available for reaction with further molecules of ethylene oxide to form polyethylene glycol. Therefore whenever any non-ionic detergent of the ethylene oxide condensate type is formed a certain amount of polyethylene glycol of varying molecular weight is formed from the trace quantities of moisture

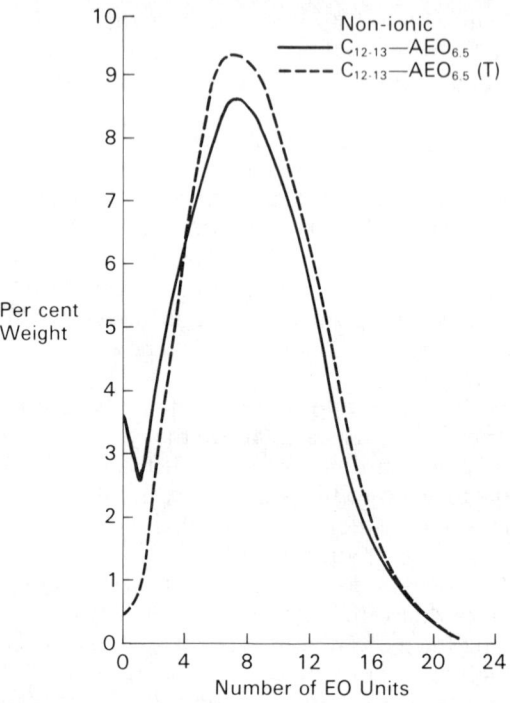

Fig 2.2 EOD of conventional versus narrow range ethoxylate

which are invariably present. For this reason although two different non-ionic detergents may have the same nominal composition or formula, they might vary in some of their physical or physico-chemical properties because their spread of ethylene oxide units will be different and both the content and type of polyethylene glycol will not be the same owing to variations in raw materials and manufacturing conditions.

2. Fatty Acid Condensates

Another type of non-ionic can be prepared by reacting fatty acids with ethylene oxide. This is the same type of reaction

$$R-COOH + n(CH_2OCH_2) \rightarrow$$

$$R-\overset{O}{\overset{\|}{C}}-O-CH_2-CH_2-(OCH_2CH_2)_{n-2}OCH_2-CH_2OH$$

A second method of manufacture is by the esterification of a fatty acid with a polyethylene glycol, $HOCH_2(CH_2OCH_2)_nCH_2OH$. The difference in these two methods is that the polyethylene glycol has two —OH groups, so that unless very specific catalytic control is taken, non-ionic esters made by esterification will always have a certain proportion of di-esters. This is not necessarily a disadvantage, but makes the esterification ester different from the ethoxylation ester.

Being esters, these materials have one drawback that non-ionics in general do not suffer from. Non-ionic detergents are not in general affected by metallic ions, acids or alkalis and their functions as detergents are only marred by the way these extraneous ions in the water affect the physical properties of the water itself. However, the fatty acid ethylene oxide condensates are readily hydrolysed by acids or alkaline solutions to their component fatty acid and to polyethylene glycol. Thus in strong alkaline solutions these materials are no better than the soap of the fatty acid from which they were made, and in strong acid solution they are ineffectual. However, as components in household detergent powders, they do perform quite well.

3. Condensation of Ethylene Oxide with an Amine

Two reaction products are possible:

$$R-NH_2 + n(CH_2OCH_2) \rightarrow$$
$$R-NH-CH_2-CH_2(OCH_2CH_2)_{n-2}OCH_2-CH_2OH$$

or

$$R-NH_2 + 2n(CH_2OCH_2) \rightarrow$$

$$R-N \begin{cases} CH_2CH_2(OCH_2CH_2)_{n-2}OCH_2-CH_2OH \\ CH_2-CH_2(OCH_2CH_2)_{n-2}OCH_2-CH_2OH \end{cases}$$

In the case of the second reaction, the end product is shown as being symmetrical, but it need not be, and can again be a mixture of many homologues and isomers. Other than specialized items, this class of non-ionic has not been used to a large extent in cleaning compositions. It should be noted, however, that these materials in acid solution can exhibit cationic characteristics, whereas in neutral or alkaline solutions they are non-ionic.

4. Condensation of Ethylene Oxide with an Amide

We shall later describe the alkylolamide type of products. These are basically amides of an alkylolamine and a fatty acid. Certain members of this class exhibit detergency and others do not. To convert the non-detergent materials into detergents, they can be condensed with ethylene oxide:

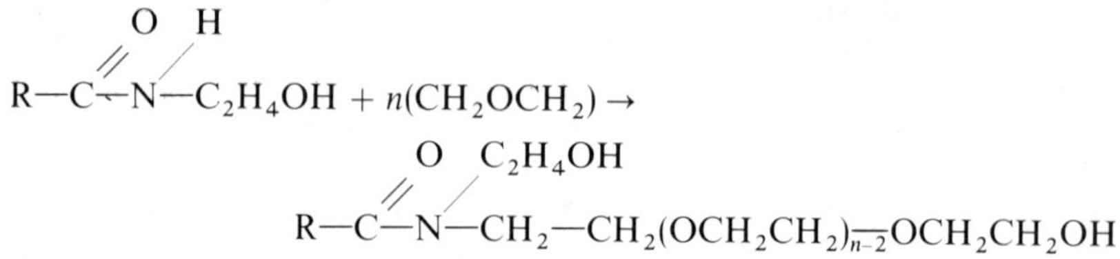

$$R-\overset{O}{\overset{\|}{C}}-\overset{H}{\overset{|}{N}}-C_2H_4OH + n(CH_2OCH_2) \rightarrow$$

$$R-\overset{O}{\overset{\|}{C}}-\overset{C_2H_4OH}{\overset{|}{N}}-CH_2-CH_2(OCH_2CH_2)_{\overline{n-2}}OCH_2CH_2OH$$

or

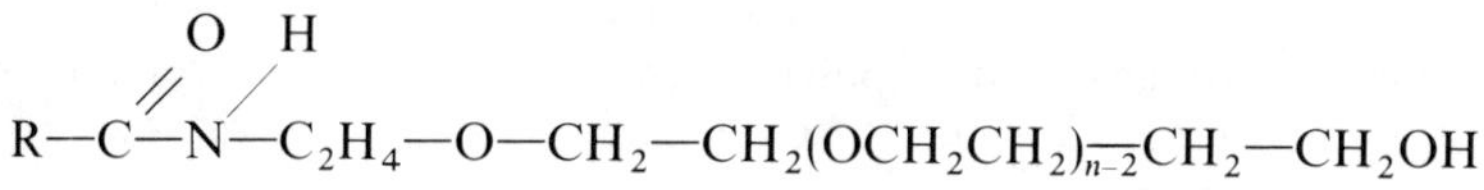

$$R-\overset{O}{\overset{\|}{C}}-\overset{H}{\overset{|}{N}}-C_2H_4-O-CH_2-CH_2(OCH_2CH_2)_{\overline{n-2}}CH_2-CH_2OH$$

or

$$R-\overset{O}{\overset{\|}{C}}-\overset{H}{\overset{|}{N}}-C_2H_4OH + 2n(CH_2OCH_2) \rightarrow$$

$$R-\overset{O}{\overset{\|}{C}}-N\begin{cases} C_2H_4-O-CH_2-CH_2-(OCH_2CH_2)_{\overline{n-2}}OCH_2-CH_2OH \\ CH_2-CH_2-(OCH_2CH_2)_{\overline{n-2}}OCH_2-CH_2OH \end{cases}$$

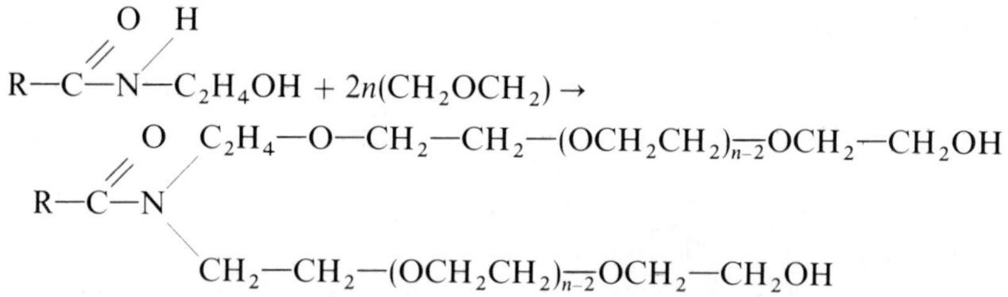

Again, the last reaction product is shown to have the ethylene oxide equally divided between the two possible directions, but there is no absolute reason for this. The ethylene oxide is divided at random in the two directions; also, any amount of either of the first two reaction products may be present. The alkylolamines are themselves complex mixtures and not as simple as shown above, so that an alkylolamide condensed with ethylene oxide consists of many components.

5. BLOCK POLYMERS

As a general rule glycols are water-soluble. Polyethylene glycols (produced by the ethoxylation of ethylene glycol) are also completely water-soluble, the lower molecular weight members of the class are described as infinitely soluble and at a molecular weight of 10,000 specifications state the solubility to be of the order of 50 per cent.

Polypropylene glycol behaves somewhat differently. The glycol itself and the lower condensation products are miscible in all proportions with water. Thus at molecular weight 400 the solubility is still infinite. Above 400 it starts falling off; polypropylene glycol of molecular weight 900 has a solubility in water of 1 per cent and this becomes even lower as the molecular weight rises.

These high-molecular-weight polypropylene glycols can be classed and serve as hydrophobes (they are soluble in or mix with oils, aromatic and

NON-IONIC DETERGENTS

paraffinic solvents) and in addition have two terminal —OH groups:

$$\text{HO-CH(CH}_3\text{)-CH}_2\text{-O-CH}_2\text{-CH(CH}_3\text{)-O-}\cdots$$

$$\cdots\text{O-CH(CH}_3\text{)-CH}_2\text{-O-CH(CH}_3\text{)-OH}$$

Ethylene oxide, added to this molecule, will attach itself at random to either —OH group and then again at random on to the new —OH groups formed. The ethylene oxide chains serve as the hydrophilic portion. Surface activity is attained in the resultant product, called a block polymer, when the ethylene oxide added has reached a definite ratio to the starting hydrophobe.

Block polymers can be and are made with a wide range of hydrophobic/hydrophilic ratios and molecular weights to give surfactants 'tailor-made' to specific physico-chemical requirements.

Several modifications of the basic block have been developed. One is the 'inverse' type in which the propylene oxide is condensed on to the pre-formed polyethylene glycol. Here the hydrophilic portion is in the centre of the molecule with hydrophobic tails emerging.

All these materials need an initiator—a core on to which the initial charge of either propylene or ethylene oxide can be attached. In the two examples given above the initiators can be propylene or dipropylene glycol or ethylene or diethylene glycol, all four of which have two reactive —OH groups. Initiators with more reactive hydrogens are also used. On occasion these initiators are condensed with formaldehyde to give a bigger, more compact core with several hydrogen nuclei. One of the more interesting products uses ethylene diamine to which four molecules of propylene (sometimes ethylene) oxide are condensed

$$\text{H}_2\text{N-CH}_2\text{-CH}_2\text{-NH}_2 + 4\text{CH}_2\text{-CH-CH}_3 \text{ (epoxide)} \rightarrow$$

$$(\text{HO-CH(CH}_3\text{)-CH}_2)_2\text{-N-CH}_2\text{-CH}_2\text{-N-(CH}_2\text{-CH(CH}_3\text{)-OH})_2$$

This serves as a base for further growth reactions with propylene oxide to form the hydrophope and then ethylene oxide to make the surfactant. The molecule has four hydrophilic tails spreading out from the hydrophobe in the shape of an X.

Further variations are possible. In one the propylene and ethylene oxides are pre-mixed before condensation. The reactivity of ethylene oxide is nine times that of propylene oxide, but for all that random distribution of the two is obtained. In another, propylene oxide and ethylene oxide are added alternately to the chains.

It can be seen that the possible permutations and combinations are enormous to give an extremely wide spectrum of properties. Block polymers are manufactured for use in a large variety of processes and products from defoamers and de-emulsifiers through wetting agents, emulsifiers and cosmetics to laundry detergents. One particular use where they have proved very successful is in automatic dishwashing powders where foam is unacceptable.

6. Sucrose Esters

These are products which have had their ups and downs in the detergent world. A future is being predicted for sucrose esters, particularly when production costs can be brought into line with other detergent raw materials.

Production is based on the original Foster D. Snell method which reacts sucrose with a fatty acid ester (methyl stearate) in dimethyl formamide solution. Mono- or di-esters can be formed, depending on the original ratio of reactants. The mono-esters are suited to detergent use.

The Japanese firm of Dai-Nippon has modified the technique in that it works with a large excess of sugar, and uses potassium carbonate as a catalyst, operating at 90°C at a pressure between 9·3 and 13·3 kPa. Unreacted sugar is precipitated from the reaction solution by the addition of toluene.

Foster D. Snell[17] has developed a technique where the sucrose is dissolved in propylene glycol and the methyl ester emulsified into this solution. The glycol is distilled off and in the distillation process transesterification takes place. This process eliminates the use of expensive and poisonous solvents used hitherto.

7. Sorbitan Esters

Sorbitol and mannitol (its stereo-isomer), both hexahydric alcohols, can react with fatty acids (usually in the form of fatty acyl chlorides) to form the mono-, di-, tri-, ..., hexa- esters, but these products have not attained great commercial importance.

When sorbitol (or mannitol) is heated with a maximum of four molecules of a fatty acid to a temperature of close to 200°C in the presence of an acid catalyst, esters and an internal ether are formed. In the case of the monoester the product could be

```
        H
        |
    H—C
        \
   HO—C—H
        |            \
    H—C—OH            O
        |            /
    H—C
        /
   HO—C—H
        |
    H—C—O—C—R
        |      ‖
        H      O
```

called sorbitan ester. These are water-insoluble, oil-soluble and useful as water-in-oil emulsifiers in the food, pharmaceutical and cosmetic industries. Except for sorbitol hexa-ester or sorbitan tetra-ester they all have at least one reactive hydrogen derived from an —OH group. Ethylene oxide is coupled to these hydrogens to make the material more hydrophilic and in the ultimate water-soluble.

Ethoxylated sorbitan esters are available in varying degrees of esterification, ethoxylation and with different fatty acid tails. They are valuable emulsifiers for cosmetics and polishes, in the food industry, and for specialized treating purposes in the paper, textile and plastics industries. As emulsifiers they behave differently from the conventional non-ionics because of the distribution of the hydrophilic chains. Very rarely is the ethylene oxide attached to one —OH group only. The molecule has thus several (relatively) short ethylene oxide chains rather than one long one, making it more compact.

8. Alkylolamides

Another group of synthetic detergents which falls into the non-ionic category is the fatty acid alkylolamides. These were first discovered by Kritchevsky[18] and are made by reacting fatty acids with alkylolamines:

$$R-\underset{\underset{O}{\|}}{C}-OH + H-\underset{H}{\overset{}{N}}-C_2H_4OH \rightarrow R-\underset{\underset{O}{\|}}{C}-\underset{H}{\overset{}{N}}-C_2H_4OH + H_2O$$

The reaction outlined above is between a fatty acid and a monoethanolamine. This has no detergent properties nor is it soluble to any great extent in water. The uses of this type of compound will be dealt with below.

However, if certain fatty acids are reacted with a di-alkylolamine as above, and if the amide formed is then reacted with a further molecule of the di-alkylolamine, the resultant product becomes water-soluble and has detergent properties.

The materials commonly used for this condensation are lauric acid, myristic acid, coconut fatty acids, palm-kernel fatty acids, and occasionally the fatty acids of vegetable oils with higher molecular weights. The alkylolamines used are generally monoethanolamine and diethanolamine, but the isopropanolamines are also employed.

The alkylolamides are rarely used by themselves as detergents, but as additives to other detergent materials, for example, with dodecyl benzene sulphonates. They act as foam boosters and increase the detergency synergistically, and in liquid products they increase viscosity without increasing the cloud point. They also have the valuable property of acting as skin-protecting agents: whereas detergents in general tend to de-fat the skin, the emollient effect of these alkylolamides helps to overcome this tendency.

A further use of alkanolamides is as an intermediate for the manufacture of sulphosuccinates. The preference is for the monoethanolamide or isopropanolamide of coconut fatty acid.

The thickening effect of alkanolamides on detergents has been demonstrated to be dependent on inorganic salts present as an effect of the inorganic salt on the micelles.[19-21] Salts need not necessarily be added, they can be present as the neutralization products of the excess sulphonating agent.

More recently monoethanolamides, especially those derived by reacting a fatty acid methyl ester with monoethanolamine, are further condensed with ethylene oxide (see p 29). Depending on the type of fatty acid methyl ester and the number of molecules of ethylene oxide, non-ionic surfactants with excellent detergency power may be obtained.

Table 2.1 shows the change in viscosity of a 12 per cent neutralized dodecyl benzene sulphonate solution with the addition of 2 per cent alkylolamide and the foam height of a 0·1 per cent active solution in hard water of 300 ppm at 45°C.

TABLE 2.1

Effect of the Addition of Alkylolamides to Dodecyl Benzene Sulphonate Solution

Alkylolamide	Viscosity at 25°C (Pa s)	Foam height	
		Immediate	After 5 minutes
None	0·075	16	15
Coconut monoethanolamide	0·305	18	17
Coconut monoisopropanolamide	0·305	16·5	16
Coconut diglycolamide	0·960	18·5	17
Coconut diethanolamide	0·650	17·5	16
Oleic diethanolamide	0·840	17	17
Coconut di-isopropanolamide	0·360	17	16

NON-IONIC DETERGENTS

TABLE 2.2

Effect of the Addition of Ethanolamide on the Foaming Power of Various Detergents

Type of detergent	Concentration of detergent %	Detergent alone		Detergent with ethanolamide*	
		Foam height at 45°C in hard water			
		Immediate cm	After 5 minutes cm	Immediate cm	After 5 minutes cm
Lauryl alcohol triethanolamine salt	0·02	3·8	1·5	11	9
Lauryl alcohol monoethanolamine salt	0·02	3·0	1·2	10·5	10
Lauryl ether sulphate	0·02	16·0	15·0	15·0	14·5
Lauryl alcohol sodium salt	0·02	5·5	5·0	13·5	12·5
C_{10-14} alcohol sulphate sodium salt	0·02	2·0	1·0	4·0	3·5
Octyl phenol, 10 mols ethylene oxide	0·12	14·0	9·0	17·0	15·5
Nonyl phenol, 10 mols ethylene oxide	0·12	13·5	12·5	15·5	14·5
Octyl phenol, 14 mols ethylene oxide	0·12	15·5	14·5	15·5	15·0
Nonyl phenol, 14 mols ethylene oxide	0·12	16·0	14·5	14·5	14·0
Low foam non-ionic†	0·12	1·5	1·5	2·0	2·0

* The ethanolamide was added at the rate of 20 per cent of the active matter.
† See p 228.

Table 2.2 indicates the increase in foaming power of various detergents at low concentrations when coconut diethanolamide is added. It will be noted that the effect is very marked for alcohol sulphates, the ether sulphates are hardly affected, and the effect on non-ionic detergents varies.

9. Fatty Amine Oxides

A further type of non-ionic detergent is the group known as fatty amine oxides. These again offer many different possibilities, but basically they are made by treating tertiary amines with hydrogen peroxide. For our purposes, those amines which contain at least one long-chain group, usually from a fat source, are detergents. Two typical reactions are shown:

$$C_{12}H_{25}(CH_3)_2N + H_2O_2 \rightarrow C_{12}H_{25}-\underset{\underset{CH_3}{|}}{\overset{\overset{CH_3}{|}}{N}}\rightarrow O + H_2O$$

$$\underset{\underset{C_{12}H_{25}}{|}}{\underset{N}{\bigcirc}}^{O} + H_2O_2 \rightarrow \underset{\underset{C_{12}H_{25}}{|}}{\underset{N\rightarrow O}{\bigcirc}}^{O} + H_2O$$

Again these two representative types include the dodecyl grouping.

These amine oxides are in themselves good detergents,[22] but their main use is as foam boosters, viscosity increasers, and skin-protecting agents in liquid detergents. They are sticky materials and are not suggested for incorporation into powdered materials.

Their foam-boosting ability has been compared with lauric diethanolamide by Jungerman and Ginn[23] in Table 2.3. The method was to wash soiled plates in water of 125 ppm hardness at 46–49°C till the foam disappeared, with a 0·0125 per cent solution of the following formulation:

Sodium linear alkyl benzene sulphonate	18
Sodium lauryl ether sulphate	12
Foam stabilizer	x
Water	to 100

From the table, it appears that the fatty amine oxides are as effective as the alkylolamides as foam boosters, but in smaller concentrations.

TABLE 2.3

Comparison of Foam-Boosting Ability between Amine Oxides and Lauric Diethanolamide

Foam stabilizer	x	Plates washed to foam end-point
Dimethyl cocoamide oxide	0·48	26
Dimethyl hydrogenated tallowamide oxide	0·48	26
Dimethyl hexadecylamine oxide	0·48	27
Bis (2-hydroxyethyl) stearylamine oxide	0·60	20
Lauric diethanolamide	1·5	25

Amphoterics and Zwitterionics

This class of material contains in the same molecule both an anionic and a cationic group. The amphoterics are usually manufactured so that the anionic (carboxylate or sulphate/sulphonate) group is neutralized by Na and the cationic (quaternary ammonium) has associated with it, depending on the pH, an hydroxyl or chloride anion.

Zwitterionics also contain the same groups but the positive electric charge is neutralized by the negative charge on the same or an adjacent molecule. As will be explained below amphoterics can become zwitterionics under certain conditions of pH.

The chemistry of the syntheses is smothered in trade secrets. One route to the manufacture of amphoterics is the condensation of a fatty acid with a substituted ethylene diamine:

$$R-COOH + H_2N-CH_2-CH_2-NH-CH_2CH_2OH \longrightarrow$$
fatty acid $\qquad$ aminoethylethanolamine

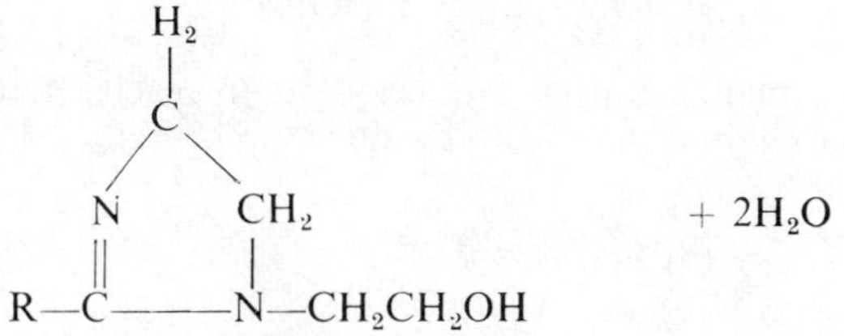

$$\underset{\text{imidazoline intermediate}}{
\begin{array}{c}
H_2 \\
| \\
C \\
\diagup \quad \diagdown \\
N \qquad CH_2 \\
\| \qquad | \\
R-C \text{———} N-CH_2CH_2OH
\end{array}
} \qquad + 2H_2O$$

Cyclozation has taken place to form the imidazoline. This is not the complete picture as some esterification can occur. This imidazoline is now both quaternarized and given an anionic group. The common reactant for this is sodium chloracetate in strongly alkaline solution to produce

either

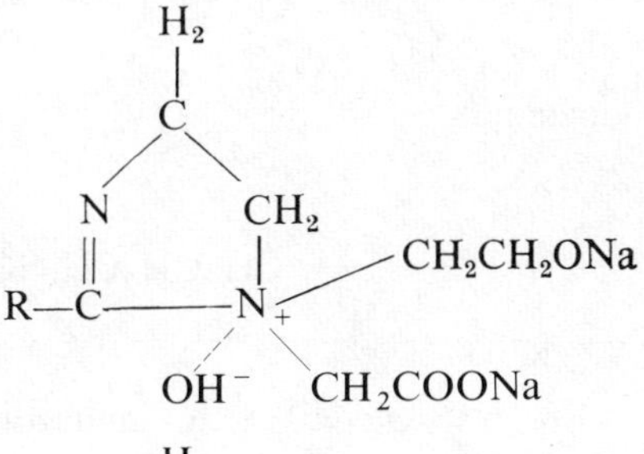

$$\begin{array}{c}
H_2 \\
| \\
C \\
\diagup \quad \diagdown \\
N \qquad CH_2 \qquad CH_2CH_2ONa \\
\| \qquad | \diagup \\
R-C \text{———} N_+ \\
\diagdown \\
OH^- \qquad CH_2COONa
\end{array}$$

or

$$\begin{array}{c}
H_2 \\
| \\
C \\
\diagup \quad \diagdown \\
N \qquad CH_2 \qquad CH_2-CH_2OCH_2COONa \\
\| \qquad | \diagup \\
R-C \text{———} N_+ \\
\diagdown \\
OH^- \qquad CH_2COONa
\end{array}$$

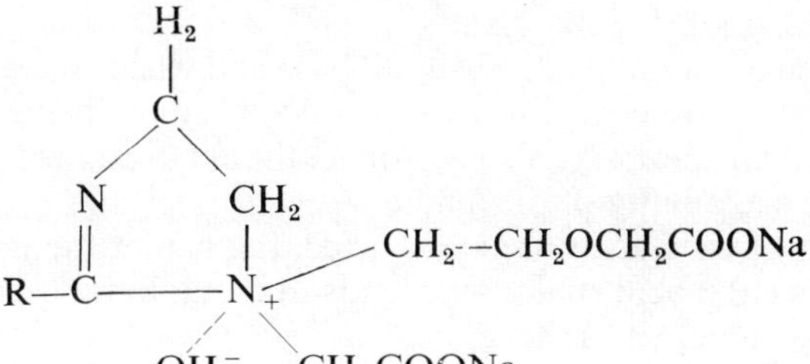

It will be noted that in the first example the caustic soda reacts with the —OH group to produce an alcoholate (in addition to the quaternarization) and in the second example further chloracetate has reacted with this group to give an ether linkage and a dicarboxylate. In both instances the quaternary N will be associated with an hydroxyl ion unless the pH is lowered.

Variations of the above are possible, mainly in the choice of the fatty acid, but an interesting product on the market has a fatty alcohol sulphate anion to neutralize the positive charge on the quaternary N:

$$C_9H_{19}-\underset{C_{12}H_{25}OSO_2^-}{\overset{\overset{\displaystyle CH_2}{\underset{\displaystyle \|}{N}}\diagdown\overset{\displaystyle CH_2}{|}}{N^+}}-CH_2CH_2OH$$
$$\phantom{C_9H_{19}-N^+-}CH_2COONa$$

One method of manufacturing zwitterions is to react a tertiary fatty amine with either chloracetic acid to produce the betaine

$$R-\underset{R''}{\overset{R'}{\underset{|}{\overset{|}{N^+}}}}-CH_2COO^-$$

or with propane sultone

$$\begin{array}{c} CH_2\!-\!-\!CH_2 \\ || \\ CH_2O \\ \diagdown\diagup \\ SO_2 \end{array}$$

to form the sulphobetaine:

$$R-\underset{R''}{\overset{R'}{\underset{|}{\overset{|}{N^+}}}}-CH_2CH_2CH_2SO_3^-$$

Sultones can also be reacted with imidazolines to form the corresponding sulphonate.

Ammonia is a relatively weak base and when converted to an amine the basicity is weakened further, a primary amine being however a stronger base than a tertiary. Also the longer the chain on the N nucleus, the weaker the basic quality of the amine.

Carboxylic acids are weak acids, the higher the molecular weight, the weaker the acid. Sulphonic acids and sulphuric esters are strong acids, comparable with mineral acids.

AMPHOTERICS AND ZWITTERIONICS

If the pH of the amphoteric is altered, either the anionic or cationic portion is partially neutralized. When the residual anionic (or cationic) radical is exactly balanced in its acidity (basicity) by the cationic (anionic) portion, the iso-electric point is reached and the resultant becomes a zwitterion.

To summarize, in acid solution amphoterics behave as cationics, in alkaline solution as anionics and at the iso-electric point the behaviour is between the two extremes. If one of the functions of detergency, wetting, foaming, surface tension or detergency itself is plotted against pH, all the curves will have the shape:

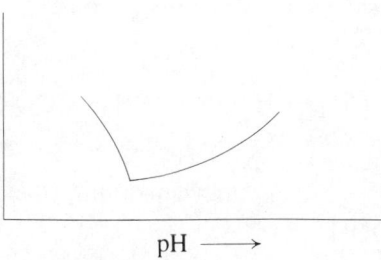

pH ⟶

The dip in the curve indicates the iso-electric point which is rarely in the neighbourhood of 7 and will vary depending on the various components of the molecule. At the iso-electric point the solubility is also at its lowest but because of extraneous salts etc, invariably present, solubility cannot be used as an indicator.

It was from work done on the iso-electric point that the development of compatibility of anionic and cationic detergents came about.

Amphoterics are used in cosmetics, textile and metal cleaners and as fabric softeners. They have excellent compatibility with inorganic salts, acids and bases. Advantage is taken of the zwitterion/pH effect to promote a desired characteristic. For example, at the iso-electric point they are almost completely non-irritating to the skin and eyes, indicating the use in baby shampoos and skin cleaners.[24]

We list below some amphoterics which are available on the world markets:

$$C_{11}H_{23}-\underset{\underset{OH^-}{\overset{\overset{CH_2}{\underset{}{N}}}{\underset{\|}{C}}}}{}-\underset{CH_2COONa}{\overset{\overset{CH_2}{|}}{N^+}}-CH_2CH_2ONa \qquad \text{Miranol CM}$$

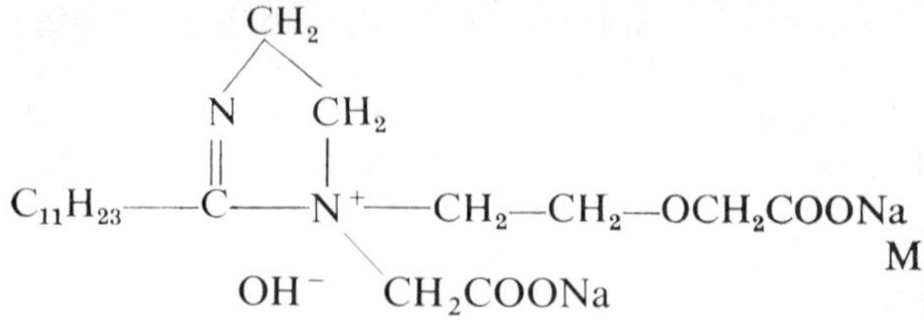

PRINCIPAL GROUPS OF SYNTHETIC DETERGENTS

$$C_{11}H_{23}-\underset{\underset{OH^-}{\|}}{C}-\underset{\underset{CH_2COONa}{|}}{\overset{\overset{CH_2}{\diagup\diagdown}}{N^+}}{}\overset{CH_2}{}-CH_2-CH_2-OCH_2COONa$$

Miranol C2M

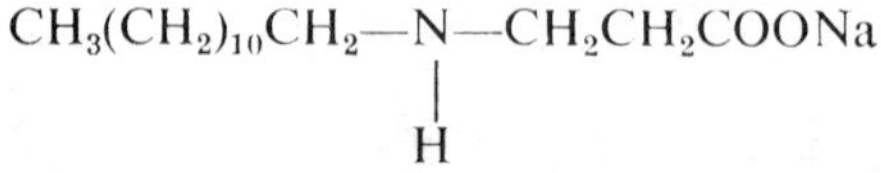

$$CH_3(CH_2)_{10}CH_2-\underset{\underset{H}{|}}{N}-CH_2CH_2COONa \qquad \text{Deriphat}$$

Derivatives of the basic betaine structure[25] can be:

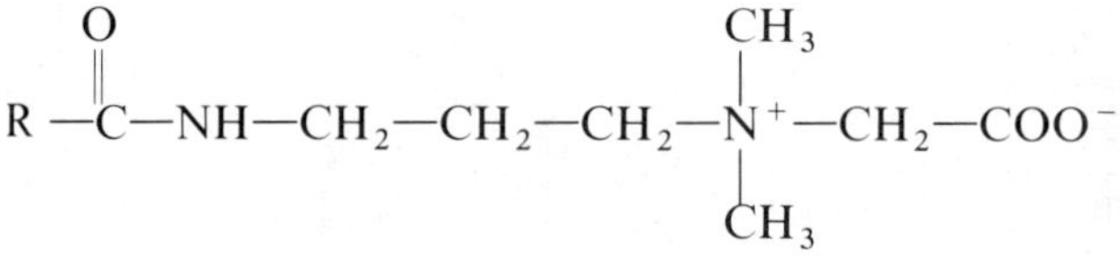

$$R-\overset{\overset{O}{\|}}{C}-NH-CH_2-CH_2-CH_2-\underset{\underset{CH_3}{|}}{\overset{\overset{CH_3}{|}}{N^+}}-CH_2-COO^-$$

Alkylamidopropylbetaine

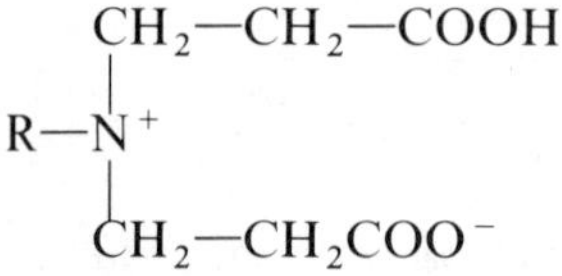

$$R-\underset{\underset{CH_2-CH_2COO^-}{|}}{\overset{\overset{CH_2-CH_2-COOH}{|}}{N^+}}$$

Alkylimidodipropionate

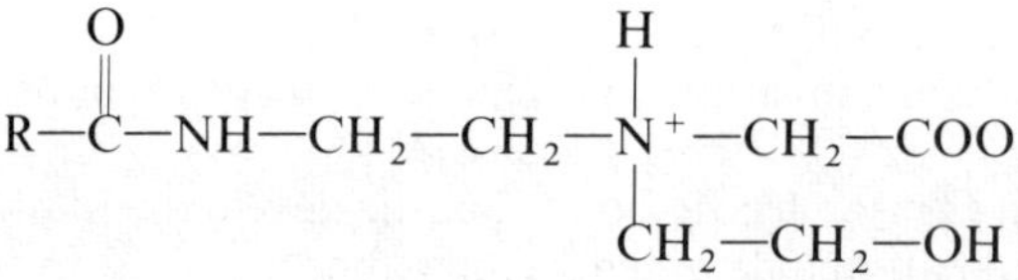

$$R-\overset{\overset{O}{\|}}{C}-NH-CH_2-CH_2-\underset{\underset{CH_2-CH_2-OH}{|}}{\overset{\overset{H}{|}}{N^+}}-CH_2-COO^-$$

Hydroxyethyl-alkylamidoethylglycinate

The chain length of the R group being of course dependent on the fatty amine or acid used in the synthesis.

Linfield and his associates[26] state that optimum detergency is obtained with only one long-chain component and at a pH of 7. At a pH of 9–10 there is a considerable drop in detergency. In acid solutions these amphoteric compounds are mainly cationic and in alkaline solutions they behave as anionic detergents. Depending on the type of 'cationic' radical being used they may be incompatible with anionic detergents in acid solution but they are all compatible with cationic and non-ionic detergents.

Some of the amphoterics give very excellent foam, which makes them suitable for foam cleaners which are to work at a neutral pH, such as hair shampoos and rug shampoos.

A very useful property of the amphoterics is that they are stable in highly acid solutions, and have thus found use in acid cleaners based on hydrofluoric acid.

BIODEGRADABILITY

With legislation or agreements having appeared or appearing in many countries, it is important for the chemist to know whether a material he is using or intends using is biodegradable. A test for biodegradability is usually beyond the means of even a well-equipped laboratory as found in most detergent factories, and reliance must be placed on accepted principles or manufacturers' advice. The degree of biodegradability is also to a large extent dependent on the method of test.

As a general rule we can, however, state that it has been found that straight-chain materials are biodegradable. Hence soaps, linear alkyl benzene sulphonates, olefin sulphonates, alkane sulphonates, fatty alcohol and ALFOL sulphates, ether sulphates, sucrose esters, alcohol phosphates and fatty alkanolamides are all sufficiently biodegradable for all standards laid down by legislation or agreements. The position of non-ionic detergents of the ethylene oxide condensate type is obscure. Various claims for the biodegradability of alkyl phenol ethylene oxide condensates have been made but not substantiated. It appears that fatty alcohol and ZIEGLER ethoxylates are biodegradable but this falls somewhat with increasing amounts of ethylene oxide. Some OXO alcohol sulphates and ethoxylates appear to be biodegradable but this again is dependent on the degree of branching.

Most manufacturers of either the raw or processed material supply adequate information on the degree of biodegradability of their products, but these figures must be read in conjunction with the method of test.

Results of practical biodegradation tests, undertaken by the German firms Sunlicht and Henkel in conjunction with the Hamburg authorities, are illuminating.[27] Before the introduction of biodegradable detergents (1963–64), degradation at the Hamburg sewage plant (using activated sludge) was between 15·4 and 22·2 per cent. In the summer of 1965 the degradation was between 39·2 and 49·6 per cent; and in the summer of 1967 it reached 71·3–80·3 per cent. Before the introduction of biodegradable detergents the water discharged from the plant contained 10–11 mg/litre detergent and in 1967 2–3 mg/litre.

References

1. Sisley, J. P., *Encyclopaedia of Surface-Active Agents*, Chemical Publishing Co, Inc, New York (1964).
2. Schwartz, A. M. and Perry, J. W., *Surface-Active Agents*, Vol I, p 116, Interscience Publishers, New York (1949).
3. Rubinfield, J. and Cross, H. D., *Soap*, **43**, 3, 41 (1967).
4. Holtzman, S. and Milwidsky, B. M., *Soap*, **41**, 10, 64 (1965).
5. Haupt, D. E., Moffet, J. G. and von Hennig, D. H., *Soap*, 42 (Feb 1975).
6. *US Patent* 3,896,057, Ethyl Corporation.
7. Richter, J., *Proceedings of International Symposium on Detergency*, Bled, Yugoslavia (1969).

8. Davidsohn, A., Moretti, G. and Adami, I., Paper read at International Symposium on Natural Base Cleaning Agents, Marseille (April 1985).
9. Boy, A., Brárd, R. and Passedroit, H., *Chem. Eng.*, 84–5 (13 Oct 1975).
10. *US Patent* 3,888,917, Marathon Oil.
11. *US Patent* 4,171,291, Lever Bros.
12. Kharasch et al., *J. Org. Chem.*, **3,** 175 (1938).
13. Hutchinson, J., Talk presented on the occasion of the H. P. Kaufman Prize at the DGF meeting, Berlin (Oct 1973).
14. *US Patent* 2,870,220.
15. Rosen, M. J., Current Developments in Surfactants, Presented at CSMA Seminar (June 1985). Cahn, A., New Products on the Market, Presented at CSMA Seminar (June 1985).
16. Kravetz, L., The Changing Role of Surfactants in Liquid Detergents, Presented at CSMA Seminar (June 1985).
17. *Chem Eng News*, 56 (12 Feb 1968).
18. *US Patent* 2,089,212, W. Kritchevsky (10 Aug 1937).
19. Hunting, A. L. L., *Cosmetics & Toiletries*, **97,** 3, 53 (1982).
20. Milwidsky, B. and Holtzman, S., VIth Congress of Surface Active Substances, Zurich (Sept 1972).
21. Donaldson, B. R. and Messenger, E. T., *Int. Cosm. Science*, **1,** 4 (1979).
22. *CISP* 2, 169,976.
23. Jungerman, E. and Ginn, M. E., *Soap*, **40**, 9, 61 (1964).
24 Bass, D., *Mfg Chemist* (Aug 1970), p 30.
25. Ploog, U., *Seifen-Oele-Fette-Wachse*, **108**, 373–376 (1982).
26 Linfield, E. and Ginn, M. E., *Soap*, **40,** 9, 61 (1964).
27. Spohn, H. and Fischer, K., *Tenside*, **4,** 241 (1967).

3. Inorganic Components of Detergents, Builders and Other Additives

If ready-for-use detergents were to be prepared only from the items mentioned in Chapter 2, they would work quite satisfactorily for certain operations, but they would be so concentrated that the consumer would have difficulty in measuring out the small quantities involved. For this reason the active material is diluted in such a way that it can easily be measured for use. In the case of liquids or pastes this diluent can be water, and in the case of powders it is generally sodium sulphate.

It must be emphasized that these two inert materials are not added merely to cheapen the product. They are added to make the end product easily measurable, and to convert the detergent into a form which can be handled conveniently. For example, sodium dodecyl benzene sulphonate in a highly concentrated form of 70 per cent or more is a very stiff paste. If this were to be marketed as a dishwashing product, only a few drops would be needed in a sink of water, and the paste would take a long time to dissolve. But if diluted to 10 or 12 per cent concentration, a tablespoon of solution would be required. This is much more manageable, and it disperses instantly. It is also more attractive in appearance. The same dilution effect can be had by adding sodium sulphate and then spray-drying to give a dry, free-flowing powder of attractive appearance and quick solubility.

Historically, sodium sulphate appeared as the inert filler in detergent powders because the original methods of sulphonation were such that large amounts of free sulphuric acid were present in the sulphonate prior to neutralization. This acid was naturally converted to sodium sulphate in the neutralization process, and when the material was to be dried to flakes in a drum-drier, or to beads in a spray-drier, extra sodium sulphate was added to the slurry to give the finished powder its free-flowing properties.

Sodium sulphate occurs naturally in many parts of the world, often mixed with sodium or magnesium chloride. It is also made as a by-product in the manufacture of hydrochloric acid from salt:

$$2NaCl + H_2SO_4 \rightarrow 2HCl + Na_2SO_4$$

and as a by-product wherever excess sulphuric acid needs to be neutralized with an alkali, or excess sodium hydroxide or sodium carbonate must be neutralized with sulphuric acid. Hitherto sodium sulphate had been a comparatively coarse powder with a bulk density of 1200 g/litre. A spray-dried type of sodium sulphate with a bulk density of 650 g/litre has now come on to the market in Europe. This sodium sulphate is preferred by manufacturers of dry-mixed powders for obvious reasons and in the spray-drying slurry process this has the added advantage of helping to produce a finer slurry. This makes it relatively cheap and a completely inert material, but with fatty alcohol sulphates and dodecyl benzene sulphonates it tends to increase the foaming properties slightly. It is also said to have a negative effect on detergency; this, however, can be counteracted by the addition of CMC.

In addition to water and sodium sulphate, all detergents other than those meant for very light-duty work require 'builders' to enhance the effect of the detergent raw material. By thus improving the overall detersive qualities, it is possible to use less of the more expensive active material. These 'builders' may be inorganic materials (generally alkaline) and organic materials, which are used in relatively small amounts compared to the inorganic materials, but their function in these small quantities is no less important.

The inorganic constituents fall into four main categories, each performing a definite function in the cleaning process. Much has been written on the physico-chemical aspects of the work of the builders, and we shall therefore merely confine ourselves to a statement of the properties of the builders which fall into the following groups:

1. phosphates;
2. silicates;
3. carbonates;
4. oxygen-releasing materials;
5. sundry builders.

1. Phosphates

The phosphates are again subdivided into two classes—orthophosphates and condensed or complex phosphates.

The orthophosphates used in the detergent industry are trisodium phosphate, Na_3PO_4, available as the anhydrous salt, or the crystalline $Na_3PO_4 \cdot 12H_2O$, and disodium phosphate, available both in the anhydrous form, Na_2HPO_4 and the crystalline $Na_2HPO_4 \cdot 12H_2O$.

Trisodium phosphate was in the past used as a soap-builder, but it is seldom used in detergent formulations nowadays other than in speciality products where the high alkalinity it produces is required. Nevertheless, it has the property of softening water by precipitating the metallic ions present in the water as a gelatinous precipitate; it also helps in dispersing the soil, and can dissolve fatty acids by saponification.

Disodium phosphate is even more rarely used in detergent mixtures and only when the lower alkalinity which it produces is desired.

Trisodium phosphate has the property of forming co-crystals and advantage is taken of this fact in the production of the material called commercially 'chlorinated trisodium phosphate' which has a formula approximating to $(Na_3PO_4.12H_2O)_5NaOCl$. This is used in admixture with cleaning materials which require the bleaching and sterilizing effect of sodium hypochlorite together with the high alkalinity, water-softening and dirt-suspending properties of trisodium phosphate.

Of great importance to the detergent industry, however, are the so-called condensed phosphates. If phosphates are considered to be a chemical combination of sodium oxide Na_2O, and phosphorus pentoxide P_2O_5, trisodium phosphate can be written:

$$1\tfrac{1}{2}Na_2O : \tfrac{1}{2}P_2O_5 \text{ or } (3Na_2O : 1P_2O_5)_{\tfrac{1}{2}}$$

The condensed phosphates are materials which have a higher proportion of P_2O_5 in the molecule and conversely, a lower proportion of Na_2O, giving them a lower alkalinity. The condensed or complex phosphates which are in use or have been used for cleaning materials are:

Tetrasodium pyrophosphate	$Na_4P_2O_7$ or $2Na_2O : 1P_2O_5$
Sodium tripolyphosphate	$Na_5P_3O_{10}$ or $2\tfrac{1}{2}Na_2O : 1\tfrac{1}{2}P_2O_5$
Sodium tetraphosphate	$Na_6P_4O_{13}$ or $3Na_2O : 2P_2O_5$
Sodium hexametaphosphate	$(NaPO_3)_6$ or $3Na_2O : 3P_2O_5$

Both sodium tetraphosphate and sodium hexametaphosphate are hygroscopic and are unsuitable for formulation into dry powders. They are now used only in very special circumstances.

The above phosphates are mentioned for historical reasons only to explain phosphate condensation. Sodium tetraphosphate has virtually disappeared from the world market, sodium hexametaphosphate is both hygroscopic and subject to reversion in solution. Tetrasodium pyrophosphate is now used for rather specialized purposes only but its potassium analogue is used in liquids.

These phosphates are produced by molecular dehydration, at an elevated temperature, of mixtures of mono- and di-sodium phosphates, according to the equations:

$2Na_2HPO_4$	$\rightarrow Na_4P_2O_7 + H_2O$
	Tetrasodium pyrophosphate
$2Na_2HPO_4 + NaH_2PO_4$	$\rightarrow Na_5P_3O_{10} + 2H_2O$
	Sodium tripolyphosphate
$2Na_2HPO_4 + 2NaH_2PO_4$	$\rightarrow Na_6P_4O_{13} + 3H_2O$
	Sodium tetraphosphate
$6NaH_2PO_4$	$\rightarrow (NaPO_3)_6 + 6H_2O$
	Sodium hexametaphosphate

The reactions are more complex than shown and unless the ingredients are intimately mixed and conditions carefully controlled, small amounts of the other phosphates will always be produced. More will be said about this in the manufacture of spray-dried powders (Chapter 6, p 213).

The four phosphates are listed above in descending order of alkalinity.

They all have the power of softening water by sequestering the polyvalent metal ions which cause the hardness. These metallic ions react by double decomposition with two or more of the sodium ions from the phosphate molecule and the metallic ions enter into the anion of the molecule. This can be written ionically:

$$Na_2{}^{2+}(Na_2P_2O_7)^{2-} + Mg^{2+}Cl_2{}^- \rightarrow Na_2{}^{2+}(MgP_2O_7)^{2-} + 2NaCl$$

Thus in this instance the magnesium ion is inactivated or sequestered, and is not available to enter into further ionic reactions. Anions of this type do not precipitate as salts from solution, which means that the water-softening effect of condensed phosphates does not result in the formation of any precipitate.

It has been found that sodium hexametaphosphate has the greatest power to sequester calcium ions and that tetrasodium pyrophosphate is best for magnesium ions. The power of sodium tripolyphosphate lies between the two in both instances.

The actual amounts of phosphate required to prevent precipitation of unit weight of metallic soap are shown in Table 3.1.

TABLE 3.1

Amounts of Phosphates Necessary to Prevent Precipitation of Unit Weight of Metallic Soap

	Mg soap	Ca soap
Sodium hexametaphosphate	16	16
Sodium tripolyphosphate	14	30
Tetrasodium pyrophosphate	11	130

A comparison of the water-softening properties of phosphates with organic chelates is given at the end of the section on organic chelates.

In addition to the water-softening power of the phosphates, they have the power of redissolving insoluble salts of these metals. Thus, if a piece of cloth has embedded in its fibres an insoluble calcium soap from a previous washing or treatment, or soil consisting of a metallic soap, the phosphate will dissolve this soap and also sequester the calcium ion. As sodium ions have been released from the phosphate molecule, they will now combine with the fatty-acid anion from the metallic soap, and soap is regenerated. This regenerated soap is now available for work and in this instance the presence of the condensed phosphate has enhanced the detergent power of the solution.

All the condensed polyphosphates, but particularly sodium tripolyphosphate in themselves synergistically aid the detergent. The term 'synergism' has already been used in this volume and we shall now explain it.

If standard soiled clothes are washed separately in the following solutions:

(i) 1 g/litre dodecyl benzene sulphonate
(ii) 1 g/litre sodium tripolyphosphate
(iii) 0·5 g/litre dodecyl benzene sulphonate and 0·5 g/litre sodium tripolyphosphate,

it will be found that the piece washed in half the quantities of both is considerably cleaner than the pieces washed in one or either of the materials separately.

This illustrates synergism: in this instance sodium tripolyphosphate is present in approximately the same concentration as the actual detergent, but where the sum total of the two is the same or less than the individual materials, it will considerably enhance the performance of the detergent, compared with the use of detergent on its own. This can be shown graphically, as in Fig. 3.1.

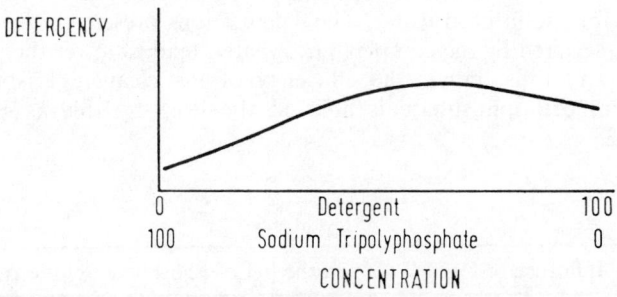

Fig 3.1   Synergism of sodium tripolyphosphate with detergents

Finally, the condensed phosphates are able to deflocculate and keep in suspension insoluble materials like clay, and to emulsify oily materials. Related to this property of deflocculation, that is, the breaking up of large masses into small ones, is the property of peptization, that is, the ability to keep finely divided solids in suspension and to prevent their coagulation. All the complex phosphates have this property. Deflocculation, emulsification of oily materials (which is a deflocculation process) and peptization all add to the very complicated washing process.

The complex phosphates, however, suffer from one defect. In solution they tend to revert to orthophosphates, a reaction which is the reverse of the one by which they were manufactured. This reversion is quite fast for hexametaphosphates, and in the cold is very slow for pyrophosphates. Polyphosphates are again between the two extremes. For this reason heavy-duty liquid detergents are not usually made with polyphosphates, but the rate of reversion of pyrophosphates is such that they are suitable for incorporation into these liquid detergents. For reasons of solubility, the pyrophosphate used is mostly tetrapotassium pyrophosphate, $K_4P_2O_7$.

Tetrapotassium pyrophosphate has a solubility at 20°C in water of 60 per cent while, under the same conditions, tetrasodium pyrophosphate dissolves to the extent of 5 per cent. It is a very hygroscopic material, so much so, that more often than not it is sold as a concentrated solution rather than the salt, and for this reason is only suitable for liquid products. All the other properties of tetrapotassium pyrophosphate are identical with those of tetrasodium pyrophosphate.

Sodium tripolyphosphate is however reasonably stable if the pH of the solution is held between 8 and 10, and as its solubility in pure water is 15 per cent it is finding a place in the production of medium- or light-duty liquid detergents.

Potassium tripolyphosphate is now appearing on the market and it has greatly increased solubility.

Many of the new alkaline cleaners such as machine dishwashing compounds, both solid and liquid, and hard surface cleaners, usually liquid, are using a mixture of trisodium phosphate and sodium tripolyphosphate for powders or tetrapotassium pyrophosphate for liquids particularly for use in hard water. The calcium ions present in the water, in being sequestered by the complex phosphates, tend to lower the pH of the solution and thus impair the efficiency of the cleaner. Trisodium (or potassium) orthophosphate is affected the least by this as shown in Table 3.2.

TABLE 3.2

Influence of Hard Water on the pH of Phosphate Solutions

Phosphate	pH of 1% sol. in distilled water	pH of 1% sol. in hard water (400 ppm)
$K_4P_2O_7$	10·1	9·4
$Na_4P_2O_7$	10·3	9·5
$Na_5P_3O_{10}$	9·4	8·9
Na_3PO_4	11·9	11·7

Many specifications call for a minimum P_2O_5 content without specifying the type of phosphate required. Table 3.3 lists the various phosphates encountered, their molecular weights and their P_2O_5 content.

It is worth recording that for liquid detergents the potassium orthophosphates can be produced quite easily in solution by neutralizing phosphoric acid to the required degree with caustic potash.

Sodium tripolyphosphate or rather detergent phosphates in general have been blamed for eutrophication* in certain inland lakes and ponds (see

* Eutrophic waters may be defined as those with a high concentration and growth rate of plant life. The opposite are called oligotrophic waters which have little or no primary productivity.

Table 3.3

Phosphates Used in Detergents

Type	Formula	Molecular weight	Theoretical P_2O_5 content
Disodium phosphate crystalline	$Na_2HPO_4 \cdot 12H_2O$	358	19·8
Disodium phosphate	Na_2HPO_4	142	50·0
Trisodium phosphate crystalline	$Na_3PO_4 \cdot 12H_2O$	380	18·7
Trisodium phosphate	Na_3PO_4	160	43·3
Dipotassium phosphate	K_2HPO_4	174	40·8
Tripotassium phosphate	K_3PO_4	212	33·5
Tetrasodium pyrophosphate crystalline	$Na_4P_2O_7 \cdot 10H_2O$	446	31·8
Tetrasodium pyrophosphate	$Na_4P_2O_7$	266	53·4
Tetrapotassium pyrophosphate	$K_4P_2O_7$	330	43·0*
Sodium tripolyphosphate crystalline	$Na_5P_3O_{10} \cdot 6H_2O$	476	44·7
Sodium tripolyphosphate	$Na_5P_3O_{10}$	368	57·9*
Potassium tripolyphosphate	$K_5P_3O_{10}$	448	47·5*
Sodium hexametaphosphate	$(NaPO_3)_6$	612	69·6*

* These figures are rarely achieved in practice.

p 6) and much work has been and is being done on this subject, both from the point of view of the ecology of the waters and from the detergent side. Ways and means are being sought to replace or reduce the phosphate content of detergents. A case in point is the work done at Lake Constance (Bodensee).[1] This is of particular interest as this lake has no outlet and there is a continual build-up of materials in the water. It is considered that the lake has been in existence for some 10,000 to 15,000 years. In the early 1930s the waters contained 0·2 mg of phosphorus per cubic metre. In 1953 the figure rose to 3–4 mg and today the amount is 37 mg. Similarly in the 1920s the unicellular life was determined at 440 thousand per litre whereas today the population is 40–60 million per litre, an increase of more than one-hundredfold. This increase in both phosphorus content and protozoa runs exactly concurrent with the great upsurge in the use of household detergents; but in fairness it must be pointed out that there was a simultaneous increase in the use of chemical fertilizers.

Both Numan[1] and Hudson & Marson[2] (together with many other authorities) are of the opinion that detergents contribute only some 25 per cent of the phosphate influx into these lakes, the balance coming from fertilizers and human excrement. Banning or lowering the phosphate content of detergents will not solve the problem to any great extent but will cause other side effects such as increases in costs and lowering of cleaning standards. It is suggested that a better approach to the problem would be to pre-treat water being discharged into the inland lakes, to remove phosphates. This, of course, would require huge capital outlay, but the long-term results are likely to be better.

One method of treating municipal effluents to precipitate phosphates[3] has been started in Sweden. The effluent has 16–18 g of iron per cubic metre (as ferrous salts) added. This reduces, by precipitation, the phosphorus content of the effluent to 0·3–0·5 ppm, a level sufficiently low to minimize eutrophication in lakes.

Attempts have been made by manufacturers to replace phosphates, in certain instances with more expensive ingredients and in others, with cheaper materials. There has been a singular lack of success with these alternatives. Werdelmann[4] has written the requirements for a phosphate substitute: 'it seems that the more intensely we search for a suitable substitute builder, the clearer it becomes how minimal are the chances of finding one which is nearly as good as phosphate. A suitable substitute must meet the requirements as shown in Table 3.4.'

2. Silicates

Soapmakers used to consider that sodium silicates were inferior ingredients, mainly added to soaps to 'fill' and cheapen them. In fact, silicates play a

TABLE 3.4

Requirements for Phosphate Substitutes

Technical requirements	Ecological requirements
Detergency	Biological degradability
Synergistic effect with surfactants	Non-toxic to humans, animals and water organisms
Good complexing and softening action	No harmful interactions with other substances (directly or by degradation products)
Emulsify fatty soil deposits	
Compatibility with perborate	No interaction with heavy metals
Break up soil particles	Safe for drinking water obtained from surface waters
Disperse and suspend soil particles	
Reduce soil deposition	No eutrophication factor:
Alkalinity and buffering action	Free from phosphorus, preferably free of phosphorus and nitrogen
No damage to fibres	
Non-corrosive	
Availability (economical and technological)	
Consumer acceptability	

very important role in the formulation of washing materials, and particularly those based on synthetic detergents.

Sodium silicate is made usually by the fusion of sand, containing a high proportion of silica, with soda ash in an electric furnace according to the equation:

$$Na_2CO_3 + SiO_2 = Na_2SiO_3 + CO_2$$

sodium metasilicate and carbon dioxide being produced.

The sodium metasilicate is the salt of (possibly hypothetical) silicic acid, which although a weaker acid than carbonic acid, displaces the carbon dioxide because the sodium carbonate decomposes to Na_2O and CO_2 at the furnace temperature.

This is not the whole story by any means. The ratio of silica sand to soda ash is varied to give a variety of alkalinities. It has become common to write the molecular formula of silicates according to the methof Berzelius, thus metasilicate is written: $Na_2O:SiO_2$, and the molecular ratio (sometimes called the modulus) is 1:1. Sodium metasilicate is considered to be a true chemical compound as the components are combined in molecular proportions. Many other ratios are available and some can be considered mixtures or solid solutions. The crystalline forms of sodium silicate available commercially are:

	Formula	Ratio
Sodium orthosilicate	Na_4SiO_4 or $2Na_2O:SiO_2$	2:1
Sodium sesquisilicate	$(Na_6Si_2O_7)$ or $1\frac{1}{2}Na_2O:SiO_2$	$1\frac{1}{2}$:1
Sodium metasilicate	Na_2SiO_3 or $Na_2O:SiO_2$	1:1

All these forms are available as the anhydrous or hydrated forms. It is questionable whether the ortho- and sesquisilicates are manufactured as such or whether they are physical blends of metasilicate and powdered caustic soda. For example, orthosilicate is now offered as $Na_4SiO_4 \cdot H_2O$ which indicates that it might be a physical mixture. However, true crystalline forms of orthosilicate, $Na_4SiO_4 \cdot 6H_2O$ and sesquisilicate $Na_6Si_2O \cdot 11H_2O$ are offered. These are manufactured by adding the requisite amount of caustic soda to a metasilicate solution and crystallizing the resultant mix. Metasilicate is also available as the pentahydrate, $5H_2O$, and used to be sold as the nonahydrate, $9H_2O$, but this has been withdrawn from the market as it tended to become liquid or pasty if stored at somewhat high temperatures.

Where the silica ratio is above 1 (the molecular weight of silica is 60 and that of sodium oxide is 62 so the molecular and weight ratios for sodium silicates are nearly identical), the sodium silicates are no longer crystalline and are termed colloidal. These are available in three forms:

1. *Soluble glass.* After being fused in the furnace the melt is cooled and broken up mechanically into small lumps or platelets. This form cannot be used as such, but must be converted into one of the two following forms.

2. *Water glass.* The lumps described above are dissolved by heat and pressure to form a concentrated solution. This solution is often, although not always, filtered to remove insolubles present in the original sand.
3. *Soluble powders.* The solution is then spray-dried to form hollow walled beads with 18–20 per cent retained water. One manufacturer adds sodium sulphate and retains 10–15 per cent water. Either of these two forms is readily soluble in cold water and can be used in dry mixing operations.

Soluble glass and soluble powders are normally made in the ratios 1:2 and 1:3·3, the former being called alkaline, the latter neutral glass. Other ratios from 1:1·6 to 1:1·38 are available as 'water glass' for special purposes. A common ratio used in detergent powders is 1:2·4 made by blending neutral and alkaline glasses or adding caustic soda to neutral glass.

Some detergent manufacturers use the so-called 'wet method' to produce a solution of up to 40 per cent disilicate starting from fine sand (glass grade) and caustic soda solution (30–50 per cent) under pressure 10 bar. The chemical reaction is as follows:

$$2SiO_2 + 2NaOH \rightarrow Na_2O \cdot 2SiO_2 + H_2O$$

Typical material balance is as follows:

To produce 1000 kg sodium disilicate (calc. as 100 per cent equal to 2500 kg 40 per cent solution):

sand (granulometry: at least 99 per cent 0·5 mm, at least 80 per cent 0·125 mm)	690 kg
NaOH (calc. as 100 per cent)	450 kg
process water (soft water)	1000 litres
steam	800 kg
electric power installed	30 kW

Investment for such units is much lower than for production units based on the fusion process with soda ash.

Potassium silicate, available commercially in colloidal ratios only, traditionally used for non-detergent purposes, welding rods, cathode ray tubes, etc, is being used more and more for specialized liquid detergents. It is available in weight ratios of 1:1·5, to 1:2·5, molecular ratios of 1:1·25 to 1:1·35.

The properties of the various silicates used in detergents are summarized in Table 3.5.

Sodium silicate was first added to laundry soap over a century ago, initially to cheapen the soap which was then known as 'filled' or 'silicated' soap, but the advantages the silicate imparted to the soap were soon perceived and they became an integral part of laundry soaps and powders. With the development of the synthetic detergents, silicates rapidly found their niche and are included in virtually all forms of cleaning materials

TABLE 3.5

Properties of Common Silicates Used in Detergents (on an anhydrous basis)

Silicate	pH 1% solution	Active alkalinity as Na_2O	Total alkalinity as Na_2O	Molecular ratio
Sodium orthosilicate	13·0	58·5	60·5	2:1
Sodium sesquisilicate	12·6	54·3	56·0	1½:1
Sodium metasilicate	12·5	49·0	51·5	1:1
Alkaline glass	11·3	29·5	34·0	1:2
Potassium silicate	11·2	22·2*	25·6*	1:2·5
Neutral glass	10·5	21·6	28·2	1:3·3
Potassium silicate	10·1	15·5*	20·3*	1:3·5
Sodium disilicate†	11·2	23·5	27·5	1:2
Sodium trisilicate†	10·4	16·5	21·5	1:3·3

* Calculated as the sodium salt.
† These products appear under various commercial names and include 15–20 per cent water or 10–15 per cent sodium sulphate and 10–15 per cent water of hydration. Special spray-dried sodium disilicate with a bulk density of only 80 g/litre and a water content of c. 18 per cent is on the market now. It is especially useful when producing detergent powder by spray-mixing, dry neutralization, and other non-spray-drying methods (p 202). It serves also to reduce bulk density of otherwise heavy powders.

whether based on synthetics or pure soap with the exception of toilet articles.

Both portions of the molecule contribute to the performance of enhancement of the cleaning process.[5] The functions can be summarized:

Contribution by the alkaline radical (Na_2O or K_2O)
1. Alkalinity.
2. Buffering capacity. The pH is sustained at 9·5 or higher until the silicate is exhausted. (This property is developed in conjunction with the SiO_2 radical.)
3. Saponification (neutralization) of acidic soils.
4. Emulsification of oils.

Contribution by the SiO_2 radical
1. Deflocculation of soils.
2. Anti-redeposition.
3. Water softening.
4. Corrosion inhibition, both on metals (ferrous and non-ferrous) and on the glaze on china or porcelain. The mechanism is assumed to be the formation of a monomolecular layer on the surface and, in the case of metals, the formation of an insoluble monomolecular layer of the metal silicate.
5. For spray-dried powders the silicate gives crispness to the bead.
6. For synthetic laundry bars the silicate plasticizes the mass giving a uniform appearance and easing the extrusion process.

The chemistry of the silicates is continually being investigated.

Weldes[6] has stated that when the silica ratio is higher than one the silicate ions polymerize to form polyelectrolytes with multiple charges, thus the term colloidal. This distinguishes the silicates from the common alkalis and an analogy can be taken from the phosphates in that the condensed phosphates perform in a vastly superior fashion to the simple orthophosphates.

In addition to being used as pure additives or builders, sodium ortho-, sesqui-, or metasilicates can be used as neutralization agents for alkylbenzene sulphonic acids and fatty acids. Only a portion of the Na_2O is used for the neutralization, the balance remains reacted with the SiO_2 to give a product with a higher silica ratio. Thus colloidal silicate can be introduced into a powder, paste or liquid without the intermediate energy intensive solution process or without the necessity for transporting water (if water glass is used).

When used in admixture with condensed phosphates, as builders in synthetic detergent mixtures, silicates provide a further synergistic action over and above the synergism between the phosphate and the detergent.[7]

In the manufacture of spray-dried powders, the silicate used is almost invariably of the colloidal type. The source (if the above mentioned neutralization process is not used), will be water glass, whether purchased as such or dissolved in the plant. Concentrations of these solutions are easily measured by hydrometer and the density/concentration relationship is shown in Table 3.6.

3. Zeolites

The second most abundant element in the earth's crust is silicon after oxygen and before aluminium.

Silicon is never found free (nor is aluminium). Silicon is invariably associated with oxygen (silica, quartz) and with one or more metallic oxides. Aluminium in kaolin, magnesium in talc are only two examples of the hundreds of metallo-silicates available naturally.

In the middle of the eighteenth century a Swedish mineralogist, Baron Cronstedt, found that on rapidly heating certain classes of rocks, they vibrated as the bound water evaporated. He coined the name 'zeolite' for this type of mineral from the Greek words meaning 'boiling stone'.

Zeolites, also known as molecular sieves, are now made synthetically and have for many decades been used as water softeners, as catalysts or constituents of catalysts and for gas purification. With the search for a phosphate replacement, their water-softening ability suggested their use.

Type A zeolites are hydrated sodium aluminium silicates of the generic formula $Na_2O \cdot Al_2O_3 \cdot xSiO_2 \cdot yH_2O$. The structure of the crystalline material is a three-dimensional lattice which, written in a simplified form in

TABLE 3.6

Concentration of Alkaline and Neutral Silicate Solutions

Specific gravity	Degrees Baumé	Degrees Twaddell	Alkaline silicate 1:2			Neutral silicate 1:3·3		
			% Na_2O	% SiO_2	% Sodium silicate	% Na_2O	% SiO_2	% Sodium silicate
1·014	2	2·8	0·5	1·0	1·5	0·4	1·3	1·7
1·029	4	5·8	1·05	2·05	3·1	0·77	2·53	3·3
1·045	6	9	1·55	3·15	4·7	1·15	3·75	4·9
1·060	8	12	2·1	4·2	6·3	1·55	5·15	6·7
1·075	10	15	2·6	5·3	7·9	1·95	6·45	8·4
1·091	12	18·2	3·15	6·35	9·5	2·4	7·8	10·2
1·108	14	21·6	3·7	7·4	11·1	2·8	9·2	12·0
1·125	16	25	4·25	8·55	12·8	3·2	10·6	13·8
1·142	18	28·4	4·8	9·6	14·4	3·65	12·05	15·7
1·162	20	32·4	5·35	10·65	16	4·05	13·35	17·4
1·180	22	36	5·9	11·8	17·7	4·45	14·75	19·2
1·200	24	40	6·4	12·9	19·3	4·9	16·2	21·1
1·220	26	44	7·0	14	21·0	5·3	17·6	22·9
1·241	28	48·2	7·6	15	22·6	5·75	18·95	24·7
1·263	30	52·6	8·15	16·35	24·5	6·15	20·30	26·5
1·285	32	57	8·75	17·45	26·2	6·6	21·7	28·3
1·308	34	61·6	9·3	18·7	28·0	7·0	23·1	30·1
1·332	36	66·4	9·9	19·9	29·8	7·4	24·5	31·9
1·357	38	71·4	10·6	21·1	31·7	7·85	25·85	33·7
1·383	40	76·6	11·2	22·4	33·6	8·25	27·25	35·5
1·410	42	82	11·8	23·7	35·5			
1·438	44	87·6	12·5	25·0	37·5			
1·468	46	93·6	13·2	26·4	39·6			
1·498	48	99·6	13·8	27·8	41·6			
1·530	50	106	14·6	29·1	43·7			
1·563	52	112·6	15·3	30·6	45·9			
1·592	54	119·4	16·0	32·1	48·1			
1·635	56	127	16·8	33·6	50·4			
1·672	58	134·2	17·6	35·1	52·7			
1·710	60	142	18·3	36·7	55			

one plane can be

```
                           —Al—
                            |
       —O      O—    —O     O      O—
         \    /        \    |     /
   —O—Al—O—Si—O—Al—O—Si—O—Al—O—
        /    \        /    |     \
       O—     —O     O—    O      —O
                            |
                           —Al—
```

The silicon atom being tetravalent, is electrically neutral but the trivalent aluminium, sharing four oxygen atoms, has a surplus negative charge associated with it. This requires a cation to make the molecular crystal electrically neutral. The lattice structure results in a pore in the centre of the three-dimensional crystal. The associated cation is found at the opening of this pore and it is this cation which gives the zeolite its peculiar properties.

Type 4A zeolites have a sodium ion partially blocking the pores to give an opening of about 4 ångström units. Type 3A zeolites have potassium as the cation and, as the K ion is larger, the opening is roughly 3 ångström units. Type 5A zeolites have calcium as the cation and as two pores are associated with only one cation, the effective opening is 5 ångström units.

These zeolites are made commercially by two routes, from aluminia, which is first converted to sodium aluminate:

$$Al_2O_3 \cdot 3H_2O + 2NaOH \rightarrow 2NaAl(OH)_4$$

by heating at moderate temperature. This sodium aluminate is then reacted with sodium silicate, again at moderate temperatures

$$12NaAl(OH)_4 + 4Na_2O:3SiO_2 + 7H_2O \rightarrow$$
$$N_{12}(AlO_2 \cdot SiO_2)_{12} \cdot 27H_2O + 8H_2O$$

The second method is to use good quality kaolin (china clay) which can be written $2Al_2Si_2O_5(OH)_4$. This is calcined at a temperature in excess of 500°C, becoming dehydrated to form metakaolin $Al_2Si_2O_7$. Metakaolin is then reacted at moderate temperature with water and caustic soda:

$$6Al_2Si_2O_7 + 12NaOH + 21H_2O \rightarrow Na_{12}(AlO_2 \cdot SiO_2)_{12}27H_2O$$

In both instances the final reaction is in an aqueous phase and the insoluble zeolite produced is filtered out, dried and ground.

The particle size has been optimized by one of the manufacturers to be in the range of 4–6 microns,[8] their arguments being that a smaller particle would adversely affect the bulk handling properties of the zeolite, a larger particle would be slower in reacting with the cations it is required to sequestrate and would also be entrapped on the cloth in the wash. A 5 micron particle has been proved to pass through the weave of all cloth.

A tendency has developed to use the zeolites in the form of a slurry, thus eliminating the drying stage in the production of the zeolite as described above, saving a considerable amount of energy. The next development will be for the detergent manufacturer to produce his own zeolite by one of the above methods, and to add the slurry formed to the detergent slurry production system. It is claimed that optimum activity of the zeolites is thus obtained.

Two of the sodium ions associated with the cage or lattice of the crystal are readily and rapidly exchanged for one calcium ion present in solution but the speed of reaction with magnesium ions is rather slower. (See p 95 for a comparison of sequestering properties of various chelates.) Experiments conducted by the PQ Corporation indicate that in 120 ppm hardness water with a 2:1 Ca:Mg ratio, 98 per cent of the calcium was removed from solution within the first minute, but the magnesium concentration is only halved in ten minutes.

No claim is made that zeolites are a complete replacement for STP as they can only perform the water-softening function of the polyphosphate and are really effective only on calcium.

Powders formulated with zeolites, when phosphate is completely disallowed require additional ingredients to complement the missing functions supplied by the STP and also an additional sequestering agent for magnesium. This latter condition is only really necessary if soap is present or if the alkalinity is high, because magnesium soaps, carbonate or hydroxide are all very insoluble in water and these are likely to precipitate unless strong chelating agents are present. In relatively neutral solutions, the magnesium salts of anionic syndets function quite well and are used for specialized purposes.

Where phosphates are allowed or limited, the zeolites can act as a partial replacement and here, at the present time, there is a cost advantage if zeolites replace 50 per cent of the STP, taking into account the fact that in sequestering power, a zeolite on a weight basis, 1·2 parts of zeolite hydrate is equivalent to 1 part of anhydrous STP.

An extra advantage claimed for the zeolites is that they readily and rapidly remove from solution obnoxious heavy metal ions such as manganese and iron which can be present in small quantities.

One of the zeolite manufacturers has developed a co-crystal of sodium silicate and zeolite containing 54 per cent anhydrous zeolite, 21 per cent sodium silicate, $Na_2O:2\cdot4SiO_2$, and 25 per cent moisture.

This material is in a granular form specially for formulation in processes other than spray-drying, allowing the incorporation of the correct ratio of silicate without using the more expensive di- or trisilicates. All the effects available from zeolites and silicate separately are available and in addition it has an adsorption capacity of 20 per cent of its own weight making it ideal as a component in dry mixed and agglomeration processes.

4. Carbonates

Four carbonates are available to the manufacturer:

(a) *Sodium Carbonate or Soda Ash*—Na_2CO_3: Soda ash provides high alkalinity and softens water by precipitation of calcium and magnesium carbonates, provided that the pH of the solution is over 9 and remains so after the precipitation has occurred. It is usually available in two physical forms, dense and light, depending on the conditions of manufacture.

Soda ash is mined in certain areas of the world, notably at Lake Magadi in Kenya and in California, but this natural soda is usually unsuitable for detergent manufacture, as it contains traces of organic matter and iron, giving its solutions a dirty colour, which can be transmitted to the articles to be washed. In addition, the soda ash from Lake Magadi contains 1 per cent of sodium fluoride which, being poisonous, is an undesirable ingredient of household cleaning materials. Recently processes for soda ash from natural soda have been improved to give detergent grade soda ash.

The bulk of soda ash of commerce is produced by the Solvay process which, in brief, uses as raw materials sodium chloride and carbon dioxide (obtained from burning limestone to lime).

The sodium chloride as a brine is saturated with ammonia and then pumped down a tower up which carbon dioxide is passed counter-currently. The carbon dioxide dissolves in the water, forming carbonic acid:

$$CO_2 + H_2O \rightarrow H_2CO_3$$

and because of the buffering effect of the ammonia present, this carbonic acid is converted to the bicarbonate ion. There are now present in the original brine solution: sodium ions, chloride ions, ammonium ions, and bicarbonate ions. Under these conditions sodium bicarbonate, which is only sparingly soluble in the solution, precipitates and is separated and roasted to sodium carbonate.

The equations for the various reactions are:

$$CaCO_3 \rightarrow CaO + CO_2$$
$$CO_2 + NaCl + NH_4OH \rightarrow NaHCO_3 + NH_4Cl$$
$$2NaHCO_3 \rightarrow Na_2CO_3 + CO_2 + H_2O$$

The carbon dioxide released from the conversion of the bicarbonate to carbonate is re-used in the absorption tower (in addition to the carbon dioxide obtained from the burning of the lime). The lime obtained from this burning is slaked and then added to the ammonium chloride solution obtained after the sodium bicarbonate has been separated. The ammonia is then regenerated:

$$2NH_4Cl + Ca(OH)_2 \rightarrow CaCl_2 + 2NH_4OH$$

and on heating this solution gives off ammonia gas which is used again. The full process, therefore, uses as raw materials only limestone (calcium carbonate) and salt (sodium chloride) and produces sodium carbonate and calcium chloride, ammonia being needed only for replacement of losses. Since both limestone and salt are cheap raw materials, the soda ash produced by the Solvay process is cheap. This soda ash is called, in commerce, light soda ash since its bulk density is 600–700 g/litre.

On this score it should be noted that anhydrous soda ash is hygroscopic. If exposed to the atmosphere it can absorb moisture slowly to form the monohydrate (and conversely any of the other crystalline forms deliquesce down to the monohydrate). Theoretically, therefore, soda ash on storage can increase in weight by some 18 per cent. This seldom happens, as the outer layer, after having become hydrated, forms a semi-impervious layer preventing further moisture penetration. It is, however, quite common to find in bagged soda ash an increase of 10 per cent over the 'when packed' weight. This should be taken into account when bagged soda ash is used in batch processing.

Where caustic soda is cheap, and where there is a surplus of carbon dioxide, soda ash can be manufactured by absorbing carbon dioxide in a caustic soda solution:

$$2NaOH + CO_2 \rightarrow Na_2CO_3 + H_2O$$

Besides being a cheap source of alkalinity in detergent powders, soda ash is also used in the preparation of dry products by processes other than by spray-drying, since particularly in the light form, it can absorb large quantities of liquid material on to its surface and still remain dry to the touch and free-flowing. In this instance, more often than not, it is used both as an absorbent and neutralizing agent, in addition to the use being made of it as an alkaline builder.

(b) Sodium Bicarbonate—$NaHCO_3$: Sodium bicarbonate is normally produced en route to soda ash but this grade is heavily contaminated with ammonia.

Bicarbonate is produced commercially by saturating a solution of soda ash with CO_2. The sodium bicarbonate being relatively insoluble in water precipitates out in a very pure form, so much so that it is usually of BP quality.

The material has no water-softening properties, but it is sometimes added to detergent compositions to lower the pH of the material. Also, if there is any danger of free caustic alkalinity being present, the addition of bicarbonate will neutralize this alkalinity:

$$NaOH + NaHCO_3 \rightarrow Na_2CO_3 + H_2O$$

Thus, by the addition of bicarbonate the caustic alkalinity has been destroyed and the pH of the solution of the material will have been brought down to that of soda ash. If excess of bicarbonate is added, some sodium sesquicarbonate will be formed, bringing the pH down further.

(c) Sodium Sesquicarbonate, or Modified Soda: Sodium sesquicarbonate has the formula $Na_2CO_3 \cdot NaHCO_3 \cdot 2H_2O$, which makes it half-way in properties between soda ash and sodium bicarbonate. Modified sodas are physical mixtures of the two, usually in molecular proportions and usually without the water of crystallization. Sodium sesquicarbonate and modified sodas, therefore, have water-softening properties, mild alkalinity, and have the effect of neutralizing any unwanted alkalinity. Because sesquicarbonate already has water of crystallization in its molecule, it cannot absorb water or liquids to any extent, but, if this property is required, modified sodas will serve.

(d) Potassium Carbonate: Potassium carbonate, sometimes called pearl ash or potash (from 'pot ashes') although rarely used in detergents, is often used in polish manufacture as a source of alkali. As potassium salts are in

general more soluble in water than the corresponding sodium salts, potassium carbonate is not manufactured by the Solvay process, because the potassium bicarbonate will not precipitate from the solution. It may be produced by the Leblanc process, which uses potassium chloride as a starting material. It is converted into potassium sulphate and hydrochloric acid by heating with sulphuric acid in a furnace:

$$2KCl + H_2SO_4 \rightarrow K_2SO_4 + 2HCl$$

The potassium sulphate thus produced is next roasted with limestone and coke:

$$K_2SO_4 + CaCO_3 + 2C \rightarrow K_2CO_3 + CaS + 2CO_2$$

The black solid mass (called black ash) is dispersed in water and the potassium carbonate dissolves. The surplus coke and the calcium sulphide are separated from the solution by filtration and the solution is then evaporated to dryness.

Another method is to react potassium chloride with magnesium carbonate in the presence of excess CO_2. A precipitate of magnesium potassium hydrogen carbonate forms according to the equation:

$$2KCl + 3MgCO_3 + 9H_2O + CO_2 \rightarrow 2[MgKH(CO_3)_2 . 4H_2O]$$

This precipitate is filtered off and treated with hot water. The bicarbonate grouping, which in general is unstable on heating, is destroyed giving:

$$2[MgKH(CO_3)_2 4H_2O] \rightarrow 2MgCO_3 + K_2CO_3 + CO_2 + 9H_2O$$

The magnesium carbonate is filtered off and again the potassium carbonate is obtained in solution.

A third method is the reaction of electrolytic caustic potash with carbon dioxide exactly as in the production of soda ash (see p 63).

Potassium carbonate has a molecular weight of 138·2 and specific gravity of 2·04. For industrial purposes it is marketed as 80–85 per cent or 95–96 per cent pure. It is strongly hygroscopic and, unlike soda, deliquesces by attracting moisture from the atmosphere. This is a property that will help the practical man not to confuse soda and potash.

5. Oxygen-releasing Compounds

In many parts of the world oxygen-releasing compounds are added to detergent powders. The chief material used is sodium perborate. For years there was a controversy as to the exact structural formula of sodium perborate, either $NaBO_2H_2O_2 . 3H_2O$ or $NaBO_3 . 4H_2O$. It has now been shown that the structure of what was called the tetrahydrate is actually

$$2Na^+ \begin{bmatrix} HO & O{-}{-}{-}O & OH \\ & \diagdown \diagup \diagdown \diagup & \\ & B \quad\quad B & \\ & \diagup \diagdown \diagup \diagdown & \\ HO & O{-}{-}{-}O & OH \end{bmatrix}^{2-} 6H_2O$$

thus containing a theoretical 10·4 per cent active oxygen, and the monohydrate is structurally

$$2Na^+ \begin{bmatrix} HO & O\!-\!O & OH \\ & \diagdown\!B\!\diagup & \diagdown\!B\!\diagup & \\ HO & O\!-\!O & OH \end{bmatrix}^{2-} H_2O$$

containing 16·03 per cent active oxygen.

Both when dissolved in water react

$$(NaBO_2H_2O_2)_2 \cdot xH_2O \rightarrow 2NaBO_2 + 2H_2O_2 + xH_2O$$

and they are not, as is sometimes thought, a mixture of borax with hydrogen peroxide. It is manufactured by reacting borax with caustic soda to form the metaborate:

$$Na_2B_4O_7 + 2NaOH \rightarrow 4NaBO_2 + H_2O$$

This solution is now electrolysed under careful conditions so that the anodic oxidation of the water takes place and the sodium perborate crystallizes from the solution.

In solution, the action of sodium perborate is similar to that of hydrogen peroxide: at elevated temperatures nascent oxygen is released which has a bleaching effect but does not affect animal, vegetable and synthetic fibres. Hence sodium perborate can be used as the universal laundry bleach and has, in effect, been used as such in European countries for the better part of forty years.

The action of peroxide was considered to be the release of nascent oxygen. Present ideas consider that peroxide dissociates

$$H_2O_2 \rightleftharpoons H^+ + OOH^-$$

and it is the perhydroxyl ion which is responsible for the bleaching. The reaction is reversible and will be shifted to the right if alkali is present to absorb the hydrogen ion, ie, if the pH is relatively high.

The perhydroxyl ion does not affect animal, vegetable or synthetic fibres and perborate can be used as the universal laundry bleach and has been used as such in European countries for the better part of this century.

However, sodium perborate also adds to the alkalinity and buffering properties of the detergent, and is therefore a detergent builder in its own right. Although the alkalinity of the salt itself is high (the pH of a 1 per cent solution is 10·2), it is not adversely affected in its bleaching properties if materials are added which bring the pH down. For example, if bicarbonate is added to give a lower pH specifically for the bleaching of alkali-sensitive fabrics, the bleaching effect will not be reduced.

Sodium perborate is a relatively stable material when dry-mixed with other dry ingredients. Traces of water and certain heavy metals will catalyse the decomposition of the perborate, but the addition of magnesium sulphate or silicate, or tetrasodium pyrophosphate, will prolong the storage life of these powders.

This stability is sometimes a defect, however, as the bleaching action only takes place at elevated temperatures. Niewenhuis[9] has indicated that the addition of $N'N'$tetra-acetyl ethylene diamine (TAED):

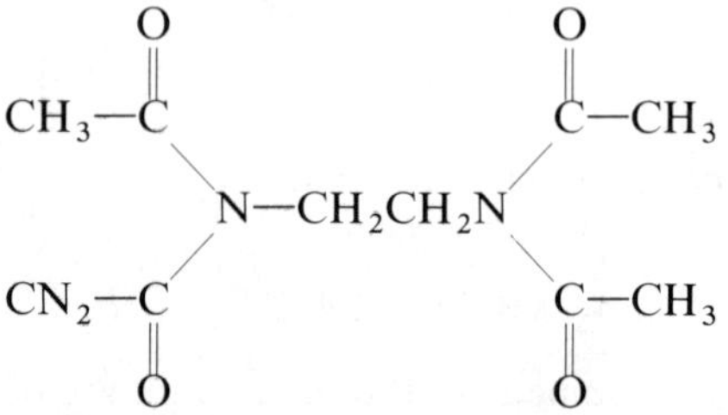

$$\begin{array}{ccc} CH_3-\overset{O}{\underset{\|}{C}} & & \overset{O}{\underset{\|}{C}}-CH_3 \\ & N-CH_2CH_2N & \\ CN_2-\overset{}{\underset{\|}{C}} & & \overset{}{\underset{\|}{C}}-CH_3 \\ O & & O \end{array}$$

can activate the perborate at a lower temperature. Other materials suggested for activating peroxy compounds are para-acetoxybenzene sulphonate,[10] tricetyl cyanurate and tetra acetyl glycol uril[4] (TAGU). TAED is now a common additive in Europe, although the others are also used.

All these materials cause the bleaching action to take place at a lower temperature. Figure 3.2 shows the reaction of TAGU.

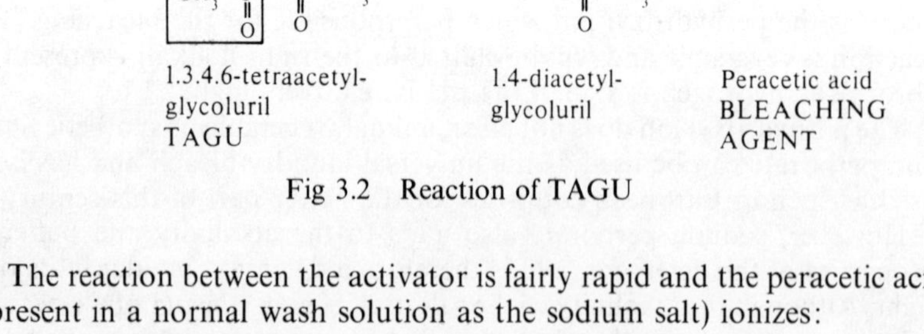

Sodium perborate
$+H_2O$

$2H-O-OH$ + [TAGU structure] ⟶ [1,4-diacetylglycoluril structure] + $2\,CH_3-\overset{O}{\underset{\|}{C}}-O-OH$

| 1.3.4.6-tetraacetyl-glycoluril TAGU | 1.4-diacetyl-glycoluril | Peracetic acid BLEACHING AGENT |

Fig 3.2   Reaction of TAGU

The reaction between the activator is fairly rapid and the peracetic acid (present in a normal wash solution as the sodium salt) ionizes:

$$CH_3-\underset{\underset{O}{\|}}{C}OOH \rightleftarrows H^+ + CH_3\underset{\underset{O}{\|}}{C}OO^-$$

in the same way as hydrogen peroxide, alkali being necessary to drive the reaction to the right. The peracetic ion is available for action at a much lower temperature than the perhydroxyl ion, thus we have the activation.

These activators, however, caused their own problems. Traces of moisture in the powder, or a relatively high storage temperature or even time caused the activator to react with the perborate in the carton with both loss of activity due to decomposition of the peracetic acid, and in the extreme, the production of a smell of acetic acid.

This was overcome by encapsulation of the activator, thus only when the powder was dissolved did the two components of the bleaching system react.

Although the costs of the activator are high, from information available, a large firm has been able to lower the perborate level from 20 per cent to 12 per cent with an activator, and to maintain the bleaching power in this powder at a lower temperature.

Another problem, more likely to be encountered in the USA, with shorter washing times and lower temperatures, is that the reaction between the activator and the perborate does not go to completion. This problem can be overcome by the use of stable pre-formed peroxyacids, to be discussed in the next chapter.

In certain parts of the world the use of perborate is restricted because the boron salts which remain in the effluent water have adverse effects on agriculture when this water is used for irrigation. In these instances, sodium percarbonate, $2Na_2CO_3 \cdot 3H_2O_2$, which is only slightly less stable than sodium perborate, is suggested as an alternative. Sodium percarbonate is sodium carbonate with 'hydrogen peroxide of crystallization', instead of water of crystallization. It can be considered as a mixture of the two and in solution, as if both sodium carbonate and hydrogen peroxide had been added separately.

Where dry bleaches are manufactured as such, sodium percarbonate is being considered. It has more active oxygen than perborate and is cheaper to manufacture, offering a double cost advantage. It is now being made by Interox in a coated form, increasing its stability.

6. Sundry Inorganic Builders

Borax: Chemically, borax is sodium tetraborate, available commercially as the decahydrate—$Na_2B_4O_7 \cdot 10H_2O$—or the pentahydrate—$Na_2B_4O_7 \cdot 5H_2O$. It is little used in detergents, although powders containing borax have appeared on the market. Although it gives a good buffering effect at a moderately low pH, all the other alkaline builders, which will be present anyway, can supply the same properties. However, borax serves a useful purpose where water softening and detergent synergism is required at a low pH when the high alkalinity of other builders might irritate the skin. In this case, borax and sodium tripolyphosphate combined will give optimum results. The pentahydrate has also the advantage of absorbing stickiness and can be used for improving the freeflowing properties of powders when stickiness is a problem.[11] The anhydrous salt, which is also available commercially can also absorb large quantities of water and still remain free-flowing.

It might be apt to correct a common misapprehension. Borax is not the salt of commercial boric acid. Boric acid is chemically orthoboric acid, H_3BO_3, whereas borax is the salt of tetraboric acid, $H_2B_4O_7$ which is not met with in industry (see also p 65).

The objection mentioned above about the discharge of perborate into water which might be used for irrigation also applies equally to borax.

Sodium Chloride: Sulphonation is sometimes carried out by using chlorosulphonic acid. Hydrochloric acid gas is then liberated from the reaction, for example in the sulphation of fatty alcohols:

$$R\text{—}OH + HSO_3Cl \rightarrow R\text{—}OSO_3H + HCl$$

When the sulphonate is neutralized with caustic soda, residual traces of this hydrochloric acid are converted into sodium chloride. In other processes the raw material is sometimes bleached with sodium hypochlorite solution and here again sodium chloride eventually appears as an adulterant in the raw material. In both types of reaction, if no special extraction procedures are used, the sodium chloride remains as a 'filler'.

Sodium chloride has been found to be the most cost-effective viscosity increasing agent for anionic detergent solutions, be they alkyl benzene sulphonates, fatty alcohol sulphates or ether sulphates, and is widely used for this purpose, particularly in the presence of alkylolamides.[12]

In the past, sodium chloride had been rarely used as a filler in powders for two reasons. In spray-dried powders unless the purest grades of sodium chloride are used it can render the powder hygroscopic. It was also considered that sodium chloride corroded the metallic parts of washing machines.

Extensive tests in the authors' laboratories have determined that about 10 per cent of sodium chloride in a well-formulated heavy-duty washing powder does not corrode the stainless steel parts that are used in modern washing machines.

Sodium chloride has an added advantage if incorporated into non-spray-dried automatic machine washing powders. It prevents caking of the powder in the dispensing chamber on contact with the incoming water.

It has also been found that in certain instances, notably the washing of garments stained with blood and also woollen garments, the addition of salt has a positive affect on detergency.

Magnesium Sulphate: Although the addition of magnesium sulphate (commercially available as Epsom salts: $MgSO_4 . 7H_2O$) to soap products is most undesirable because of the precipitation of insoluble magnesium soaps, it is often added to detergents, particularly powders based on alkyl benzene sulphonate containing sodium perborate.

Magnesium sulphate acts as a stabilizer for the perborate while still in the powder form. Also, in solution a certain portion of magnesium dodecyl benzene sulphonate in admixture with sodium dodecyl benzene sulphonate emphasizes the detergency. This conversion to magnesium sulphonate (by double decomposition) is not always achieved, as the magnesium might first be precipitated from the solution as magnesium silicate by the sodium silicate invariably present, or might be complexed with the condensed phosphate. It is not clear what happens to the magnesium ion when a powder containing magnesium sulphate is dissolved in water. Suffice it to say that the addition of magnesium sulphate is in no way detrimental to the detergent action.

With the switchover to linear alkyl benzene sulphonate it was found that the addition of magnesium sulphate to the slurry for the manufacture of spray-dried powders helped somewhat to eliminate the stickiness inherent in powders using LABS. It is considered that in this instance, magnesium sulphate is converted by reaction with sodium silicate to magnesium silicate and this type of material is often used as an 'anti-stick' agent.

Insoluble Inorganic Fillers: In this sub-grouping are included abrasive materials such as silica, quartz, marble dust and kieselguhr, which are used in the manufacture of scouring powders and abrasive soaps and pastes. Their action is purely abrasive and they do not in any way 'build' the detergent. They are discussed in Chapter 6 in the section on abrasive materials.

However, one natural insoluble material has intrinsic detergent properties, namely bentonite. Many countries have standard specifications limiting the amount of insoluble material that may be present in soap products, and where these rules apply, bentonite, being insoluble in water, cannot be used. Bentonite is a natural clay product containing complex silicates, mainly aluminium, with some iron and magnesium. Its peculiar property is that it is formed in a lattice-like crystal structure and has the property of swelling in the presence of water. It is able to emulsify oily materials, and to peptize fine particles of soil, keeping them in suspension. In addition, the bentonite molecule contains a small proportion of alkali ions and bentonite is thus able to soften water by ion-exchange. Subject to the limitation imposed by national standards, bentonite is therefore useful as a low-cost builder in detergent compounds (see also p 84).

Caustic Alkalis: Sodium hydroxide, sodium orthosilicate and sodium sesquisilicate are very alkaline, highly corrosive and hygroscopic materials and are rarely used in cleaning materials. The exceptions are for bottle, glass and metal cleaning by machine when a very high pH is desired.

Caustic soda, and to a lesser extent caustic potash, are used in very large quantities for neutralizing the anionic active matter. This neutralization may be done by the sulphonator, who will then supply a concentrated liquid, paste or powder for further compounding. Alternatively, the detergent manufacturer can buy an alkyl benzene sulphonic acid, and carry out his own neutralization process to suit his particular requirements. In the instances where the caustic alkalis are used for neutralization, they cannot be considered to be builders.

Both caustic soda, and caustic potash are available in powder, flake, solid and liquid forms. The liquid forms are solutions of between 40 and 50 per cent strengths, and if the extra freight cost on the liquid is not too high, the use of caustic solutions is preferable, since they can easily be handled by pumping and the need for special dissolving arrangements is obviated.

If the manufacturer buys the caustic soda (or potash) in one of the solid forms, it is necessary for him to dissolve it, as, for neutralizing purposes, the material must be introduced in a liquid form. The handling of these caustic alkalis requires special precautions, as they are highly corrosive to the human skin. Elaborate solution arrangements are available in some factories, but if the caustic soda is bought in the solid form, a simple procedure to dissolve this is to use a castiron tank of a volume sufficient for about one month's supply of caustic soda lye (solution). This tank is fitted with a false bottom or grating to raise the solid material from the bottom of the tank. This false bottom is situated more than the diameter of a drum of solid caustic soda from the top of the tank and should be slightly inclined to one end. The top and bottom of the drums of solid caustic soda are now knocked out with a hammer and cold chisel, but the cylindrical side of the drums can be left in place. The drums are raised by a chain hoist into the solution tank and rolled into place on their sides. When enough drums are in the tank, water is allowed to run in until the drums are covered. The heat of solution of the caustic soda is such that if a solution of 40 per cent or over is being produced, the temperature of the solution in the tank will approach boiling-point. As the solid material is dissolved its solution will sink, displacing the more dilute water from below the false bottom to the upper portion of the tank.

This in turn will dissolve more caustic soda and natural circulation will take place. In twenty-four hours, all the soda will dissolve and a homogeneous solution will be obtained. The cylindrical portion of the drums can be left in the tank until sufficient liquid has been pumped away and these drums can then be rinsed with a jet of water and lifted out. During the actual solution process workers should take care not to stand close to the tank, as the caustic soda solution might boil and cause dangerous splashing. Caustic potash can be handled in a similar way, but in this case the heat of solution is greater and it is not advisable to make a solution of over 40 per cent.

To dissolve powdered, flaked or lump materials, the best procedure is to fill the tank with water, pour the contents of one drum (only) into a castiron basket, and lower the basket into the water by a hoist. When all the material has been dissolved, the hoist is raised, recharged with fresh material and lowered again. This process is continued until a sufficiently strong solution is obtained. It is essential that workers should be provided with rubber gloves, goggles, protective aprons and respirator masks.

The strengths of both caustic-soda and caustic-potash solutions are best determined by measuring their density *when cold* with a hydrometer (note that after the solution has been completed there is a contraction in volume). The percentage of material in solution can then be read off directly from Tables 3.7 and 3.8.

SUNDRY INORGANIC BUILDERS

TABLE 3.7

Baumé Concentration Table for Caustic Soda Solutions at 15·6°C (60°F)

Baumé	Sp gr 60/60°F	Twaddell	Na_2O %	NaOH %	NaOH g/l	NaOH lb/US gal	NaOH lb/imp gal	NaOH lb/ft^3
1	1·007	1·4	0·47	0·603	6·06	0·05	0·06	0·38
2	1·014	2·8	0·95	1·22	12·35	0·10	0·12	0·77
3	1·021	4·2	1·43	1·85	18·87	0·16	0·19	1·18
4	1·028	5·6	1·93	2·49	25·61	0·21	0·26	1·60
5	1·036	7·1	2·44	3·15	32·57	0·27	0·33	2·03
6	1·043	8·6	2·96	3·81	39·75	0·33	0·40	2·48
7	1·051	10·1	3·48	4·49	47·15	0·39	0·47	2·94
8	1·058	11·7	4·02	5·18	54·79	0·46	0·55	3·42
9	1·066	13·2	4·56	5·88	62·65	0·52	0·61	3·91
10	1·074	14·8	5·11	6·59	70·73	0·59	0·71	4·42
11	1·082	16·4	5·67	7·31	79·06	0·66	0·79	4·94
12	1·090	18·0	6·24	8·05	87·65	0·73	0·88	5·47
13	1·098	19·7	6·81	8·79	96·47	0·81	0·96	6·02
14	1·107	21·4	7·40	9·55	105·50	0·88	1·06	6·59
15	1·115	23·1	7·99	10·31	114·90	0·96	1·15	7·17
16	1·124	24·8	8·59	11·09	124·50	1·04	1·25	7·78
17	1·133	26·6	9·21	11·88	134·40	1·12	1·34	8·40
18	1·142	28·3	9·83	12·69	144·70	1·21	1·45	9·03
19	1·151	30·2	10·46	13·50	155·20	1·30	1·55	9·69
20	1·160	32·0	11·11	14·33	166·10	1·39	1·66	10·37
21	1·169	33·9	11·76	15·18	177·30	1·48	1·77	11·07
22	1,179	35·8	12·43	16·04	188·80	1·58	1·89	11·79
23	1·189	37·7	13·10	16·91	200·70	1·68	2·01	12·53
24	1·198	39·7	13·79	17·79	213·00	1·78	2·13	13·30
25	1·208	41·7	14·49	18·70	225·60	1·88	2·26	14·09
26	1·218	43·7	15·20	19·62	238·80	1·99	2·39	14·91
27	1·229	45·8	15·92	20·55	252·30	2·11	2·52	15·75
28	1·239	47·9	16·66	21·51	266·20	2·22	2·66	16·62
29	1·250	50·0	17·42	22·48	280·60	2·34	2·81	17·52
30	1·261	52·2	18·18	23·47	295·60	2·47	2·96	18·45
31	1·272	55·4	18·96	24·47	310·90	2·60	3·11	19·41
32	1·283	56·6	19·76	25·50	326·90	2·73	3·27	20·41
33	1·295	58·9	20·56	26·54	343·20	2·86	3·43	21·43
34	1·306	61·3	21·40	27·62	360·40	3·01	3·60	22·50
35	1·318	63·6	22·25	28·72	378·20	3·16	3·78	23·61
36	1·330	66·1	23·13	29·85	395·70	3·30	3·96	24·77
37	1·343	68·5	24·03	31·01	415·90	3·47	4·16	25·97
38	1·355	71·0	24·95	32·20	435·90	3·64	4·36	27·21
39	1·368	73·6	25·91	33·44	456·90	3·81	4·57	28·53
40	1·381	76·2	26·89	34·71	478·80	4·00	4·79	29·89
41	1·394	78·8	27·89	36·00	501·40	4·18	5·01	31·30
42	1·408	81·6	28·93	37·34	525·10	4·38	5·25	32·78

INORGANIC COMPONENTS OF DETERGENTS, BUILDERS AND OTHER ADDITIVES

TABLE 3.7 (continued)

Baumé	Sp gr 60/60°F	Twad-dell	Na$_2$O %	NaOH %	NaOH g/l	NaOH lb/US gal	NaOH lb/imp gal	NaOH lb/ft^3
43	1·422	84·3	29·99	38·71	549·70	4·59	5·50	34·32
44	1·436	87·1	31·09	40·12	575·40	4·80	5·75	35·92
45	1·450	90·0	32·23	41·60	602·50	5·03	6·03	37·63
46	1·465	92·9	33·41	43·13	631·00	5·27	6·31	39·40
47	1·480	95·9	34·63	44·69	660·50	5·51	6·61	41·24
48	1·495	99·0	35·86	46·28	691·10	5·77	6·91	43·15
49	1·510	102·1	37·12	47·91	722·90	6·03	7·23	45·13
50	1·526	105·3	38·43	49·60	756·20	6·31	7·56	47·21
51	1·543	108·5	39·77	51·33	791·00	6·60	7·91	49·39

* Based on tables from International Critical Tables.

TABLE 3.8

Potassium Hydroxide—Density of Aqueous Solutions at 15°C (59°F)

Baumé	Sp gr 60/60°F	Twad-dell	KOH %	KOH g/l	KOH lb/US gal	KOH lb/imp gal	KOH lb/ft^3
1·2	1·0083	1·66	1	10·80	0·10	0·08	0·63
2·5	1·0175	3·50	2	20·35	0·17	0·20	1·27
3·8	1·0267	5·34	3	30·80	0·26	0·31	1·92
5·0	1·0359	7·78	4	41·44	0·35	0·41	2·57
6·3	1·0452	9·04	5	52·26	0·44	0·52	3·26
7·5	1·0544	10·88	6	63·26	0·53	0·63	3·95
8·7	1·0637	12·74	7	74·46	0·62	0·74	4·65
9·9	1·0730	14·60	8	85·84	0·72	0·86	5·36
11·0	1·0824	16·48	9	97·42	0·81	0·97	6·08
12·2	1·0918	18·36	10	109·20	0·91	1·09	6·82
13·3	1·1013	20·26	11	121·10	1·01	1·21	7·56
14·5	1·1108	22·16	12	133·30	1·11	1·33	8·32
15·6	1·1203	24·60	13	145·60	1·22	1·46	9·09
16·7	1·1299	25·98	14	158·20	1·32	1·58	9·87
17·8	1·1396	27·92	15	170·90	1·43	1·71	10·67
18·8	1·1493	29·86	16	183·90	1·54	1·84	11·48
19·9	1·1590	31·80	17	197·00	1·64	1·97	12·30
20·9	1·1688	33·76	18	210·40	1·76	2·10	13·13
22·0	1·1786	35·72	19	223·90	1·87	2·34	13·98
23·0	1·1884	37·68	20	237·70	1·98	2·38	14·84
24·0	1·1984	39·68	21	241·70	2·10	2·42	15·71
25·0	1·2083	41·66	22	265·80	2·22	2·66	16·60
26·0	1·2184	43·68	23	280·20	2·34	2·80	17·49
27·0	1·2285	45·70	24	294·80	2·46	2·95	18·41
27·9	1·2387	47·74	25	309·70	2·58	3·10	19·33
28·9	1·2489	49·78	26	324·70	2·71	3·25	20·27

SUNDRY INORGANIC BUILDERS

TABLE 3.8 (continued)

Baumé	Sp gr 60/60°F	Twaddell	KOH %	KOH g/l	KOH lb/US gal	KOH lb/imp gal	KOH lb/ft^3
29·8	1·2592	51·84	27	340·00	2·84	3·40	21·23
30·8	1·2695	53·90	28	355·50	2·97	3·56	22·19
31·7	1·2800	56·00	29	371·20	3·10	3·71	23·17
32·6	1·2905	58·10	30	387·20	3·23	3·87	24·17
33·6	1·3010	60·20	31	403·30	3·37	4·03	25·18
34·5	1·3117	62·34	32	419·70	3·50	4·20	26·20
35·4	1·3224	64·48	33	436·40	3·64	4·36	27·24
36·2	1·3331	66·62	34	453·30	3·78	4·53	28·30
37·1	1·3440	68·80	35	470·40	3·93	4·70	29·37
38·0	1·3549	70·98	36	487·80	4·07	4·88	30·45
38·8	1·3659	73·18	37	505·40	4·22	5·05	31·55
39·7	1·3769	75·38	38	523·20	4·37	5·23	32·66
40·5	1·3879	77·58	39	541·30	4·52	5·41	33·79
41·4	1·3991	79·82	40	559·60	4·67	5·60	34·94
42·2	1·4103	82·06	41	578·20	4·83	5·78	36·10
43·0	1·4215	84·30	42	597·00	4·98	5·97	37·27
43·8	1·4329	86·58	43	616·10	5·14	6·16	38·47
44·6	1·4443	88·86	44	635·50	5·30	6·36	39·67
45·4	1·4558	91·16	45	655·10	5·47	6·55	40·90
46·2	1·4673	93·46	46	675·00	5·63	6·75	42·14
47·0	1·4790	95·80	47	695·10	5·80	6·95	43·40
47·7	1·4907	98·14	48	715·50	5·97	7·16	44·67
48·5	1·5025	100·50	49	736·20	6·14	7·36	45·96
49·2	1·5143	102·86	50	757·20	6·32	7·57	47·27
50·0	1·5262	105·24	51	778·40	6·50	7·78	48·59
50·7	1·5382	107·64	52	799·90	6·68	8·00	49·94

Computed from values given in International Critical Tables

Ammonia: Ammonia is finding favour nowadays as an additive in 'ammoniated' liquid cleaners and is also occasionally used as a neutralizing agent.

Pure ammonia is a suffocating and pungently smelling gas. Aqueous solutions have a strongly alkaline reaction. Commercially, ammonia is usually a 25 per cent solution in water (density 0·9).

Let us now summarize the various requirements of the inorganic builders, and compare the properties of the various builders commonly used.

When using soap it is essential that the solution should be kept alkaline. The soap must contain sufficient alkali both to neutralize the acidity invariably present in soils and to maintain the pH of the soap solution above its hydrolysis point, when it will split off fatty acids and thus reduce the amount of effective soap present, and also increase the amount of fatty acid to be cleaned. Detergents in general are not hydrolysed as easily as soap, but it is still necessary to have sufficient alkali present to neutralize the acidity of the soil, and the presence of builders tends to reduce both the surface and interfacial tensions of their solutions.

The above statement although true for washing and cleaning operations, is not strictly true for personal cleansing. The human skin has a natural pH of the order of 5·5 and shampoos and cleansing lotions are now being manufactured at that pH or even lower. The manufacturers are using the claim, rightly so, of 'no alkali added'.

In Table 3.9 we show the effect of the addition of various builders to sodium dodecyl benzene sulphonate solutions.

From the table it is seen that with the addition of builders one can obtain the same or lower surface and interfacial tensions (which both play an important part in the washing process) as with the use of builders and less active material.

Caustic soda is in itself a source of high alkalinity, but the disadvantage in its use is that all its alkalinity is available at once. The ideal detergent-builder is one that releases its alkalinity only on reaction, but this alkalinity will still be available at a relatively high level (pH).

TABLE 3.9

Influence of Builders on Surface and Interfacial Tensions of Dodecyl Benzene Sulphonate

Concentration DDBS %	Concentration builder %		Surface tension	Interfacial tension
1·0	nil		32·0	5·2
0·4	NaCl	0·6	28·4	2·5
0·4	Na_2SO_4	0·6	29·2	2·9
0·4	Na_2CO_3	0·6	28·5	2·7
0·4	$Na_4P_2O_7$	0·6	29·4	2·7
0·25	nil		31·2	3·5
0·1	NaCl	0·15	29·7	2·5
0·1	Na_2SO_4	0·15	30·2	2·8
0·1	Na_2CO_3	0·15	30·3	3·6
0·1	$Na_4P_2O_7$	0·15	30·6	4·3
0·0625	nil		30·3	3·1
0·0250	NaCl	0·0375	31·6	3·5
0·0250	Na_2SO_4	0·0375	30·4	3·3
0·0250	Na_2CO_3	0·0375	30·9	5·1
0·0250	$Na_4P_2O_7$	0·0375	34·8	8·4

In Table 3.10 we show the pH and available alkalinity of all the alkalis commonly used. In this table the total, active and inactive, alkalis are shown separately. The total alkali is that which is titratable with acids to methyl orange, and the active alkali is that which does work in the washing process, titratable only with phenolphthalein.

Usually, to indicate the buffering capacity of alkalis, an electrometric

TABLE 3.10

Characteristics of Builders used in Detergent Mixtures

Alkaline builder	Formula	Molecular weight	pH of 1% solution at 20°C	Total titratable Na_2O %	Active alkali titratable with phenolphthalein as Na_2O %	Inactive alkali titratable with methyl orange as Na_2O %
Sodium hydroxide	NaOH	40·01	13·1	76·00	76·00	—
Sodium metasilicate	$Na_2SiO_3 \cdot 5H_2O$	212·13	12·3	29·37	27·99	1·38
Alkaline sodium silicate	$Na_2O:2SiO_2$	182·00	11·2	34·00	29·50	4·50
Neutral sodium silicate	$Na_2O:3SiO_2$	242·00	10·5	28·20	21·60	6·60
Sodium carbonate	Na_2CO_3	106·00	11·3	58·00	29·00	29·00
Sodium sesquicarbonate	$Na_2CO_3 \cdot NaHCO_3 \cdot 2H_2O$	226·03	10·0	41·27	13·76	27·51
Sodium bicarbonate	$NaHCO_3$	84·01	8·4	36·92	—	36·92
Borax	$Na_2B_4O_7 \cdot 10H_2O$	381·44	9·2	16·40	11·50	4·81
Trisodium phosphate	$Na_3PO_4 \cdot 12H_2O$	380·16	12·0	18·88	10·00	8·80
Tetrasodium pyrophosphate	$Na_4P_2O_7$	265·95	10·5	22·91	6·70	16·21
Sodium tripolyphosphate	$Na_5P_3O_{10}$	367·93	9·5	16·90	1·60	15·30
Sodium hexametaphosphate	$(NaPO_3)_6$	612·10	6·9	2·95	—	2·95
Sodium tetraphosphate	$Na_6P_4O_{13}$	469·92	8·5	8·80	—	8·80
Sodium perborate	$NaBO_2 \cdot H_2O_2 \cdot 3H_2O$	153·87	10·2	21·60	17·00	4·60

titration of the alkali is done against standard acid and a graph plotted of the results. We have departed from this procedure and in Table 3.11 we show the residual pH of solutions of the alkalis after discrete amounts of acid have been added. This table shows the 'rate of release' of alkalinity, in that the pH does not rise to the level of caustic soda initially; it is maintained at a relatively steady figure, while the acid is reacting with it. In every case, 1 gram of the anhydrous material was titrated with 0·5 mol/l acid.

In some cases the figure in the first column, that is the natural pH of the substance, differs from that stated in Table 3.10. This is because the pH in Table 3.10 is measured on a 1 per cent solution, whereas the initial concentration in Table 3.11 is 0·4 per cent.

TABLE 3.11

Residual pH of Solution after Addition of 0·5 mol/litre Acid to 100 ml of 0·4 per cent Solution

Material	pH 0·5 mol/litre HCl added					
	0 ml	5 ml	10 ml	15 ml	20 ml	25 ml
Sodium hydroxide	12·7	12·6	12·5	12·5	12·4	12·3
Sodium metasilicate	12·3	12·2	12·0	11·7	11·0	10·4
Trisodium phosphate	11·9	11·5	10·4	7·1	6·3	3·0
Alkaline silicate 1:2	11·2	10·7	10·2	9·7	8·1	2·3
Sodium carbonate	11·0	10·6	9·9	9·3	7·0	6·4
Neutral silicate 1:3	10·5	10·1	9·3	2·8	—	—
Sodium perborate	10·2	9·3	8·6	2·6	—	—
Tetrasodium pyrophosphate	9·7	8·5	7·0	5·9	2·7	—
Sodium tripolyphosphate	9·4	7·6	5·6	2·6	—	—
Borax	9·2	9·0	8·6	8·2	6·6	2·2
Sodium hexametaphosphate	6·9	2·8	—	—	—	—

As can be seen from the first column of Table 3.11 the alkalis are set out in descending order of alkalinity (pH), but the alkalinity does not fall in the same order. The alkalinity of condensed phosphates drops very rapidly, indicating that their buffering capacity is not good. Metasilicate and alkaline silicate maintain an alkaline pH almost until they are exhausted.

To give an idea of the water-softening power of the alkalis we show in Table 3.12 the residual hardness of water after the addition of 0·2 per cent of each of the materials.

It is evident that, in softer waters, sodium tripolyphosphate and sodium tetraphosphate have the best softening ability, but as the water becomes harder, soda ash overtakes them. This softening of water by carbonate ions is via the precipitation of the insoluble calcium and magnesium carbonates, whereas the phosphates act by maintaining these metallic ions in solution

Table 3.12

Residual Hardness after Addition of 0·2 per cent Alkaline Builder

Original hardness* ppm	Sodium tripoly-phosphate	Sodium tetra-phosphate	Tetra-sodium pyro phosphate	Tri-sodium phosphate	Sodium metasilicate	Borax	Soda ash	Sodium sesquicarbonate
60	0	0	16	28	34	54	12	20
120	0	0	36	40	68	110	24	40
180	12	0	44	64	110	146	30	53
300	52	24	60	108	164	232	36	72
400	112	60	76	128	253	336	46	92

* The hardness consisted of two parts calcium and one part magnesium. The test was conducted at 49°C and hardness was measured by the amount of soap required to give a stable foam.

and thus not adding to the amount of physical foreign matter to be flushed away.

This, however, is not the whole picture. The actual softening of water for synthetic detergents is not as important as it is for soap, but Porter[13] has shown the effect of builders on the action of non-ionic detergent (which in no way reacts with the metallic ions present in the water) in removing clay and grease.

The results are summarized in Table 3.13, which shows the amount of grease and clay removed from swatches of cloth which had been im-

Table 3.13

Effect of Builders on Clay and Grease Removal Action of Non-ionic Detergents

Calcium chloride present in water mmol/litre	Builder added mmol/litre	Percentage clay removed	Percentage grease removed
0	none	92	92
2	none	66	54
2	2 Na_3PO_4	78	18
2	2 $Na_4P_2O_7$	92	72
2	2 Metaphosphate	90	73
2	2 EDTA	87	72
2	2 $Na_5P_3O_{10}$	92	78
2	10 Na_2CO_3	60	56
2	20 Na_2CO_3	79	60
2	50 Na_2CO_3	80	68
2	20 Na_2SiO_3	88	80

pregnated with a standard amount of these two soils and then washed under standard and uniform conditions with a non-ionic detergent with and without additions of calcium chloride and builders.

From the table it appears that in distilled water the non-ionic detergent does the best removal job. The presence of the calcium ion, however, radically interferes with this work, and EDTA (which is described on p 89) or large amounts of sodium carbonate, both of which should theoretically remove the hardness completely, do not restore to the non-ionic detergent the power it had with no calcium ions present. Sodium tripolyphosphate gives the best all-round help to the non-ionic, but it should be noted that for the removal of grease the very best effect was obtained from sodium metasilicate.

As foam, or rather the lack of foam, is playing a large part in detergent formulations (see p 254), the effect of alkaline builders on the foaming power of typical non-ionic detergents (nonyl phenol with 9 or 10 mol of ethylene oxide) and the same detergent modified to a low foamer (p 190) is shown in Table 3.14.

It will be noted that the low foam detergent gives practically no foam in warm water, the presence of alkali increases the foaming power slightly for normal detergents and fairly extensively in the case of the low foamer. Surprisingly enough, the mixture of builders gives less foam than do individual alkalis.

Finally, one aspect of the alkalis as powders that should be taken into account is the ability to adsorb liquids on to the surface and still remain free-flowing, which is important when making 'dry mixes'.

This ability is given in Table 3.15.

Colloidal Silica: Not to be confused with colloidal silicates, colloidal silica is very finely divided silica produced in such a way that its bulk density is of the order of 100 g/litre. Being so bulky, the solid particles have an enormous surface area and their structure is such that they can absorb liquid materials on to their surface. Colloidal silica is often added to powders, made by processes other than spray-drying, to render them free-flowing and to reduce any stickiness.

A variation of colloidal silica is colloidal calcium silicate, with the same physical properties and used for the same purposes.

Sodium Hypochlorite: Sodium hypochlorite very rarely appears in formulations with detergents, but is often used to bleach detergents during the manufacturing process (see pp 204, 211, 244, 268).

It can be made by a variety of methods[15] but usually by the interaction of chlorine and caustic soda according to the equation:

$$Cl_2 + 2NaOH \rightarrow NaOCl + NaCl + H_2O$$

and is usually sold as a solution containing 10 per cent available chlorine. Sodium chloride is always present in equimolecular proportions with the sodium hypochlorite and when the sodium hypochlorite has completed its

TABLE 3.14

Effect of Alkaline Builders on Foaming Power of Non-ionic Detergents

Detergent	Builder	Foam height in cm							
		Distilled water				Hard water			
		25°C		55°C		25°C		55°C	
	Nonyl phenol with 10 mols ethylene oxide	Immed	5 min	Immed	5 min	5 min	Immed	Immed	5 min
As such	Nil	10½	10	9	8½	9½	9	9	9
Modified	Nil	5½	5	2	1½	5½	5	1½	1½
As such	Silicate	11	11	10½	9½	12	11½	11½	11½
Modified	Silicate	7½	7½	7½	7	8	8	7½	7
As such	Tripolyphosphate	11½	11	11½	10	10	9½	9	8½
Modified	Tripolyphosphate	7½	7	7	6½	5½	4½	2½	2
As such	Borax	12	11½	11	10	10	10	9½	9
Modified	Borax	7	6½	6½	6	7	6	7	6
As such	Silicate, Tripolyphosphate, Borax	9	8½	6½	6	10½	9½	8	7½
Modified	Silicate, Tripolyphosphate, Borax	8	7½	4½	4	7½	6½	4½	4
	Nonyl phenol with 9 mols ethylene oxide								
As such	Nil	—	—	—	—	—	—	10	9
Modified	Nil	—	—	—	—	—	—	1½	—
As such	Soda ash*	—	—	—	—	—	—	10	9
Modified	Soda ash*	—	—	—	—	—	—	9	7

Builder concentration was 0·1 per cent except for those marked * where the concentration was 0·2 per cent

TABLE 3.15[14]

Surface Absorptive Capacity of Builders

Builder	Gram liquid absorbed per 100 g powder
Anhydrous sodium metasilicate	5
Low-density anhydrous sodium metasilicate	12·5
Trisodium phosphate	4
Tetrasodium pyrophosphate	12·5
Sodium tripolyphosphate	13
Dense soda ash	12
Light soda ash	25

bleaching action it remains in the solution as a further molecule of sodium chloride. Account should therefore be taken of this in bleaching operations when the presence of inorganic salts can affect the final product, either in viscosity or concentration.

Only very rarely are liquid formulations containing both active chlorine with organic wetting agents manufactured, as the organic matter and the sodium hypochlorite are mutually incompatible.

Certain of the lower molecular weight phosphate esters have been found not to react with hypochlorite rapidly and can be added to a bleach solution to increase its wetting power, with a reasonable stability of the active chlorine. This, of course, is to be considered only as a compromise and shelf-life is never longer than six months at temperatures in the high 20's centigrade.

For blending into chlorine based powders, whether for bleaching or sterilizing, powdered chlorine releasing compounds are available, and in the absence of water, or with very little moisture, the stability is normally good, shelf-lives of over a year easily being attained.

There are a variety of powders that release active chlorine when brought into solution; bleaching powder (chloride of lime, $CaCl.OCl$) containing 32–36 per cent available chlorine and calcium hypochlorite (HTH), $(Ca(OCl)_2)$, containing 72 per cent available chlorine. Neither of these has found great favour with detergent manufacturers because of the soluble calcium ions introduced into the working solution. The use of dry bleaching powder or calcium hypochlorite in scouring powder is not to be recommended. Stability, especially in the presence of active detergent matter, is limited. Lithium hypochlorite, available as a dry free-flowing powder containing 35 per cent available chlorine, has none of the disadvantages of the other inorganic chlorine bearing agents.

There are a variety of organic materials which liberate hypochlorite ions in solution—Chloramine-T, Halane to name a couple—but these have all been superseded by the chloroisocyanurates (chloro-triazine triones). These have in the past been available in a variety of forms with differing chlorine contents. The choice has now settled down to two materials: trichloro-isocyanuric acid (trichloro-s-triazinetrione) $(ClNCO)_3$, containing 90 per cent active chlorine, with a solubility of 1·2 per cent in water (25°C) and yielding a pH of 2·0–3·5 in a 1 per cent solution and sodium dichloro-isocyanurate (sodium dichlor-s-triazinetrione) available in two forms, as the anhydrous salt $(NaCl_2(NCO)_3)$ containing 61 per cent available chlorine, soluble at the rate of 25 per cent in water and giving a pH (1 per cent) of 6–7, and the hydrate containing two molecules of water and 55–56 per cent available chlorine. Surprising as it may seem the hydrate is more stable than the other two varieties when compounded.

References

1. Nüman, W., *Tenside*, **8,** 82 (1971).
2. Hudson, E. J. and Marson, H. W., *Chem & Ind* pl 449 (14 Nov 1970).
3. Dahlqvist, Hall and Bergman, *Wasser Luft und Betrieb*, 107–112 (1976).
4. Werdelmann, B. W., *Soap*, **40** (March 1974).
5. Schweiker, G. C., *JAOCS*, **55,** 1, 36 (1978).
6. Weldes, H., *Soap* (May 1972).
7. Philadelphia Quartz Co Technical Publication.
8. Rock, S. L. and Coffey, R. T., Paper presented at CSMA Seminar (June 1985).
9. Niewenhuis. *Proceedings of the Fourth International Congress on Surface Activity*, Vol III, p 321.
10. Harris, S. C., *Oil and Soap*, **23,** 101 (1946).
11. Kali, V. M., Private communication.
12. Hunting, A. L. L., *Cosmetics & Toiletries*, **97,** 3, 53 (1982).
13. Porter, *Proceedings of Fourth International Congress on Surface Activity*, Vol III, p 187.
14. *Cowles Silicates*, Cowles Chemical Company.
15. Milwidsky, B. M., *Mfg Chemist*, **33,** 400 (1962).

4. Sundry Organic Builders

In addition to the inorganic builders described in Chapter 3 various organic agents are also added. Although these are generally employed in much smaller amounts than the inorganic builders, they play a no less important part in the final cleaning process.

Anti-redeposition Agents: When cellulose, which is insoluble in water, is treated with chloracetic acid, carboxy-methyl cellulose is formed. When neutralized to the sodium salt, this is known as CMC, which has already been mentioned in Chapter 1 as an anti-redeposition agent.

The structure of cellulose is:

[cellulose structural formula showing linked anhydroglucose units]

that is, linked anhydroglucose units, with three —OH groups per unit, one primary, the other two secondary.

When reacted with sodium chloracetate (chloracetic acid in strongly alkaline solution) the reaction is as shown:

$$-\overset{|}{\underset{|}{C}}-OH + CH_2ClCOONa + NaOH \rightarrow$$

$$-\overset{|}{\underset{|}{C}}-O-CH_3COONa + NaCl + H_2O$$

Whalley has indicated[1] that the relative reactivity, in this reaction is 2·5 for the primary alcohol, 2 for the —OH next to the ether linkages and 1 for the other secondary alcohol group. This means that the reaction will go preferentially to the normal alcohol with the second —OH group not far behind.

Theoretically each of —OH can be etherified but this almost never happens. The number of OH groups reacted per anhydroglucose unit is called the degrees of substitution (DS).

This can be three but rarely exceeds one, and more often than not is fractional.

Thus, if a particular CMC has a degree of substitution of 0·5, this means that one half of a molecule of —CH_2COOH on average is attached to each glucose unit; in other words, every second glucose unit has one carboxy-methyl group added.

On this degree of substitution are dependent all the properties of the CMC and for use in detergents a value of between 0·4 and 0·6 is suggested. The figure is only an average; when a degree of substitution of 0·5 is obtained, it might in fact mean that every second glucose unit has one carboxy-methyl group attached to it, but it more probably means that some glucose units have one, some have two and many have none. Some CMCs have a fairly narrow random distribution and others a very wide one. Niewenhuis[2] is of the opinion that a narrow distribution, with a degree of substitution of 0·5, gives the best results.

The ratio of CMC present in a washing powder is also important, and this is best taken as a definite proportion of the active detergent material present. One-twelfth to one-tenth of the active matter present is a usual figure for CMC in washing powders for use on cotton materials.

CMC with a degree of substitution of the order of 0·5 is not very soluble in water, so much so that a 10 per cent solution is a very viscous gel. The presence of inorganic salts lowers the solubility considerably and if CMC were to be included in a heavy-duty liquid formulation, the CMC would be thrown out of solution. Various methods have been devised to overcome this problem, but the most promising is the use of the sodium salt of carboxy-methyl-hydroxyethyl cellulose.

Besides its main use as an anti-redeposition agent, CMC is often used in detergents and cosmetic pastes as a thickening agent, mainly to increase the consistency of pastes and to prevent separation into phases. In this case the important characteristic of the CMC is the apparent viscosity. CMC solutions or dispersions display non-Newtonian flow and the viscosity is dependent on the stresses applied to the solutions and therefore the term apparent viscosity is used. This viscosity, assuming standard and constant methods of determination, is dependent on the degree of substitution, on the presence of inorganic salts, on the presence of non-solvents, on the pH, on the uniformity of the distribution of the carboxyl group and finally on the actual molecular weight (degree of polymerization) of the cellulose. For these reasons CMC when used for viscosity control can only be judged by making trial batches under conditions identical to the practical ones.

Of recent years the position of CMC as the sole or universal anti-redeposition agent is being challenged by PVP (polyvinyl-pyrrolidone). This material has for a considerable period been used in heavy-duty liquid detergents[3] where its greater solubility and compatibility with inorganic salts has eliminated many of the problems associated with the incorporation of CMC into solutions.

The mechanism of the anti-redeposition action of CMC is not fully understood, but it appears that CMC has a preferential attraction towards cellulose (to which it is related structurally) over normal dirt particles and thus CMC seems to be most effective on cotton. The effectiveness of CMC on synthetic and wool fibres is questionable and PVP does give better anti-redeposition characteristics both on synthetic and resin treated cotton fabrics.

A new development competing with CMC is a solid plastic containing multiple carboxyl groups. The plastic is put into the washing machine with the clothes. The carboxyl groups attract the dirt and hold it in the plastic, preventing redeposition. The plastic with the dirt is discarded after the wash.[4]

Thickening Agents: CMC is often used as a relatively cheap thickening agent, one of the considerations being that it is already in stock in most detergent factories. It is a useful thickening agent for paste products where clarity of solution is not necessary. Other thickening agents of the inorganic type also suitable for pastes of all sorts are the fumed colloidal silicas mentioned on p 78, special varieties of which can also be used for the thickening of non-aqueous solutions, and the montmorillonites and hectorites, available both as the natural and modified products. Modified montmorillonite is available as Veegum and Vangell (Vanderbilt Corporation) and the Bentones (NL Chemicals). The Bentones are modified bentonites which in turn are part of the montmorillonite family of aluminium silicates. It is interesting to note that as long ago as 1938, in the October issue of *Soap*, an article describing bentonites stated:

> Base Exchange: Bentonite acts somewhat like a mild zeolite in that it softens water by taking the soluble lime and magnesium out of solution and replacing them by sodium and potassium. Such water-hardening materials become part of the bentonite and are carried in suspension by it rather than being precipitated as insoluble deposits on fabrics, as occurs with alkalis.

This, before the large-scale advent of synthetics and condensed phosphates, ante-dates the present-day ideas on the use of (synthetic) zeolites.

The hectorites are magnesium silicates and are available both as natural and synthetic materials (Laponite, Laporte). They have in their lattice structure some lithium ions and outside the cage are some sodium ions. The montmorillonite materials give opaque 'solutions' whereas the hectorites, especially the synthetic grades, produce crystal clear ringing gels (if the concentration is sufficient).

The organic materials usually give clear or only slightly turbid solutions or suspensions and these are quite numerous.

Carbopol and Carbomer (Goodrich Corp.) and the Acrysols (Rohm & Haas) need to be neutralized with a base to achieve maximum thickening. The Acrysols are available as a thin emulsion and can be easily dispersed with all the other ingredients before the necessary alkali is added to give the gelling effect. The Carbopols are powders which need to be pre-dissolved before they can come into contact with the base. The Carbopols have an added advantage in that they have in their molecule several carboxylic radicals. If these are neutralized by both water-soluble and water-insoluble (higher amine) bases, the neutralized polymer can act as a bridge between oil and water phases, producing emulsions.

Modified celluloses of the non-ionic type, methyl cellulose, hydroxyethyl cellulose, methylhydroxypropyl cellulose are all available in varying viscosities.

All these modified celluloses are soluble in cold water and insoluble in hot water and most organic solvents. They are best brought into solution by dispersing them, with stirring, in a non-solvent (hot water or alcohol) and then adding cold water while stirring.

Increasing interest is now being shown in Guar gum and Locust Bean gum and their modifications. These can act as combined thickeners with other built-in properties such as conditioners for hair shampoos.

From the above it appears that if it is necessary to thicken a preparation with one of the above organic materials, the thickening agent needs to be dissolved or dispersed in water prior to its addition to the preparation except for the Acrysols. As solutions of these materials must of necessity be of the order of 5 per cent, the preparation must first be made in a more concentrated form before being diluted and thickened by the thickening agent dispersion. This does not always pose a problem but on occasion it is not always feasible to hold back water for the thickening agent from the basic preparation. In these instances the inorganic types of thickening agent should be considered. They can normally be blended directly into the finished product with little trouble.

Inorganic salts and pH affect the properties of these thickening agents and formulae should be submitted to exhaustive storage tests before deciding on a particular type.

Finally the Xanthan gums maintain their viscosity characteristics over the whole pH range and are used as thickeners for oven cleaners (which are basically caustic soda solutions), a property which very few other thickeners can claim.

Optical Brighteners: Wrongly called optical bleaches (no bleaching action takes place), optical brighteners are now an integral part of all washing powders. They are dyestuffs which are absorbed by textile fibres from solution, but not subsequently removed in rinsing, and they have the property of converting invisible ultra-violet light into visible light on the blue side of the spectrum. As the fibre will then reflect a greater proportion of visible light than if it had not been treated with the dyestuff, it will appear brighter, and as the tone of the extra light being reflected is on the blue side of the spectrum, this blue-violet tinge will complement any yellowishness present on the fibre, to make it look whiter as well as brighter. This action is similar to, but not to be confused with, the old method of blueing washing. In this instance, a blue dye or pigment is adsorbed on to the fibre, and this blue tends to absorb any yellow light falling on it, again reflecting light richer in blue, but in this case there is an absorption of light, and less light is reflected than is passed on to it, and although the fibre looks whiter, it is not brighter.

The structure of optical brightening agents is complicated and the subject of many trade secrets. They are usually derivatives of coumarin or stilbene. Initially, they were absorbed only by cotton fibres, but types are now available for all the new synthetic fibres on the market. There are also the 'general purpose' types which are substantive (sometimes to a lesser degree) to two or even more different families of fibres.

There are many manufacturers of optical brightening agents, and each one produces some tens of types. Hence, the problem of deciding which type to use becomes acute. From the point of view of the detergent manufacturer the basic variations are:

(i) Substantivity to various fibres (see above).
(ii) Stability. Most optical brighteners are compatible with oxygen bleaches and very few are stable to chlorine bleaches. Some, however, are stable to these bleaching agents after they have been adsorbed on to the fibre.
(iii) Solubility and dispersibility. Not all optical brighteners are soluble in water under the conditions of manufacture of, for example, a slurry prior to spray-drying, while others are not sufficiently soluble and stable for incorporation into liquid detergents.

The detergent manufacturer is advised to work in close co-operation with the supplier of his optical brightening agents, as to quantities to be added and we suggest the following points be considered before deciding on a particular type:

(*a*) Is the washing material under consideration to be used for one type of fibre only? Even though a powder is to be marketed for, say, cotton only, it is advisable to use a brightener which will work on resin-treated cotton and synthetic fibres as well, as no amount of printing of detailed instructions will prevent the consumer from deciding to use the washing powder for other materials, with adverse results.
(*b*) Is the washing powder to be sold with a chemical bleach incorporated in it, and if so will the brightener be compatible with the bleach on storage and stable when dissolved in water? If no chemical bleach is to be added to the powder, are washing practices in the area to add a chemical bleach after the washing process (or even during it) and if so which bleach, and will the optical brightener still be effective under those conditions?
(*c*) Is the optical brightener to be incorporated into a liquid, or paste, a spray-dried, drum-dried, or dry-mixed powder? Can the optical brightener withstand the manufacturing conditions of the detergent and will it then be stable on storage?
(*d*) As optical brighteners are usually anionic, is there any possibility of any material in the formulation being incompatible with the brightener?

SUNDRY ORGANIC BUILDERS

This point is rather important where cationic softeners are incorporated into a washing powder or where it is desired to add an optical brightener to a softening formulation. Cationic materials can quench the fluorescence of the whitening agent.[5]

(e) These optical brighteners when converting the ultra-violet light into visible light emit this light at a definite frequency. The colour tones of this frequency vary from violet to green. Is any particular colour shade favoured?

(f) Account must be taken of the particular washing practices or the use to which the detergent is to be put, ie, hand or machine washing, cold, warm or hot water. For example, the practice for machine washing in the USA is at a temperature of 54·5°C with a cycle of 5–15 minutes. In Europe cotton washing is done at a temperature of between 60°C and 93°C for 15–30 minutes. Different optical brighteners will be absorbed at different rates under these sets of conditions.

(g) Will the brightener have any effect on the colour of the finished formulation? This is best tried out in practice as actual manufacturing conditions, apart from the constitution of the brightener, play their part.

(h) Is the brightener light fast (ie, not attacked by light) once it is adsorbed on to the fibre? If not, minimum standards for light stability need to be determined.

No hard-and-fast rules can be given for the choice of these materials, as much depends on individual preferences, manufacturing procedures and local washing practices. It is now becoming common practice to add considerably more than the minimum quantity of optical brighteners to household powders. In general, brighteners after ten washes build up on the fibres to an optimum. If double (say) the normal quantity of the brightener is incorporated, the brightening effect will be enhanced after only one wash and the brightening effect of this powder will be visible to the naked eye. This is now being taken advantage of particularly in the introduction of new brands on the market. In addition it is now becoming common practice to incorporate mixtures of optical brighteners for different fibres.

Although the concentration of these brighteners is small they are expensive ingredients and an accurate appraisal of all the above figures is necessary.

Different from the optical brighteners are the photoactivated bleaches. These are sulphonated aluminium or zinc phthalocyanines.[6] These photoactivators can convert oxygen (from the air or that dissolved naturally in water) to its active form. Thus if the photoactivator is present in the wash water, either as a constituent of the powder or added separately, it will be adsorbed on to the fibre and after the washing process, when the clothes are hung up to dry, in the presence of light (preferably sunlight) the photoactivator comes into play, bleaching any bleachable stains. This is a true chemical bleaching action, not a brightening or whitening effect.

It needs to be emphasized that the action takes place on wet clothes only. The requirements for the action are therefore moisture (wet clothes), light (preferably sunlight), oxygen (either from the air or dissolved in the water) and of course the photoactivator adsorbed on to the cloth.

At the time of going to press powders containing these photoactivators are being test marketed in Spain.

Interox of the USA has developed another novel approach to bleaching.[7] Instead of using perborate and an expensive activator to yield an organic bleaching compound in solution, and to hope that the reaction between the two will have gone to completion (not always guaranteed), the company suggests the use of a pre-formed peroxyacid and is already offering commercially magnesium monoperoxyphthallate hexahydrate:

$$\left[\begin{array}{c} \text{C}_6\text{H}_4 \begin{array}{c} \text{CO}_3\text{H} \\ \text{CO}_2^- \end{array} \end{array} \right]_2 \text{Mg}^{2+} \cdot 6\text{H}_2\text{O}$$

Interox H48

It is possibly early to forecast whether this will have an impact on washing powder formulations but should be borne in mind as the possible low temperature bleach additive of the future.

Chelating Agents: In Table 3.13 we mentioned EDTA, and NTA has been mentioned in discussing eutrophication. EDTA is an abbreviation for ethylene diamine tetra-acetic acid and NTA is the abbreviation for nitrilo triacetic acid, both of which are members of the group of chelating agents which have come into prominence in the last few years.

These chelating agents have been used in minor quantities as additives to detergents for many years; but as has been explained, their use is now increasing.

Their action is to 'lock up' (the word chelate comes from the Greek κελος—claw) polyvalent ions in their molecules and make these ions undetectable and ineffective. As an illustration let us take the case of EDTA and calcium ions.

The sodium salt of EDTA has the structure:

$$\begin{array}{c} \text{NaOOC}-\text{CH}_2 \\ \phantom{\text{NaOOC}-}\diagdown \\ \phantom{\text{NaOOC}-\text{CH}_2}\text{N}-\text{CH}_2-\text{CH}_2-\text{N} \\ \phantom{\text{NaOOC}-}\diagup \phantom{\text{CH}_2-\text{CH}_2-\text{N}}\diagdown \\ \text{NaOOC}-\text{CH}_2 \text{CH}_2-\text{COONa} \end{array}$$

(with CH$_2$—COONa branches on both nitrogens)

and this reacts with calcium ions to give the complex:

$$\begin{array}{c} \text{NaOOC—CH}_2 \qquad\qquad\qquad \text{CH}_2\text{—COONa} \\ \diagdown\qquad\qquad\qquad\diagup \\ \text{N—CH}_2\text{—CH}_2\text{—N} \\ | \qquad\qquad\qquad\qquad | \\ \text{CH}_2 \qquad\qquad\quad \text{CH}_2 \\ | \qquad\qquad\qquad\qquad | \\ \text{O}=\text{C—O—Ca—O—C}=\text{O} \end{array}$$

releasing two sodium ions. The calcium ion, joined to the EDTA molecule by two ionic and two co-ordinate bonds, is effectively removed as an ion.

The chelating agents fall into three groups:

(a) Aminocarboxylic acids of which there are three representatives:

EDTA (ethylene diamine tetra-acetic acid)

$$\begin{array}{c} \text{HOOC—CH}_2 \qquad\qquad\qquad \text{CH}_2\text{COOH} \\ \diagdown\qquad\qquad\qquad\diagup \\ \text{N—CH}_2\text{—CH}_2\text{—N} \\ \diagup\qquad\qquad\qquad\diagdown \\ \text{HOOC—CH}_2 \qquad\qquad\qquad \text{CH}_2\text{COOH} \end{array}$$

NTA (nitrilo triacetic acid)

$$\begin{array}{c} \qquad\qquad\qquad \text{CH}_2\text{COOH} \\ \qquad\qquad\qquad\diagup \\ \text{HOOC—CH}_2\text{—N} \\ \qquad\qquad\qquad\diagdown \\ \qquad\qquad\qquad \text{CH}_2\text{COOH} \end{array}$$

DTPA (diethylene triamine pentacetic acid)

$$\begin{array}{c} \text{HOOC—CH}_2 \qquad\qquad\qquad\qquad\qquad \text{CH}_2\text{—COOH} \\ \diagdown\qquad\qquad\qquad\qquad\qquad\diagup \\ \text{N—CH}_2\text{CH}_2\text{—N—CH}_2\text{—CH}_2\text{—N} \\ \diagup\qquad\qquad\quad |\qquad\qquad\quad \diagdown \\ \text{HOOC—CH}_2 \qquad\quad \text{CH}_2 \qquad\qquad \text{CH}_2\text{—COOH} \\ \qquad\qquad\qquad | \\ \qquad\qquad\qquad \text{COOH} \end{array}$$

The differences between the three are that NTA has one, EDTA has two and DTPA has three amino groups, NTA has three, EDTA has four and DTPA has five carboxylic units.

The reaction between the chelate and the ion to be chelated is on a mol-to-mol basis and therefore theoretically the chelate with the lowest molecular weight will sequestrate the largest quantity of metallic ions on a weight-for-weight basis. Thus, NTA will sequestrate more calcium ions per unit weight than will EDTA. Similarly, both will sequester more calcium ions than magnesium because the molecular weight of magnesium is smaller than that of calcium. In addition to the molecular weight ratio there is the fact of the stability of the complex. As the bonding is to both the carboxylic and amino groups, the molecule with the greatest number of these will give the most stable chelate.

For practical purposes, the complexes formed by NTA are sufficiently stable and as this has the greatest sequestering action per unit weight and as its price has fallen considerably in recent years, NTA is being used more and more in detergents, especially as the use of phosphates is being limited somewhat.

The complex EDTA forms with calcium is shown above. The NTA complex with calcium can be shown pictorially:

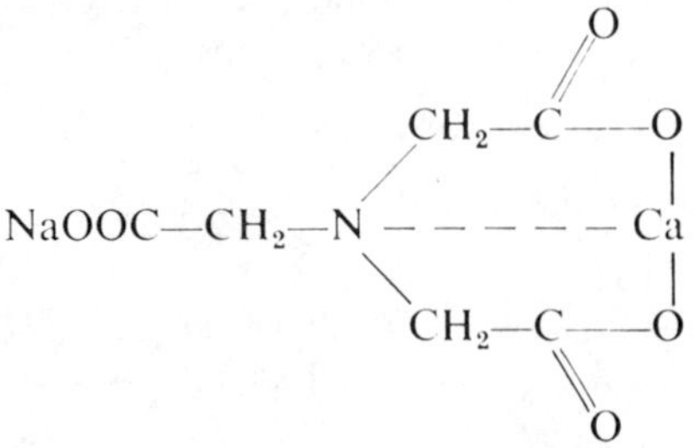

$$\text{NaOOC-CH}_2-\text{N} \begin{array}{c} \diagup \text{CH}_2-\text{C} \overset{\displaystyle \diagup \text{O}}{-\!-\!-\text{O}} \\ \diagdown \text{CH}_2-\text{C} \underset{\displaystyle \diagdown \text{O}}{-\!-\!-\text{O}} \end{array} \text{Ca}$$

These complexes are formed best in an alkaline pH, ie, where the carboxyl groups are completely neutralized and therefore if the acid or partially neutralized material is used in a formulation, sufficient external alkali needs to be added for maximum effect. As an example the complex of EDTA shown above is formed at pH 11. At a lower pH the complex:

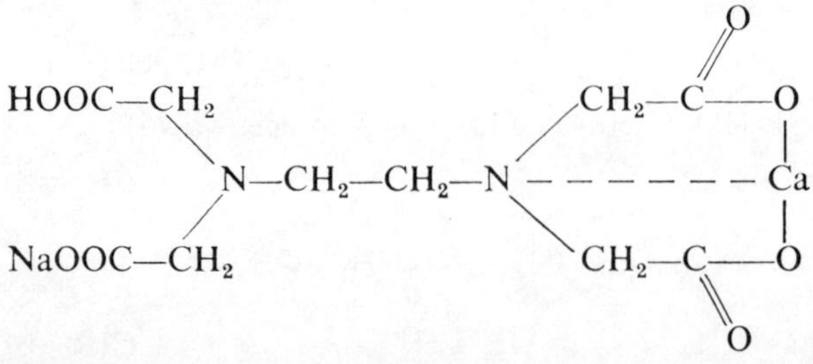

$$\begin{array}{c} \text{HOOC-CH}_2 \\ \diagdown \\ \text{N-CH}_2\text{-CH}_2\text{-N} \\ \diagup \\ \text{NaOOC-CH}_2 \end{array} \begin{array}{c} \diagup \text{CH}_2-\text{C} \overset{\displaystyle \diagup \text{O}}{-\!-\!-\text{O}} \\ \diagdown \text{CH}_2-\text{C} \underset{\displaystyle \diagdown \text{O}}{-\!-\!-\text{O}} \end{array} \text{Ca}$$

is formed. This is of the same structure and basic stability as that formed with NTA.

In addition to the simple chelation of the ions which cause hardness in water, these amino carboxylic acids can serve other functions and EDTA has for many years been used for this purpose. Phosphates are not very efficient in sequestering trivalent ions and therefore chelating agents are added to detergents to prevent iron stains in laundering, and to complex metallic ions which might catalyse the decomposition of sodium perborate (or other persalts) or dull the effect of optical brightening agents.

Where detergents are used in hot solutions and the solution needs to remain hot for an extended time, sodium tripolyphosphate will slowly decompose and lose its effectiveness. In these instances chelating agents are indicated. For the same reason and by virtue of its limited solubility in water, sodium tripolyphosphate cannot be used in liquid detergent formulations. Here tetrapotassium pyrophosphate which is reasonably stable and very soluble is commonly used, but again the amino carboxylic acids will serve a useful purpose as a partial replacement.

It should be noted that NTA, EDTA and DTPA are all acids and each forms a series of salts. The materials are sold commercially as either the acid or with some or all of the acid group neutralized. They have different efficiencies for different metals at different pH values, but in general the formulations used in cleaning materials are on the alkaline side so that if a chelating agent which is not fully neutralized is used, part of the inherent alkalinity of the formula will be utilized in neutralizing the chelating agent.

Although the above chelating agents can sequester iron in moderately alkaline solutions, at a pH of over 9 or so ferric ions tend to precipitate from the solution and for this purpose the next class of chelating agents have been developed.

(b) Hydroxyaminocarboxylic acids of which two main examples are:
HEDTA (hydroxyethylene diamine triacetic acid)

$$\begin{array}{c} HOOC-CH_2 \\ \diagdown \\ N-CH_2-CH_2-N \\ \diagup \\ HOOC-CH_2 \end{array} \begin{array}{c} CH_2CH_2OH \\ \diagup \\ \\ \diagdown \\ CH_2COOH \end{array}$$

ie, EDTA with one of the carboxylic groups replaced by an alcohol grouping.
DEG (dihydroxyethyl glycine)

$$\begin{array}{c} HOCH_2-CH_2 \quad CH_2CH_2OH \\ \diagdown \diagup \\ N \\ | \\ CH_2COOH \end{array}$$

ie, NTA with two of the carboxylic groups replaced by alcohol groupings.

These materials have been developed specially for the chelation of ferric ions. Ferrous iron behaves like any other divalent metal but is subject to ready oxidation to the ferric state so that if only ferrous iron is present in any particular process with no reducing agent it can be considered that in the course of time ferric ions will be present.

HEDTA will sequester ferric ions at a pH of 9 and DEG will sequester them at a pH of 12. DEG will not sequester calcium or magnesium ions but does other metals. HEDTA is not as effective as the aminocarboxylic acids for calcium and magnesium. Their use is therefore rather limited.

If the formula of triethanolamine (p 103) is examined pictorially:

$$\text{HO—CH}_2\text{—CH}_2 \quad \text{CH}_2\text{—CH}_2\text{—OH}$$
$$\diagdown \quad \diagup$$
$$\text{N}$$
$$|$$
$$\text{CH}_2$$
$$|$$
$$\text{CH}_2$$
$$|$$
$$\text{OH}$$

it will be noted that it is related to DEG, except that it does not contain an acid group. Triethanolamine can serve a useful purpose for the control of ferric iron in 1–18 per cent caustic soda solution.[8]

In the years 1969 and 1970 after the public outcry against eutrophication, NTA was introduced into domestic powders as a whole or partial replacement for sodium tripolyphosphate in 'phosphate-free' or 'low phosphate' powders.

At the end of 1970 the US Surgeon General reported that NTA in the presence of large amounts of mercury and cadmium salts caused birth defects in rats. The mechanism is the production of a complex which can pass the placental barrier. The report emphasized that the quantities of the heavy metals used in the tests were far in excess of that found in actual practice as contaminants in water and that no health hazard is envisaged. Owing to this adverse publicity, the use of NTA, particularly in the USA, is being diminished. It is presumed that other aminocarboxylic acids will behave in a similar way and there is a reluctance to use these materials in large quantities.

(c) *Hydroxycarboxylic acids.* The principal members of this group are:

Gluconic acid

$$\begin{array}{c}
\text{COOH} \\
| \\
\text{H—C—OH} \\
| \\
\text{H—C—OH} \\
| \\
\text{H—C—OH} \\
| \\
\text{H—C—OH} \\
| \\
\text{H—C—OH} \\
| \\
\text{H}
\end{array}$$

Citric acid

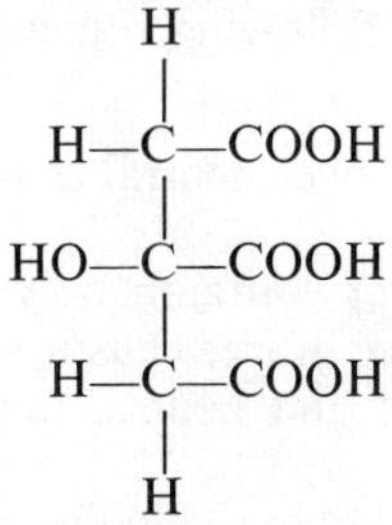

$$\begin{array}{c} H \\ | \\ H-C-COOH \\ | \\ HO-C-COOH \\ | \\ H-C-COOH \\ | \\ H \end{array}$$

Tartaric acid

$$\begin{array}{c} H \\ | \\ HO-C-COOH \\ | \\ HO-C-COOH \\ | \\ H \end{array}$$

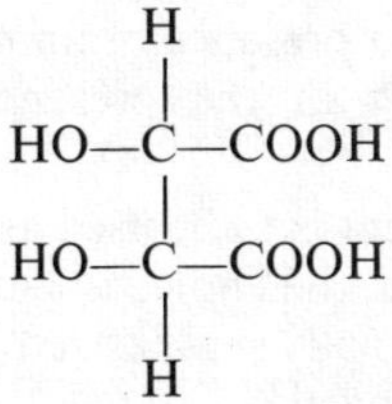

Gluconic acid is a good all-round chelating agent, but its particular usefulness appears in solutions containing free caustic soda. One gram of sodium gluconate at a pH below 11 is able to chelate 25 mg of calcium carbonate, whereas in a 3 per cent solution of caustic soda it can sequestrate 325 mg $CaCO_3$.[9] In addition it is able to chelate ferric iron over the whole pH range.

Citric and tartaric acids are effective chelating agents for most bivalent and trivalent ions except for the alkaline earths. Citric acid has the added advantage of completely sequestrating iron when ammonia is present[10] but loses its effectiveness above 60°C, an important point for European washing conditions.

An acid which is being suggested as a replacement for phosphates is oxydiacetic acid or diglycolic acid.

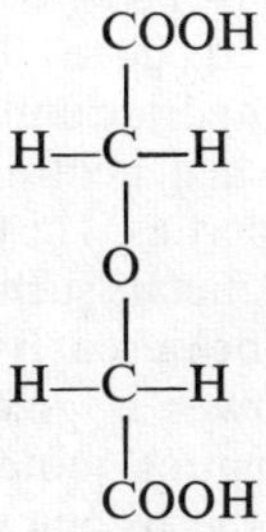

$$\begin{array}{c} COOH \\ | \\ H-C-H \\ | \\ O \\ | \\ H-C-H \\ | \\ COOH \end{array}$$

This material is under test both for its chemistry and as a detergent additive to replace phosphate. It appears to be intermediate in chelating power, for alkaline earths, between sodium tripolyphosphate and NTA. It is therefore considered that it will not give a sufficiently strong complex with mercury and cadmium to pass the placental barrier. At the moment the price is high but methods of production by the oxidation of diethylene glycol are being studied and this might bring the price into a range to make its use in detergent powders possible.

SUNDRY ORGANIC BUILDERS

To sum up we suggest the following rough guide for the selection of a chelating agent:

(a) For softening of water over neutral to mildly alkaline pH: NTA or EDTA.
(b) For the sequestering of common polyvalent ions over the whole pH range in the absence of free caustic soda: EDTA, NTA, DTPA.
(c) For iron control, acid and neutral range: EDTA, NTA, DTPA, HEDTA.
(d) For iron control, moderately alkaline range: HEDTA, DEG.
(e) For iron control in the presence of ammonia: citric acid.
(f) For general chelation in the presence of free caustic alkalis: gluconic acid.

However, if a chelating agent is to be used as a (partial) replacement for STP in a powder formulation, the choice is usually NTA and/or a zeolite. The newly developed and specially synthesized chelating agents have as yet not become commodity chemicals.

Many co-polymers have been produced specifically as detergency aids and all have had varying successes reported. One co-polymer has been available on the market for many years mooted as a thickener and stabilizer, and has been found to have water-softening properties.

We refer to Gantrex (a registered trade name of GAF Corp.), a poly(vinyl methyl ether-maleic anhydride) of the structure

$$\begin{array}{c} OCH_3 \\ | \\ -CH_2-CH-CH-CH- \\ | \quad\quad | \\ O=C \quad C=O \\ \diagdown \; \diagup \\ O \end{array}$$

of varying molecular weight from 20,000 to 80,000, depending on the grade. The molecule is a long chain of anhydride units, which hydrolyse to the acid when dissolved in water, and when neutralized with NaOH or KOH, exhibit remarkable chelating properties. The Gantrex resins do not seem to have the propensity for chelating metals such as mercury and cadmium so no environmental objections can be raised. By reference to Table 4.1, it will be noted that the chelating power is greater than that of the other commonly used materials. The figures quoted are compiled from various sources, that for Gantrex from the technical publications of GAF.[11] It should be pointed out that all the other figures, although obtained by laboratory tests, are close to the theoretical. Calculating the theoretical figure for Gantrex 119, one comes up with a figure of 641 mg $CaCO_3$/g, considerably less than the GAF figure, indicating that there are possibly some side effects as well.

The makers maintain that Gantrex AN-119 (molecular weight 20,000), the grade suggested for detergent use, can effectively replace ten times its own weight of STP. This is based on detergency trials.[12]

SUNDRY ORGANIC BUILDERS

TABLE 4.1

Chelating Ability of Various Chelates for Calcium in Alkaline Solution (pH above 10)

	Molecular weight	mg $CaCO_3$ sequestered/g chelate
EDTA	292	340
EDTA NaH_2	336	285
EDTA Na_4	380	240
NTA acid	191	525
NTA Na_3	257	375
HEDTA acid	278	173
DTPA acid	393	156
STP	368	265
TPPP	330	295
Zeolite A Hydrated	—	225
Gantrex AN-119	156*	775†

* Trade name of GAF Corporation. The molecular weight indicated is of one unit of the polymer.
† See p 94 for anomaly from theoretical value.

Being an anhydride this can react readily with —OH groups. GAF has also developed[12] another use for this material in liquid detergents. If a partial ester with a suitable non-ionic is formed, an additional effect is noted in that the ester can stabilize liquid detergents, preventing the 'salting-out' of non-ionics the grade suggested is AN-149. As mentioned this was originally used as a thickening agent. When neutralized the viscosity rises but as the un-neutralized material it can still be used as a thickener for acid cleaners. Full details for use of this material are given on p 266.

The chelating agents theoretically can sequester all the metallic cations, but in practice the reaction is preferential if more than one cation is present and also the stability of the complexes formed varies with the cation and, of course, as has become obvious, with the particular chelating agent. The main function that the materials are required to do in our industry is to soften water (and to remove the extraneous iron or manganese in water).

We detail in Table 4.1 the chelating ability of certain selected sequestrants for calcium only and in alkaline solution.

Phosphonates: Phosphonates are different from the phosphate esters in that the former have a —C—P— linkage, the latter a —C—O—P— (an analogy is the sulphate, —C—O—S—, sulphonate, —C—S—, groups, p 18). Phosphonates have for considerable time been used as chelating agents, mainly in water treatment. Typical examples are

$$\text{HO-}\underset{\underset{O}{\|}}{\overset{\overset{HO}{|}}{P}}-\underset{\underset{CH_3}{|}}{\overset{\overset{OH}{|}}{C}}-\underset{\underset{O}{\|}}{\overset{\overset{OH}{|}}{P}}-\text{OH}$$

DEQUEST 2010 Phosphonate (Monsanto), and

$$\underset{HO}{\overset{HO}{>}}P\underset{\underset{\underset{CH_2-C-OH}{\underset{\|}{O}}}{\overset{\overset{CH_2-C-OH}{\overset{\|}{O}}}{|}}}{\overset{\overset{CH_2-C-OH}{\overset{\|}{O}}}{\overset{|}{C}}}-C-OH$$

BAYHIBIT AM (Bayer).

These are good sequestering agents with good hydrolytic stability and stability against both oxidizing and reducing agents. In addition they exhibit threshold effects, the ability to sequestrate at substoichiometric levels. They have been used in boiler treatment compounds where, besides chelating metallic ions they also inhibit crystal growth, but are not of great interest to detergent manufacturers.

More recently the amino phosphonates:

$$\underset{\underset{OH}{|}}{\overset{\overset{O}{\|}}{HO-P}}-CH_2-\underset{\underset{\underset{\underset{OH}{|}}{\underset{CH_2P-OH}{|}}}{\overset{|}{O}}}{N}-CH_2-\underset{\underset{OH}{|}}{\overset{\overset{O}{\|}}{P}}-OH \qquad \text{Amino-trimethylene phosphonic acid} \\ \text{ATMP}$$

Ethylene diamine tetramethylene phosphonic acid
EDTMP

SUNDRY ORGANIC BUILDERS

$$\begin{array}{c}
\text{(HO)(OH)P(O)-CH}_2 \\
\text{(HO)(OH)P(O)-CH}_2
\end{array}\!\!\!\!\!\text{N-CH}_2\text{-CH}_2\text{-N-CH}_2\text{-CH}_2\text{-N}\!\!\!\!\!\begin{array}{c}
\text{CH}_2\text{-P(O)(OH)(OH)} \\
\text{CH}_2\text{-P(O)(OH)(OH)}
\end{array}$$

with a central $CH_2-P(O)(OH)-OH$ branch.

Diethylene triamine pentamethylene
phosphonic acid
DETMP

manufactured by Monsanto (trade name Dequest) and others, have been offered and accepted by the detergent industry.[13]

In detergent formulations the acids are neutralized to form the sodium and, particularly for liquids, their potassium, ammonium or alkanolamine salts. In experiments conducted by Nijs and his collaborators better than adequate improvements were found in detergency of liquid formulations.

The amount used in detergent formulations is generally less than 1 per cent. Used in such small quantities no ecological problems are anticipated.

Hydrotropes: In recent years, the use of liquids for heavy-duty detergents has increased considerably. These heavy-duty liquids are not only solutions of detergents, because for heavy-duty work all or some of the builders mentioned in this and the previous chapter are necessary. This poses a problem in that the solubility of detergents decreases in the presence of inorganic salts. To maintain all the materials in solution, it is then necessary to add what is called a hydrotrope.

Hydrotropes in general are very short chain alkyl aryl sulphonates, but in certain instances the use of non-ionic compounds, of the ethylene oxide condensate and of the alkylolamide types and the sodium salt of oleic acid sulphonate,[14] will give the desired effect of lowering the cloud point of detergent/salt solutions.

The most common hydrotropes are the sodium and potassium (very rarely ammonium) salts of toluene, xylene and cumene sulphonic acids and urea. To give an example of their effectiveness, ethyl acetate dissolves to the extent of 8·8 ml in 100 ml distilled water, but 42·6 ml in a 40 per cent solution of sodium xylene sulphonate.[15]

In addition to heavy-duty liquid detergents, hydrotropes are also used in light-duty liquid detergents where it is required to lower the cloud point; for example, where it is desired to use a tridecyl benzene sulphonate as a base instead of a dodecyl benzene sulphonate and to maintain the cloud point at the same figure as that obtained with DDBS. Owing to the greater solubility of linear alkyl benzene sulphonate neutralized with caustic soda and the lower cloud point of the solution, the use of hydrotropes for this purpose is less important.

Besides their use in liquid formulations, hydrotropes, when added to detergent slurries prior to spray-drying of powders, have the effect of reducing the viscosity of the slurries, which means that less water needs to be used in the preparation of the slurry and consequently a higher output is realized through the spray-drier. For this particular application, sodium toluene sulphonate is recommended, and this has the added effect of increasing the flow characteristics of the finished powder. This addition has become very necessary in the case of powders containing high concentrations of LABS as they have an inherent tendency to tackiness. In this case sodium toluene sulphonate, sodium xylene sulphonate or sodium sulphosuccinate is used.[16]

As mentioned on p 26, low-molecular-weight phosphate esters also have a hydrotropic effect. Their use in liquid hand dishwashing detergents has an added effect; the liquid produces a more compact foam, which is often desirable.

Enzymes: The most significant development in the detergent industry in recent years is the addition of enzymes to household washing powders. This had been tried before with only moderate success, but after enzymes were successfully applied to cleaning operations in the meat and fish industry, attention of the enzyme producers turned to detergents. Enzymes are of many types and sub-types, but the three which are of interest to the detergent industry at present or might be in the future are:

Proteases which act on protein to form amino acids;
Amylases which convert starches into dextrins;
Lipases which attack fats and oils.

Starch stains have never been a problem with laundering and the whole action of soap and detergent powders has been geared in the past to remove fatty stains. Proteinaceous stains have always been a problem. The addition of proteolytic enzymes to powders found immediate acceptance. The action of the enzyme is relatively slow; so this type of powder appeared initially in countries where the washing habits were to pre-soak washing prior to the wash proper. The enzymes were incorporated in concentrations of $\frac{1}{2}$–2 per cent in low-active powders and these were used for the pre-soak, cold.

The mechanism of the enzymatic action is such that it cleaves the protein into smaller peptide fractions which are water-soluble. Even if the conversion is not complete, the protein is degraded into a product which is more easily removed by the detergent. Thus proteins themselves are rendered soluble or dispersible and in addition, even though the stain might contain only small proportions of protein, the mixture of protein and normal soil can be removed more easily as hitherto the protein has bound the soil very firmly to the cloth. One important fact is that the action of the enzyme is independent of the particular type of cloth to which the dirt is adhering. Not much work has been done on the action of enzymes on woollen fibres, but in this instance caution should be exercised.

After soaking for a period of from two hours to overnight, the clothes are then washed in the normal way.

Enzymes are sensitive materials and the base powder had several *a priori* conditions. The moisture content of the powder had to be below 5 per cent, the pH of the powder was best between 8 and 9·5, and no oxidizing agents could be included. All these factors tended either to inactivate or destroy the enzyme.

Enzymes are produced by fermentation, from a strain of *Bacillus subtilis* by rather complicated processes. Better and better sub-strains were bred and enzymes are now available which are stable to oxidizing agents of the perborate type and active over a broader pH spectrum, from 7·5–10.

The development of these better strains gave a spurt to the use of enzymatic powders, and powders for handwashing and automatic washing machines were put on the market. Exact figures are not available but it is considered that at the present moment 40 per cent of all household washing powders sold in Europe contain enzymes, and in the USA, after a late start, this type of powder is forging ahead.

Enzymes are now commonly being incorporated in automatic washing machine powders containing perborate, and the action is that in the stage where the water temperature is below 50°C (which can be up to half an hour) the enzyme acts on proteinaceous stains (gravy, blood, milk, cocoa, etc.) and then when the temperature reaches 60°C and higher the perborate takes over. During this whole cycle the active matter, phosphate, CMC, etc, are acting in the normal way.

Langguth and Mecey have done considerable work on the compatibility with, and action of enzymes in, the presence of the various ingredients normally used in detergent powders.[17]

Their findings can be summarized:

(a) Although there are some subtle differences between the various alkaline protease enzymes available on the market for incorporation in detergents, their performance is quite similar.

(b) Sodium tripolyphosphate enhances the action of enzymes on stain removal up to a maximum of 30 per cent STP in the formulation. Over this figure there is a slight tailing off in enzyme action.

(c) NTA also enhances the performance of enzymes but to a lesser extent than STP. This is attributed by them to the fact that NTA gives a higher pH in solution than STP and the higher alkalinity is detrimental to the action of the enzyme. If the pH of the solution is adjusted to the natural pH of STP it was found that a formulation containing 40 per cent NTA gave the same stain removal as that obtained with 30 per cent STP.

(d) If combinations of NTA and STP are desired, 50 per cent of the STP can be replaced by NTA without seriously affecting the enzyme activity. If higher concentrations of NTA are desired, the resulting pH needs to be adjusted to that of STP.

(e) Linear alkyl benzene sulphonate up to a concentration of 20 per cent enhanced the activity slightly. Over 20 per cent there was a falling off in activity. An alcohol ethoxylate (non-ionic detergent) at a concentration of between 5 and 20 per cent did not affect the enzyme activity, and similar results were obtained for a polyoxypropylene-ethylene oxide condensate, soap and alkylolamide foam boosters, but a cationic detergent reduced the enzyme activity very quickly.

(f) Sodium sulphate and borax did not affect the activity and at concentrations of over 30 per cent borax there was a slight enhancement.

(g) Sodium carbonate and sodium sequicarbonate drastically reduced the enzyme activity. Sodium bicarbonate did not affect it at all and therefore it can be assumed that the action is due to the high pH and not to the carbonate ion *per se*. Sodium metasilicate also drastically reduced the activity but if the solution containing the metasilicate had its pH adjusted to that of STP, no effect on the enzyme activity by the silicate ions was found.

(h) CMC and PVP were tested in the formulations and it was found that at concentrations of 0·5 per cent of either the enzyme activity was enhanced by these materials.

(i) One per cent of sodium perborate had little effect on the enzyme performance but 2·5 per cent reduced the activity to 50 per cent. Over the range 2·5–7·5 per cent there was very little further reduction.

(j) One part per million of active chlorine almost completely inactivated the enzymes.

(k) Tribromosalicylanilide, trichlorocarbanilide and hexachlorophene up to concentrations of 0·25 per cent have little if any effect on the enzyme activity.

(l) Optical brightening agents in common use did not affect the enzymes.

(m) Concentrations of over 20 per cent STP or 10 per cent linear alkylbenzene sulphonate decrease the activity markedly if the enzyme–detergent mixture is left in solution for two hours or so before introducing the clothes. On the other hand, alcohol ethoxylates tend to stabilize the enzymes in solution.

The above tests were done on clothes soiled artificially with blood, milk and ink which are particularly sensitive to enzymes. Although the results must not be taken to be related directly to household washing conditions, they do indicate trends.

The use of enzymes by detergent manufacturers brought in its wake problems. Enzymes attack proteins and there was the fear that workers who have to handle the raw material in a powder form might develop skin and lung irritations. In the United Kingdom the detergent industry drew up a voluntary code of practice which has worked successfully since 1968. The aim is to reduce health hazards to the respiratory system of employees in the industry, and this is achieved by close monitoring of dust levels in the workplace and regular medical examinations, particularly lung function tests. In addition to the health hazard, dust losses and segregation of the powdery enzyme can be experienced in production. A further problem is the dosing of very small percentages of these enzymes into the ready-made powders on a continuous basis.

To overcome all these problems much work was and is being done in various countries. The enzyme is normally produced after fermentation as a liquid concentrate and one of the methods of converting it into a powder is to spray-cool this concentrate with sodium sulphate. This produces a dusty product. Another method is to mix the concentrate with fast hydrating sodium tripolyphosphate. If care is not taken and the mixing is not efficient this can produce lumps which detract from the free-flowing characteristics necessary. Still another method is to mix an enzyme powder with granular sodium tripolyphosphate and to bond the two together with a fine spray of a non-ionic detergent. This gives a friable bond, and enzyme dust can again be formed when the join is broken.

Enzymes are now being produced in modified forms in England, Germany, Italy and Denmark. One of these modifications is the incorporation of the enzyme with a non-ionic detergent containing rather a long ethylene oxide chain and a certain amount of sodium toluene sulphonate. This mixture can be either spray-cooled to a granular powder or extruded as fine needles. Another method is the encapsulating of the enzyme with a non-ionic detergent which will not destroy enzymes. In this way the enzyme is in contact only with the detergent which is the outer phase. This overcomes two of the problems of storage (shelf-life). One is the problem of humidity in the powder and the other is perborate. Although enzymes are stable on storage in the presence of perborate, the word stable is only relative in that the stability is considered satisfactory. In the case of encapsulation the enzyme does not come into contact either with the moisture or the perborate, until of course the powder is dissolved in water, when the normal reactions take place.

All the above methods tend to dilute or titurate the enzyme so that larger quantities are needed to be dosed, making the mixing more efficient. In addition the particle size of the enzyme product approaches that of the spray-dried granule and the bulk density of the products from the two latter methods approaches that of a spray-dried powder and the danger of segregation is lessened.

Contradictory claims have been made about the effect of enzyme-containing powders on the housewife (the user). The latest authoritative information taken from a report in the *British Medical Journal*[18] seems to be that no adverse effects can be found. A group of approximately 7000 women were used for an experiment running into five months. Of these women, 4119 used enzyme-containing products and 2884 were used as controls. No differences of any sort were found when the hands of all these housewives were examined. (See also p 8).

Enzymes have been developed and commercialized that are stable in liquid detergent formulations. They are either different strains or have additives added to allow their reasonable long-term shelf-life in a formulated product.

However, there are two conditions required in the ultimate liquid; the water content needs to be between 40 and 60 per cent and a certain minimum amount of calcium ions need to be present. Normal tap water will presumably supply sufficient calcium, and it is presumed that calcium ions are added by the manufacturer of the enzyme. However, this precludes the use of a chelating or sequestering agent in the liquid.

Certain manufacturers are adding chelating agents to liquids containing enzymes with water at the minimum concentration on the assumption, possibly borne out by trials, that there is not sufficient ionic mobility in the concentrated solution to sequestrate the calcium ions.

The enzymatic liquids should therefore be directed to the light-duty market, and they also have shown good performance in cold water, but a recommendation to speed the enzymatic action is to pre-spot the undiluted liquid on to problematic stains. The reaction time is thus lowered from minutes to seconds.

Interest is also being aroused in recent times in the use of lipases (enzymes which hydrolyse fat molecules) to act on grease stains. These are added to complement the action of protease.[19]

Bacteriostats: More and more cold wash detergent powders are coming on the market. Although the washing capabilities of these powders are quite satisfactory, especially now that enzymes are being added, there is a danger that clothes can be cross-infected with germs. The term germ does not necessarily mean disease carriers. Non-malignant bacteria or fungi can remain on washing and can develop objectionable odours. When high temperatures are used the germs are destroyed by a sterilization process. In cold washes bacteriostats need to be added to the powders. These are usually added in fractions of a per cent and are one of tribromosalicylanilide, dibromosalicylanilide, trichlorocarbanilide. These products do not act as antiseptic agents but prevent growth of bacteria in the concentrations used.

Due to various developments in the past, the detergent manufacturer should select a bacteriostat with care, paying attention to both health and ecological considerations. It is advisable, before making any decision on the choice of a bacteriostat, or any other disinfectant to be added to a detergent, that the manufacturer gets advice from a pharmaceutical firm specializing in this field, with adequate guarantees for safe use.

AMINES

In addition to inorganic alkaline builders, alkaline organic materials

AMINES

—the amines—are also used in large quantities for the manufacture of detergents, but they cannot be called builders as such. It is true that these materials occasionally appear as 'builders', that is, they are used to add alkalinity to a detergent, usually a liquid, but more often than not they are used purely to neutralize the sulphonic acid, or as reagents in the manufacture of alkylolamides.

These amines, all of which are derivatives of ammonia, can be divided into two classes, the alkylolamines and the alkylamines.

Alkylolamines: These are all reaction products of ethylene oxides or propylene oxide and ammonia, manufactured according to the following typical reactions:

$$NH_3 + H_2C\underset{O}{\overset{}{\diagdown\diagup}}CH_2 \rightarrow H\overset{H}{\underset{}{N}}-CH_2CH_2OH$$

Ethylene oxide   Monoethanolamine

$$NH_3 + 2H_3C-\underset{O}{\overset{H}{\underset{}{C}}\diagdown\diagup}CH_3 \rightarrow \quad \begin{array}{c} CH_3-COH-CH_3 \\ HN-COH-CH_3 \\ | \\ CH_3 \end{array}$$

Propylene oxide   Di-isopropanolamine

The important alkylolamines available to the detergent and polish manufacturer are mono-, di-, and tri-ethanolamine; mono-, di-, and tri-isopropanolamine and mixed isopropanolamine which is a mixture of the three isopropanolamines with an average molecular weight somewhat higher than that of diisopropanolamine. Triethanolamine, the most important and most widely used of all the alkylolamines, is also not a pure product, containing usually some 85 per cent pure triethanolamine, the balance being a mixture of mono- and di-ethanolamines. A pure product is available commercially, but the extra expenditure is not warranted in the manufacture of detergents and polishes.

In addition, there are available several alkyl-alkylolamines, for example diethylethanolamine of the structural formula:

$$CH_3CH_2-\underset{}{\overset{CH_2CH_3}{\underset{|}{N}}}-CH_2CH_2OH$$

and morpholine, very important in polish manufacture, less so for detergents, of the formula:

$$\begin{array}{c} CH_2-CH_2 \\ / \quad \quad \backslash \\ O \quad \quad \quad NH \\ \backslash \quad \quad / \\ CH_2-CH_2 \end{array}$$

Aminoethylethanolamine, $H_2NCH_2CH_2NHCH_2CH_2OH$, is used as the combined esterification and cyclization agent in the production of imidazolines and some amphoteric detergents.

A recent development is diglycolamine (2-(2-amino-ethoxy)-ethanol (Jefferson Chemicals, USA) of the structural formula:

$$H_2N-CH_2-CH_2-O-CH_2-CH_2OH.$$

This is isomeric with diethanolamine, but behaves rather like monoethanolamine or monoisopropanolamine when neutralizing sulphonic acids. However, the structure of diglycolamine is as if one molecule of ethylene oxide had been condensed with monoethanolamine; hence anionic detergents and alkylolamides produced with diglycolamine tend to be more soluble because of the effect of the ethylene oxide.[20]

Alkylamines: The alkylamines of the general formula:

$$R''-\underset{\underset{R}{|}}{N}-R'$$

where R' and R'' can be either hydrogen or an aliphatic radical, are used to neutralize both fatty acids and alkyl benzene sulphonic acids for both detergent and emulsification purposes. The alkyl radicals are usually propyl and butyl, although both higher and lower homologues are often used.

A cyclic alkylamine occasionally used in the manufacture of polishes is cyclohexylamine, $C_6H_{11}NH_2$.

The alkylolamines, when used to neutralize detergents, and the lower monoalkylamines, when used to neutralize detergents and fatty acids, yield products which approach liquids and are more soluble than the corresponding sodium salts in water. In addition, they are also soluble in hydrocarbon and chlorinated solvents, the isopropanolamine and the alkylamine salts being more soluble than the ethanolamine salts. The higher alkylamines in particular can be used for neutralization without the addition of any water or solvent, so that a substantially 100 per cent product can be obtained. The alkylolamines, however, produce a very viscous material and need a small amount of water or other solvent to render the final product into a form that can be easily handled.

The ethanolamines are used for the preparation of liquid detergents, emulsifiers, and solvent detergents.

The isopropanolamines can be used for the same purposes, and the alkylamines and the isopropanolamines are used for the preparation of detergents soluble in non-aqueous solvents, for example, in dry-cleaning preparations.

The important characteristics of the various alkylolamines and alkylamines are given in Table 4.2.

From the table it will be noted that diethanolamine is a solid, except in very hot climates, and that diisopropanolamine and triisopropanolamine are also solids. For this reason, triisopropanolamine is very rarely used in the detergent industry, and the di-alkylolamines are used only when their special properties are required. Triethanolamine is most commonly used for liquid detergents. The commercial variety employed normally contains only 85 per cent triethanolamine, the remainder being a mixture of mono- and diethanolamine.

For the production of solvent soluble detergents, monoethanolamine, monoisopropanolamine and isopropylamine and the higher alkylamines are commonly used. Because of the very low boiling point of isopropylamine, special precautions need to be taken when using low molecular alkylamines.

With fatty acids, morpholine forms excellent emulsifying agents where water resistance of the dried emulsion is desired. This is because morpholine soaps, when applied in a thin film (as for a polish), evaporate the morpholine quickly with the water, leaving a wax film resistant to water spotting soon after application. On occasion, however, it is necessary to remove a wax film which has become scuffed. Triethanolamine, when used as the neutralization agent for the fatty acids, gives a film with the easiest removal but the least water resistance. The use of 2-amino-2-methyl-l-propanol will give a compromise between these two extremes, that is, good water resistance and also moderate removal when required.

In agricultural emulsifiers, the calcium salts of alkylbenzene sulphonic acids play an important part. Where calcium ions cannot be tolerated, aminoethylethanolamine can be used as a divalent neutralizing agent.

Solvents

In addition many solvents are being used in modern detergents, even powders. When the soiling is either greasy or oily, solvents aid greatly in removing this soil from the article being cleaned.

(a) *Pine Oil:* A very important product, obtained chiefly in America, is the so-called pine oil, which is a terpene-rich oil produced during wood carbonization, as the fraction before pine tar; it has also been obtained by dry distillation and, more recently, by extraction of chips of resinous tree stumps. Pine oil may be regarded as the transition stage between true turpentine and resin. Its approximate composition, according to Pickett

TABLE 4.2

Properties of Alkylolamines and Alkylamines

Amine and formula	Molecular weight	Specific gravity 20/20°C	Specific gravity 60/4°C	Freezing-point °C	Boiling-point °C	pH 5% soln	Flash-point °C
Monoethanolamine $H_2NCH_2CH_2OH$	61·5	1·0113	—	10·3	170–2	11·7	90·56
Diethanolamine $HN(CH_2CH_2OH)_2$	105·1	—	1·0693	27·5	168–9 2·66 kPa	10·7	148·89
Triethanolamine $N(CH_2CH_2OH)_3$	142·0*	1·1205	—	17·9†	175–91 0·67 kPa	10·3	185·00
Monoisopropanolamine $H_2NCH_2CHOHCH_3$	75·1	0·9640	—	1·0	159–63	12·1	73·89
Diisopropanolamine $HN(CH_2CHOCH_3)_2$	133·2	—	0·9800	42·0	248–54	11·5	110·00
Triisopropanolamine $N(CH_2CHOHCH_3)_3$	191·3	—	1·0100	60·0	300–5	11·1	160·00
Mixed Isopropanolamines	140·0	1·0060	—	−26·0	—	11·6	110·00
Morpholine HNC_4H_8OH	87·1	1·0020	—	−3·0	126–30	11·8	37·78
2-amino–2-methyl–1-propanol $CH_3CNH_2CH_3CH_2OH$	89·1	0·9340	—	30–31	165	11·8	65·56
Diethylaminoethanol $(C_2H_5)_2NC_2H_4OH$	117·2	0·8850	—	−2·0	157–65	11·7	60·00
Diglycolamine	105·1	1·0572	—	−12·5	207–30	11·8	126·67
Aminoethylethanolamine $H_2NCH_2CH_2NCH_2CH_2OH$	104·2	0·9837	—	−2·0	243·8	10·7	265·00

TABLE 4.2 (continued)

Properties of Alkylolamines and Alkylamines

Amine and formula	Molecular weight	Specific gravity 20/20°C	Specific gravity 60/4°C	Freezing-point °C	Boiling-point °C	pH 5% soln	Flash-point °C
Cyclohexylamine $C_6H_{11}NH_2$	99·0	0·8647	—	—	132–6	12·5	37·78
Isopropylaminoethanol $(CH_3)_2CHNHCH_2CH_2OH$	116·0‡	0·9250	—	—	190·5	12·5	73·87
Dimethylamine $HN(CH_3)_2$	45·1§	—	—	—	—	12·5	—
Diethylamine $HN(CH_2CH_3)$	73·1	0·7000	—	—	53–59·5	12·3	<−17·78
Trimethylamine $N(CH_3)_3$	59·1§	—	—	—	—	12·0	—
Trimethylamine $N(CH_2CH_3)_3$	101·2	0·7300	—	—	85–91	11·8	−6·67
Propylamine $H_2N(CH_2)_2CH_3$	59·1	0·7180	—	—	46·5–52	12·4	<−17·78
Isopropylamine $H_2NCH(CH_3)_2$	59·1	0·6900	—	—	30·5–34·5	12·2	<−17·78
Di-n-propylamine $HN(CH_2CH_2CH_3)_2$	101·2	0·7400	—	—	105–15	11·8	7·22
Butylamine $H_2N(CH_2)_3CH_3$	73·1	0·7450	—	—	74·5–81	12·5	−12·22
Di-butylamine $HN(C_4H_9)_2$	129·2	0·7600	—	—	153–65	11·3	57·22

Notes: * Equivalent weight of the commercial product. The molecular weight of the pure material is 149·2.
† The commercial product tends to supercool considerably below this.
‡ Equivalent weight of the commercial product which is a mixture of homologues.
§ These materials are gaseous under normal conditions and are supplied commercially as water solutions containing 25 per cent of the amine.

and Schantz[21] is: terpenes 5–10 per cent, borneol 5–10 per cent, fenchyl alcohol 5–10 per cent, α-terpineol 50–60 per cent, other terpineols 15–25 per cent, esters 5–10 per cent, ketones and phenols 1–2 per cent. This solvent has a distillation range of 190–220°C.

Four types of pine oil are available:

> steam-distilled pine oil;
> destructively distilled pine oil;
> synthetic pine oil—made by partial oxidation of terpene hydrocarbons to terpene alcohols;
> sulphate pine oil.

Although pine oil is insoluble in water it has the valuable property of coupling solvents with water, particularly in the manufacture of solvent–detergent mixtures, where without pine oil there would be a tendency to separation into two phases. This property varies with the type of pine oil, which in turn varies with the amount of terpene alcohols present, the more alcohols present the greater the coupling effect.

Another important property of pine oil is its bactericidal effect. It is 1·5 to 4 times stronger than phenol[22] tested against *E. typhosa*. This property can make it an important ingredient of many types of liquid detergents (see p 206).

(*b*) *Chlorinated Solvents:* These solvents are widely used in special cleaner, paint-stripper and dry-cleaning detergents. Table 4.3 gives the main properties of the most important chlorinated solvents used in the detergent and related industry.

These solvents are all toxic to a greater or lesser degree. Carbon tetrachloride, the most toxic, is almost never used in these times. The use of trichloroethylene is being phased out. Perchlorethylene, the least toxic of these solvents still has a maximum allowable concentration in air (MAC) of 100 ppm. 1,1,1-trichorethane is considered safe with adequate ventilation, but unless a stabilized form is used, it can hydrolyse in the presence of water to form HCl. Environmental problems such as discharge into sewers should also be taken into account when formulating with these solvents. It is our considered opinion that where possible these solvents should be dispensed with (see p 290).

(*c*) *Alcohols, Glycols, Glycol Ethers, Esters:* These solvents are all of a distinctly polar nature and although not all are completely water-soluble, they all display a modicum of water solubility and are also miscible in most aromatic, paraffinic and chlorinated solvents. They thus serve as 'co-solvents' or coupling agents to combine water with solvents in special cleaning formulations.

In addition these solvents can be used to lower the viscosities of, for example, ethoxylated alcohols or the cloud points of alkyl benzene sulphonates.

Methanol is not often used as a solvent or coupling agent in detergent

SOLVENTS

TABLE 4.3

Properties of the Most Important Chlorinated Solvents

	Formula	Molecular Weight	Specific Gravity at 20°C	Boiling Point °C	Solubility of Water in Solvent at 25°C g/100 g
Carbon Tetrachloride	CCl_4	153·84	1·59	76·5	0·013
Trichlorethylene	$ClCH=CCl_2$	131·40	1·46	86·9	0·032
Perchlorethylene	$Cl_2C=CCl_2$	165·85	1·62	121·2	0·015
Methylenechloride	CH_2Cl_2	84·94	1·326	40·1	0·18
1, 1, 1-Trichlorethane	CCl_3-CH_3	133·42	1·304	74·1	—
1, 1, 2-Trichlorethane	CCl_2-CH_2Cl	133·42	1·44	113·5	0·24
Ethylene dichloride	CH_2Cl-CH_2Cl	98·95	1·256	87·1	0·16
Ortho-dichlorbenzene	$C_6H_4Cl_2$	147·01	1·306	180·4	—

With the exception of ethylene dichloride and ortho-dichlorbenzene all of these materials are non-inflammable.

formulations but is the esterification agent for the production of methyl esters. Again this is toxic, with a MAC of 200 ppm. The organoleptic threshold value for the alcohol is 2000 ppm, ten times the MAC which means that when the smell of the solvent is noted, it is far above the concentration which can be tolerated.

Particular mention must be made of the glycol ethers as they are being used more and more in all types of detergent formulations. Produced originally by Union Carbide under the names Cellosolve and Carbitol, they are now made by many other companies, one in particular being Dow under the generic name Dowanol.

The original glycol ethers were made by reacting methanol, ethanol or butanol with one or two moles of ethylene oxide, thus obtaining a large spread of both physical and chemical properties.

Dow has now extended the range by reacting the various alcohols with propylene oxide.

These solvents contain both an alcohol and at least one ether group making them both water-soluble and excellent fat and oil solvents. In addition to their solvent power these glycol ethers have an intrinsic low surface tension, therefore they aid the surfactant in one of its functions, the lowering of the surface tension of the water solution. Being alcoholic they are also excellent solvents for all types of detergent active matter. However, one drawback is that they will lower the viscosity of the finished product if it is a liquid.

TABLE 4.4

Properties of Solvents which are Also Miscible with Water

Solvents	Formula	Molecular weight	Boiling Range initial bp °C	Boiling Range end-point °C	Flash-point °C	Specific gravity at 20°C
Methanol	CH_3OH	32·03		64·5	15·6	0·792
Ethyl alcohol (pure)	C_2H_5OH	46·05	77	79	18·33	0·791
Isopropyl alcohol (pure)	$(CH_3)_2 \cdot CHOH$	60·06	82	83	19·44	0·786
Isobutyl alcohol (pure)	$(CH_3)_2CH_2CHOH$	74·08	107	111	43·89	0·803
Acetone	$CH_3 \cdot CO \cdot CH_3$	58·05	55	57	−17·78	0·793
Methyl ethyl ketone	$CH_3 \cdot COOC_2H_5$	72·06	77	82	1·11	0·809
Ethylacetate (pure)	$CH_3COOC_2H_5$	88·06	70	80	1·67	0·886
Methyl 'Cellosolve'	$CH_3OCH_2CH_2OH$	76·06	121	126	40·56	0·966
'Cellosolve'	$C_2H_5OCH_2CH_2OH$	90·08	133	137	43·87	0·931
Isopropyl 'Cellosolve'	$(CH_3)_2CHOCH_2CH_2OH$	104·09	140	143	51·67	0·906
Propylene glycol methyl ether	$CH_3OC_3H_6OH$	90·08	117	125	94	0·919
Butyl 'Cellosolve'	$C_4H_9OCH_2CH_2OH$	118·11	163	172	73·87	0·902
Methyl 'Carbitol'	$CH_3OCH_2CH_2OCH_2CH_2OH$	120·09	190	194	93·33	1·035
'Carbitol'	$C_2H_5OCH_2CH_2OCH_2CH_2OH$	134·11	189	203	96·11	1·027
Butyl 'Carbitol'	$C_4H_9OCH_2CH_2OCH_2CH_2OH$	162·14	220	231	110·00	0·955
Benzyl 'Cellosolve'	$C_6H_5CH_2OCH_2CH_2OH$	152·09	254	258	129·44	1·070
Dipropylene glycol methyl ether	$CH_3O(CH_2CHO)_2H$ $\quad\quad\quad\quad\;\; CH_3$	148·13	184	193	175·00	0·950
Ethylene glycol	$HOCH_2CH_2OH$	62·10	194	200	115·56	1·113 (25°)
Diethylene glycol	$HOCH_2CH_2OCH_2CH_2OH$	106·10	240	251	135·00	1·116 (25°)
Triethylene glycol	$HOCH_2CH_2OCH_2CH_2OCH_2CH_2HO$	150·20	275	295	154·44	1·124 (25°)
Propylene glycol industrial	$CH_3CHOHCH_2OH$	76·10	185	190	101·67	1·036 (25°)
Dipropylene glycol	$CH_3CHOHCH_2OCH_2CHOHCH_3$	134·20	220	240	121·11	1·023 (25°)
Hexylene glycol	$CH_3\!-\!\underset{\underset{CH_3}{\mid}}{\overset{\overset{\mid}{\;}}{C}}\!-\!CH_2\!-\!CHCH_3$	118·17	195	199	96–99	0·922

The most widely used glycol ether hitherto was ethylene glycol monobutyl ether, which suffered from the drawback of having a slight but penetrating odour, which was difficult to mask with a perfume. The propylene based products can give equal performance with no real appreciable odour.

Physical data of some of these solvents are given in Table 4.4.

(d) *Hydrocarbon Solvents:* Water-immiscible hydrocarbon solvents are also used in various solvent detergents. Their boiling points can range from 100°C to 200°C for use in spotting detergents and from 150°C to 300°C for metal cleaning and maintenance products. Deodorized kerosine with a boiling range of 170°C to 260°C is used for odourless products. Hydrocarbon solvents are inflammable with flash points between 15°C and 25°C for the low boiling species, around 30°C for common kerosine, and 50°C for the deodorized variety which is commonly used for such specialized products as waterless hand cleaners.

References

1. *The Chemistry and Rheology of Water Soluble Gums.* Society of Chemical Industry, Monograph No. 24, p 65.
2. Niewenhuis, *Journal of Polymer Science*, XII, 237 (1954).
3. Milwidsky, B. M., *Soap*, **39**, 4, 53 (1963).
4. Hatch, L. F., *Hydrocarbon Processing*, 79 (March 1975); Segalas, H. A., *Hydrocarbon Processing*, 74 (March 1975).
5. Findley, W. R., Paper presented at CSMA Seminar (June 1985).
6. *US Patent* 4,166,718, Ciba-Geigy Corporation.
7. Parker, J., Paper delivered at CSMA Seminar (June 1985).
8. *US Patent* 2,544,649, Dow Chemical Corp.
9. Niven, *Industrial Detergency*, p 48, Reinhold Publishing (1955).
10. *British Patent* 963,135, Unilever Ltd.
11. Technical publication, GAF Corporation.
12. *US Patent* 3,870,648, GAF Corporation.
13. Nijs, H., Godecharles, V. and May, B. H., *Seifen–Oele–Fette–Wachse*, **111**, 149–150, 203–205 (1985).
14. Getty and Streicher, *Soap*, **38**, 54 (1962).
15. Booth and Everson, *Ind & Eng Chem*, **40**, 1491 (1948).
16. Mausner, N. and Raine, E., *Soap*, **44**, 8, 34 (1968).
17. Langguth, R. P. and Mecey, L. W., *Soap*, **45**, 9, 60 (1969).
18. *New Scientist & Science Journal*, 556 (3 June 1971).
19. Andree, W. R., Mueller, W. R. and Schmid, R. D., *Journal of Applied Biochemistry*, **2**, 218 (1980).
20. Knoggs, *Soap*, **40**, 12, 79 (1964).
21. *Ind. & Eng. Chemistry*, **26**, 709 (1934).
22. McCullagh, E. C., *Disinfection and Sterilization*, 2nd edn, p 368: Lea & Ferbiger, Philadelphia (1945).

5. Synthesis of Detergents

Some information on the structure of detergents has already been given in Chapter 1. We will now give descriptions of industrial processes for the synthesis of detergents. In accordance with the aim and purpose of this book, we have selected for description in a more detailed manner those production processes which the practical man in the detergent industry is most likely to encounter. Thus sulphonation, ethoxylation, and fatty acid condensation with alkylolamines are described in detail, whereas processes for the synthesis of basic raw materials, such as the synthesis of fatty alcohols and alkyl benzene, are discussed in more general terms, to supply basic knowledge about the raw materials used for processing steps carried out in the detergent industry. For basic knowledge on the structure of a wider range of synthetic detergents the reader is referred to Chapter 1 and to the Selected Bibliography.

Raw Materials for Anionic Synthetic Detergents

Fatty Alcohols, Natural and Synthetic

Historically the first commercial synthetic detergents in our present sense of the term were derived from natural fatty materials. In the last century 'sulphonated' oils were manufactured from castor, fish and olive oils, and these still occupy a place of their own in the textile and leather industries. However, syndets as cleaning materials were based initially on spermacetti, then sperm oil, coconut oil and tallow. The first two are basically mixed esters of the cetyl palmitate type, the latter two are glycerides, coconut oil having fatty acids in the C_{12-14} range and tallow C_{16-18}.

In the later 1920s and early 1930s the fatty alcohol of the esters was obtained by saponification and subsequent distillation. The whale was not then considered to be an endangered species, but for all that attention was diverted to the much more abundant glycerides. Reduction processes were developed and coconut and tallow (also palm oil) quickly superseded sperm oil as a source of fatty alcohols.

The process used for the reduction of the glyceride to the alcohol (it also works on esters) was reduction by metallic sodium in the presence of a reducing (hydrogen donating) lower alcohol (Bouveault-Blanc), developed originally in the early part of this century but only commercialized when metallic sodium became available as a commodity in the early 1930s.

The reaction can be written:

$$RCOOR' + 4Na + 2R''OH \rightarrow RCH_2ONa + R'ONa + 2R''ONa$$

and the sodium alcoholates are then hydrolysed with water and. fractionated. The reducing alcohols can be butanol, cyclohexanol or 4-methyl-2-pentanol. The use of metallic sodium requires that all the reactants be completely anhydrous.

The reduction is only on the ester linkage and all double bonds originally present remain intact.

With the development of high-pressure hydrogenation in the organic process industry and its application to the 'oil-hardening' process for natural oils, attention was directed to the conversion of fats and oils by this means. Catalysts and process conditions were rapidly developed and the sodium reduction process fell into disfavour, only one plant having been built internationally since the early 1950s.

High-pressure hydrogenation was originally carried in batch processing units. More recently, however, continuous plants have been introduced. Such plants may use fats, fatty acids or preferably methyl esters (p 130). Pressure is high, between 30 and 32 MPa, and the temperature is between 300 and 320°C. Earlier, somewhat lower pressures but even higher temperatures had been used. However, for product quality it is better to use a higher pressure and a lower temperature for the reaction. In this way residence time is reduced to a minimum and the temperature not higher than 320°C maximum. The catalyst is usually copper chromite, sometimes promoted by cadmium or cerium compounds. The catalyst is well dispersed in the fatty acids or methyl ester, and dosed by dosing pump into the reaction system proportionally to the required hydrogen. Sometimes it is recommended to pre-mix the catalyst with a fraction of the fatty alcohol obtained in the process. As can be seen from the flow diagram (Fig 5.1), part of the spent catalyst is recycled and fresh catalyst added.

Typical process data are as follows:

For 1000 kg tallow fatty alcohol from distilled tallow fatty acids:
1060 kg distilled fatty acids
5–8 kg catalyst
hydrogen 270 m³
cooling water 50 m³
electricity 350 kWh
yield 94 per cent calculated on the fatty acids

The figures will vary slightly for coconut fatty acids or their methyl ester owing to their lower molecular weight. They would require somewhat more hydrogen and also slightly more electricity, steam and cooling water. Approximately 25–30 per cent more for coconut fatty acid, some 20 per

SYNTHESIS OF DETERGENTS

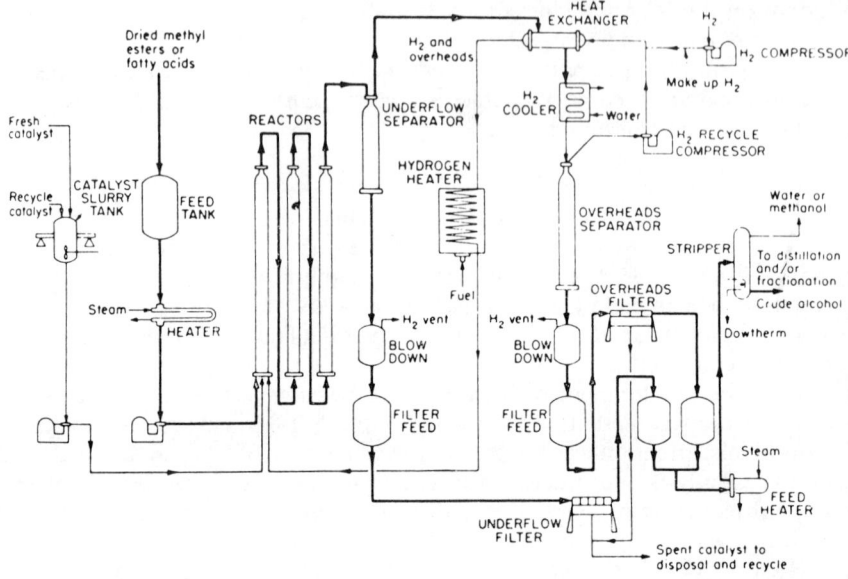

Fig 5.1 Flow chart for high-pressure hydrogenolysis to obtain fatty alcohols from natural fats

Reproduced from *Encyclopaedia of Chemical Technology*, Kirk & Othmer, by kind permission of John Wiley & Sons Inc

cent more for coconut methyl ester. The yield would be 92·5 per cent calculated on distilled coconut fatty acids.

During the hydrogenation process, the unsaturated fatty acids are transformed into saturated fatty acids of the same molecular weight, and, of course, the amount of hydrogen consumed will be correspondingly higher for unsaturated fatty acids than for saturated fatty acids or their methyl esters.

By altering the process conditions somewhat and using a selective catalyst ($Al_2O_3/CdO/Cr_2O_3$) the double bonds present can be retained in the final alcohol.

A detailed description of high-pressure hydrogenation of natural fat and oil raw materials and their transformation into fatty alcohols is given by E. F. Hill, G. R. Wilson and E. C. Steinle, Jr.[1] The chemical reaction for high-pressure reduction of fats and oils, or 'hydrogenolysis', as it is often called, is as follows:

$$R-\overset{O}{\underset{\|}{C}}-OH + 2H_2 \rightarrow RCH_2OH + H_2O$$

A more recent publication by E. Haidegger and Ll Hodossy[2] gives details of hydrogenation procedures as carried out in a pilot plant at the Institute for Highpressure Research at the University of Budapest. The authors give the equation and heat of reaction for fatty acid esters of lower alcohols and triglyceride hydrogenation. (The heat of reaction is for the hydrogenation of the COOR group only and not for the transformation of unsaturated double bounds in case unsaturated feedstock is used.)

$$R_1COOR_2 + 2H_2 \rightarrow R_1CH_2OH + R_2OH$$
$$\triangle H_R \approx -500 \text{ kJ}$$

$$\begin{array}{l} CH_2OCOR \\ | \\ CHOCOR \\ | \\ CH_2OCOR \end{array} + 8H_2 \rightarrow \begin{array}{l} CH_3 \\ | \\ CHOH \\ | \\ CH_3 \end{array} + 3RCH_2OH + 2H_2O$$
$$\triangle H_R \approx -105 \text{ kJ}$$

Virtually all plants for the production of alcohols by high-pressure hydrogenation use as starting materials the methyl esters. These have now become an important intermediate in the industry for at least two further processes. Their manufacture will be described later in this chapter.

For full process evaluation with regard to labour, raw material balance and utility requirements we refer to an article by G. R. Wilson[3] and the comprehensive book on fatty alcohols produced by Henkel.[4]

Figure 5.1 shows a detailed flow sheet for high-pressure hydrogenation.

From the late 1940s onwards plants using both OXO and Ziegler chemistry came on stream to manufacture synthetic alcohols. These were based on petroleum and with the low price of crude oil at the time, synthetically produced alcohols quickly surpassed natural alcohols in volume. With the great increase in the price in crude and with the realization that fossil fuels are a diminishing resource and that fats and oils can be obtained from self-replenishing sources, natural fatty alcohols are being favoured somewhat with the proviso that the price of the base oils fluctuates more than does the price of petroleum. A pointer is that one of the large European producers of synthetic alcohols has extended its capacity by installing an additional plant for high-pressure hydrogenation of fats and oils.

In essence, the OXO process uses an olefin, carbon monoxide and hydrogen as the raw materials to form an aldehyde with one carbon more than was originally present in the olefin[5] thus:

$$R-CH=CH_2 + CO + H_2 \xrightarrow[\text{high pressure}]{\text{catalyst}} \begin{array}{l} R-CH_2-CH_2-CHO \\ \text{or} \\ R-CH_2-CH-CHO \\ \qquad\qquad | \\ \qquad\qquad CH_3 \end{array}$$

and this aldehyde is then reduced to an alcohol.

As mentioned above, the alcohol produced contains one carbon more than the original olefin, and thus does not make OXO alcohols comparable with the alcohols produced from natural sources as these are always even-numbered carbon chains, and also the OXO alcohol need not necessarily be straight-chained.

Normal alcohols are always produced by this process, but branched olefins will give branched products, linear olefins will give a degree of branching, depending on the catalyst used. The original catalyst was an oil-soluble salt of cobalt (cobalt naphthenate for example) which was converted in the reaction conditions firstly to dicobalt octacarbonyl, $Co_2(CO)_8$, then to cobalt hydrocarbonyl, $HCo(CO)_4$. Shell uses a modified catalyst, cobalt-carbonyl-organophosphate complex, $Co_2(CO)_6[(C_4H_9)_2P]_2$, again formed *in situ*, and this catalyst, with the different reaction conditions acts also as a hydrogenation catalyst, producing alcohols rather than aldehydes, and the branching is not more than 20 per cent (2-methyl isomer), the balance being linear alcohols.

The Ziegler process, first commercialized by Continental (now Vista Chemicals) in the USA and its associate company in Germany, Condea (both using the trade name ALFOL) and shortly afterwards by Ethyl in the USA (trade name EPAL) involves reacting metallic aluminium with hydrogen gas plus aluminium triethyl to yield diethylaluminium hydride:

$$Al + 3/2H_2 + 2Et_3Al \rightarrow 3Et_2AlH$$

The hydride then reacts with ethylene to give three moles of aluminum triethyl, two of which recycle to the first step:

$$3Et_2AlH + CH_2 = CH_2 \rightarrow 3Et_3Al$$

The remaining triethyl reacts with more ethylene to produce a mixture of high-molecular-weight aluminum alkyls containing randomly distributed alkyl groups:

$$Et_3Al + nCH_2 = CH_2 \rightarrow R \cdot Al \begin{matrix} R' \\ \\ R'' \end{matrix}$$

This 'growth product' is then oxidized and hydrolysed:

$$AlR_3 + 3/2O_2 \rightarrow Al(OR)_3$$

$$Al(OR)_3 + 3H_2O \rightarrow 3ROH + Al(OH)_3$$

Ethyl uses sulphuric acid rather than water for the hydrolysis and aluminium sulphate is formed. When water is used high purity alumina is obtained.

The alcohols produced can range from C_6 to higher than C_{20}, depending on the process conditions. They are normally grouped round a peak following a Poisson distribution. Ethyl has modified the conditions to produce 'controlled peaking'. It is surmised that the company makes use of the fact that aluminium alkyls can crack at high temperatures forming dialkyl aluminium and an alpha olefin. These olefins are separated into high and low boiling fractions, the lower fraction is sent upstream for further growth and the higher fraction to another reactor for 'transalkylation', both reactors contacting the incoming streams of olefin with aluminium trialkyl. A narrower chain distribution of the final alcohol mixture is thus obtained. Table 5.1 shows a comparison of the distribution of alcohols produced by the normal Ziegler type plant and the Ethyl modification.

Reaction conditions can be controlled to shift the peak higher or lower, and the C_{14} alcohol (not very abundant in nature) can be produced at will. The alcohols in the C_{12} to C_{20} range are the important ones for the industry, the lower fractions have other uses—solvents, plasticizers, etc. Very seldom is a pure alcohol used as a detergent intermediate, blends are quite sufficient. It has been found that a blend of a (relatively) low-molecular-weight alcohol sulphate or ethoxylate with a higher molecular weight one can have synergistic effects.

In the East European countries petroleum is oxidized to fatty acids and then reduced to alcohols. These alcohols are straight-chain random secondary, not suitable for sulphonation but can be ethoxylated to non-ionic detergents.

As we shall see later in this chapter, fatty alcohols, either natural or fully synthetic, are used for three main synthetic detergent processes.[6]

1. Alcohol sulphates
    $CH_3(CH_2)_nCH_2OSO_3^-Na^+$
2. Alcohol ethoxylates
    $CH_3(CH_2)_nCH_2O(CH_2CH_2O)_mCH_2CH_2OH$
3. Alcohol ether sulphates
    $CH_3(CH_2)_nCH_2O(CH_2CH_2O)_mCH_2CH_2OSO_3^-Na^+$

Tables 5.2, 5.3, 5.4 and 5.5 give the specifications of commercial fatty alcohols produced by high-pressure hydrogenation of natural fats and oils, and of selected Ziegler alcohols.

TABLE 5.1

Comparison of Alcohol Distribution from Two Types of Ziegler Production Units

Carbon length	Normal distribution	Ethyl controlled peaking
6	9·6	1·4
8	16·9	3·2
10	20·7	7·7
12	19·4	34·5
14	15·1	26·3
16	9·8	16·7
18	5·3	8·9
20	3·2	1·3

Reproduced from *Fatty Alcohols*, Henkel KGaA, Dusseldorf, by kind permission of the publishers.

TABLE 5.2

Properties of Typical Commercial Fatty Alcohols derived from Oils and Fats

Name	Number of C-atoms	Iodine number	Solidification point °C	Boiling range °C/ 101 kPa	Hydroxyl number
Lauryl alcohol wide-range type	C_{10}—C_{18}	Less than 0·5	17–21	220–320	275–85
Lauryl alcohol	C_{12}—C_{16}	Less than 0·5	18–22	240–320	280–5
Lauryl alcohol c. 80% C_{12}	C_{12}—C_{14}	Less than 0·5	17–23	255–85	283–93
Myristyl alcohol c. 95% C_{14}	C_{14}	Less than 0·5	36–38	280–95	255–62
Cetyl alcohol c. 95% C_{16}	C_{16}	Less than 0·5	46–49	316–30	225–35
Stearyl alcohol c. 95% C_{18}	C_{18}	Less than 0·5	55–57	340–55	203–10
Tallow fatty alcohol	C_{14}—C_{18}	Less than 0·5	48–52	120–90	210–20
Oleyl-Cetyl alcohol mixture	Mainly C_{18} saturated and unsaturated	45–120*	4–35*	310–65	200–20

* The more unsaturated the mixture, the higher the iodine value and the lower the solidification point.

A new process for the conversion of alpha olefins to primary alcohols has been described by W. L. Welsh (Gulf Research and Development Co). The alpha olefins are reacted with dry hydrogen bromide in the presence of peroxides or ultra-violet light to give the primary alkyl bromides, which are then converted to an ester through metal-halide catalysed reaction with organic acid. The ester is then hydrolysed with superheated steam to yield the corresponding primary alcohols. The process has the advantage that anhydrous hydrogen bromide produced in the esterification and the organic acid can both be recycled. The Ziegler process consumes expensive compounds such as tri-butyl aluminium and boron hydride, while the OXO process for the production of secondary alcohols uses a cobalt catalyst over which olefins are passed in the presence of carbon monoxide and hydrogen.

Higher alcohols made by the Ziegler and modified OXO processes have one characteristic which make them particularly useful in detergent manufacture: both anionics and non-ionics (pp 152, 178) produced from them are easily degraded by the bacteria flora in normal sewage treatment plants.

TABLE 5.3

Properties of Pure Fatty Alcohols

Name of alcohol	Formula	Mol. weight	Melting-point °C	Boiling-point or boiling range °C	Hydroxyl number
Decanol	$C_{10}H_{21}OH$	158	7	231 (101·1 kPa)	355
Undecanol	$C_{11}H_{23}OH$	172	14	131 (1·98 kPa)	326
Dodecanol (lauryl alcohol)	$C_{12}H_{25}OH$	186	24	135–7 (1·33 kPa)	301
Tridecanol	$C_{13}G_{27}OH$	200	30	155 (1·73 kPa)	280
Tetradecanol (myristic alcohol)	$C_{14}H_{22}OH$	214	38	159–61 (1·33 kPa)	262
Pentadecanol	$C_{15}H_{31}OH$	228	44		246
Hexadecanol (palmitic alcohol or cetyl alcohol)	$C_{16}H_{31}OH$	242	49	179–82 (1·59 kPa)	232
Heptadecanol	$C_{17}H_{35}OH$	256	54		219
Octadecanol (stearyl alcohol)	$C_{18}H_{37}OH$	270	58	202 (1·33 kPa)	208
Octadecanol (oleyl alcohol)	$C_{18}H_{35}OH$	268	15–16	177–83 (0·40 kPa)	209

OLEFINS

Olefins are required as intermediates for three kinds of detergent materials, alpha olefins for the OXO process and for sulphonation, and internal olefins as alkylation reagents for alkylbenzene and alkyl phenol.

Alpha Olefins: Cracking of hydrocarbons to produce lighter fractions has been practised for many decades, both by catalytic and thermal processes. For the production of alpha olefins thermal cracking is preferred. The main reaction can be shown:

$$C_nH_{2n+2} \rightarrow C_{n'}H_{2n'} + C_{n''}H_{2n''+2}$$

where $n' + n'' = n$ and n'' is a low integer.

A linear paraffin will produce a linear olefin. Several side reactions are possible but a highly refined paraffin wax will yield 80–90 per cent alpha olefins with a small amount of branched olefins, some internal and di-olefins.

The refined wax is pre-heated, mixed with steam (to avoid coking) and fed to a cracking furnace at a temperature of about 550°C under pressure two to four times atmospheric. A low residence time is necessary (about 10

SYNTHESIS OF DETERGENTS

TABLE 5.4

Typical Composition and Properties of Selected Alfol Alcohol Blends

	Alfol 1014 CDC	Alfol 1214	Alfol 1412	Alfol 1216	Alfol 1218 DCBA	Alfol 1618
Individual alcohol content %						
C_{10}	31	0·6	0·6	0·3	0·7	—
C_{12}	37	55	38	64	39	—
C_{14}	31	44	60	24	30	1·0
C_{16}	0·8	0·9	1·9	11·	18	58
C_{18}	—	—	—	tr.	11	39
C_{20}	—	—	—	—	0·7	2·4
Colour APHA max	0	30	20	30	30	40
Density 22/22°C	0·834	0·836	0·839	0·840	0·840	0·814*
Freezing point °C	5	22	23	20	21	46
Average molecular weight	184	198	203	198	211	253
Flash point PMCC °C	121	129	132	129	135	163
Hydroxyl number	302	277	272	276	260	215
Iodine number	0·07	0·1	0·1	0·1	0·2	0·9

* Density at 60/4°C.
The name Alfol is a registered trade mark of the Continental Oil Company, Ponca City, Oklahoma, USA (now Vista Chemical Company).

seconds), this maximizes alpha olefin production, and cracking conversion is kept low (20–40 per cent per pass).

The vapour exiting from the furnace is quenched with water and the uncondensed vapour is separated from the liquid in a flash tower. These vapours are then fractionated to remove the (unwanted) lighter fraction and distilled into the various commercial cuts.

The quenched liquid is recycled to the cracker. The light fraction contains ethylene, propylene and butylene, and needs to be disposed of economically, possibly with a link to a lower olefin plant.

Published specifications for linear alpha olefins of wax origin indicate that the branched alpha olefin content is fractional, internal olefins can be from 5 to 10 per cent, di-olefins up to 4 per cent, saturates up to 4 per cent with fractional amounts of aromatics. If it is desired to sulphonate wax-cracked olefins, the exact composition should be considered as some of the extraneous constituents can adversely affect the quality of the sulphonate. The olefins will of course have both even and odd numbered carbon chains.

As mentioned in describing the Ethyl modification of the Ziegler process for the manufacture of alcohols, olefins can also be produced by this growth reaction. Alpha olefins based on Ziegler chemistry are produced by Gulf and Ethyl, each of course with its own variations. After

TABLE 5.5

Typical Composition and Properties of Selected Epal Blends

	Epal 12/85	Epal 1214	Epal 1218	Epal 1416	Epal 1418	Epal 1618
Individual alcohol content %						
C_{10}	0·1	—	0·3	—	—	—
C_{12}	86	66	48	0·3	0·7	—
C_{14}	13	27	20	62	35	1·5
C_{16}	0·01	7	17	36	40	47
C_{18}	—	—	14	1·4	23	50
C_{20}	—	—	0·3	—	0·8	2
Colour APHA max	5	5	5	5	5	5
Density	0·831*	0·831*	0·834*	0·825†	0·825†	0·819‡
Iodine value	0·02	0·02	0·34	0·2	0·4	1·0
Hydroxyl number	297	283	266	250	235	219
Average molecular weight	189	198	211	224	239	256
Flash point PMCC °C	137	137	125	143	149	202
Freezing point °C	20	22	25	37	42	46

* At 25°C.
† At 40°C.
‡ At 50°C.
The name Epal is a registered trade mark of the Ethyl Corporation, Baton Rouge, LA, USA.

the aluminium alkyl has grown to the required length, the temperature is raised and a low-molecular-weight olefin (ethylene) is added. The ethylene replaces an alkyl from the aluminium trialkyl, yielding an alpha olefin.

Olefins produced by this method are of even numbered carbon lengths. Table 5.6 indicates the typical values for detergent grade olefins produced by Ethyl. It will be noted that for the higher molecular weight products the amount of branched terminal olefins is high. Despite this the sulphonated alpha olefin produced from the 1416 olefin biodegrades at a faster rate than LAS as indicated by Ethyl.[6]

The Shell SHOP process does not use Ziegler chemistry, rather high purity ethylene is contacted with a nickel catalyst dissolved in a solvent immiscible with the finished olefin (butanediol). The process is carried out at a moderate temperature but high pressure when oligomerization, isomerization and disproportioning of undesired chain lengths causes the formation of olefins of surfactant range (C_{11}–C_{15}) and of high purity.

Linear Internal Olefins: Internal olefins are one of the preferred feedstocks for the alkylation of benzene as these are random in nature. Sweeney has pointed out[7] that the presence of 2- and 3-phenyl alkanes impairs the

Table 5.6

Typical Composition and Properties of Selected Ethyl Alpha Olefins

	Tetra decene-1	Hexa decene-1	Octa decene-1	C12–14	C14–16	C16–18
Carbon no. %						
C_{10}	—	—	—	0·3	—	—
C_{12}	2	—	—	62	1	—
C_{14}	96	2	—	36	65	1
C_{16}	2	97·5	18	1·2	33	55
C_{18}	—	0·5	81	—	1	36
C_{20}	—	—	1	—	—	8
Mono-olefin %	99·6	99·1	98·9	99·6	99·6	99·2
Paraffin %	0·4	0·9	1·1	0·4	0·4	0·8
Olefin isomers mol %						
Linear terminal	81	72	59	87	76	63
Branched terminal	14	21	35	9	19	29
Linear internal	5	7	6	4	5	8
Colour APHA	All less than 5					
Density 20°C	0·760	0·780	0·788	0·764	0·776	0·787
Flash point °C	107	128	143	81	113	135
Boiling range °C						
IBP	245	276	298	216	245	285
FBP	250	283	316	250	279	316

Reproduced by permission from Ethyl Corporation, Technical Brochure.

detersive effects of the final sulphonate, whereas by increasing the 5-, 6- and 7- isomers the detergency is increased. Internal olefins can also theoretically be used as feedstock for OXO alcohol production but here the alpha type is preferred.

Kerosene is the normal base material but the term is very loose as refineries produce to physical rather than to chemical constants. The chemical characteristics depend on the crude being used, but a kerosene with a median boiling point of 220°C (in practice IBP 190, FBP 250) will contain up to 25 per cent normal paraffins of C_{11} to C_{16} chain length, with aromatics, iso-paraffins and cyclic naphthenes. If substantially free of olefins, this kerosene is suitable for processing. The absence of olefins is necessary as the dehydrogenation step will convert to any olefin present to the undesirable di-olefin. The feedstock needs to be hydrotreated to remove any impurities which would impair the catalyst used in subsequent operations. Depending on the molecular weight required of the final linear alkylate, the hydrotreated kerosene is first fractionated to either C_{10}–C_{13} or C_{11}–C_{14} fractions or any other ranges the market calls for.

Normal paraffins are then separated by selective adsorption processes:
Urea has the property of adsorbing to itself straight-chain hydrocarbons with at least 7 carbon atoms in the molecule. It is suggested that the

mechanism is that the urea molecules wrap themselves round the hydrocarbon molecule in a hexagonal spiral and these spirals form channels which can accommodate straight-chain molecules but not the branched ones. In practice urea can adsorb some 30 per cent of its own weight of hydrocarbon.

An activator is necessary for this action to take place. Suitable activators are methanol and the lower ketones.

The procedure basically is to add the kerosene to the solution of the urea in the activator with stirring. The hydrocarbon/urea complex crystallizes (the reaction is not instantaneous), when it is filtered, washed with a short-chain paraffin solvent (less than 7 carbons) and the n-paraffin is released from the complex by hot water.

Variations of this basic method are the use of continuous processes and decomposition of the complex by non-aqueous solvents. Many patents have been granted for these variations.

On p 60 mention was made of Zeolite 5A (called also molecular sieves). This is an aluminium silicate with calcium ions having a pore diameter of about 5 ångström units. Branched chain and cyclic molecules of the feed we are considering have a cross-sectional diameter of at least 5·6 Å whereas for normal paraffins the diameter is approximately 4·9 Å. Normal paraffins can therefore enter into and be adsorbed by the lattice.

Molecular sieves are manufactured by several companies, Linde Division of Union Carbide (Iso-Siv), British Petroleum and Enjay. Their processes all involve separation in the vapour phase. The Molex process developed by Universal Oil uses molecular sieves designed to operate in the liquid phase. The desorption and regeneration of the molecular sieve is stated to be the most difficult part of the operation, it is usually accompanied by sweeping out the adsorbed material with a mixture of solvents.

These normal paraffins can now be used for alkylation (via chlorination) or be dehydrogenated to internal olefins, also as the alkylation reagent.

Dehydrogenation of the paraffin to an internal olefin is an endothermic, catalytic process. On reading the patent literature it appears that considerable work has been done by UOP and in fact the bulk of installations supplied world-wide are based on this company's technology, other than in companies who developed their own 'in house' processes.

In brief, the process involves passing the vaporized feedstock, together with recycle, through a fixed bed reactor where only part of the paraffin is dehydrogenated to linear mono-olefins. Hydrogen is produced as a by-product, and the olefin/paraffin mixture is separated from light materials by fractionation. The olefins are separated from the unreacted paraffins by selective solvents, the paraffin being fed back to recycle. If the olefin is to be used for alkylation, this final separation need not be too thorough as the paraffin can act as a diluent in the reaction and can easily be separated after the alkylation, then recycled. Alkylation units therefore normally start with an olefin unit, if not with a linear paraffin unit.

The catalyst is the important point in this operation and new and efficient ones have been developed capable of working for long periods without the need for regeneration.

Normal Paraffins

For the production of linear internal olefins, normal paraffins are required and their production is described above.

The Molex process normally achieves a separation of a minimum of 98 per cent n-paraffins, the balance being iso-paraffins and aromatics.

When paraffins are required as feedstock for the production of alkane sulphonates, the aromatic content needs to be brought down to below 50 ppm, otherwise undesirable side reactions will occur in the sulphoxidation process.

The removal of aromatics can be accomplished by two methods, hydrogenation, when they are converted to cycloparaffins, which remain in the liquid and do not increase the n-paraffin content, or by sulphonation and removal as the sulphonic acid.

The Ballestra group has developed the 'Solvex' process for the ultra-refining of n-paraffins using SO_3. The biggest plant in the world was commissioned in Sicily by this group in 1974, producing 60 tons per hour of ultra-refined n-paraffins, about one-third of the total world production.

Alkyl Benzene

Among the many different synthetic detergents the alkyl aryl sulphonates are the most important, and recent statistics show that in the United States about 50 per cent of all synthetic detergents are still of this type.

There are many important reasons for this. Not only are the alkyl aryl sulphonates outstanding in their properties as detergents, but they are also based on raw materials which are easier to obtain and cheaper than those on which the other types of detergents are based.

The main source of alkyl aryl sulphonates is the petroleum industry, but before going into details about raw materials and methods of manufacture, it will be convenient to deal with their chemical structure.

As the name implies, these products are based on aromatic compounds combined with an aliphatic chain bound to the aromatic nucleus. Thus, dodecyl benzene, which is the most important type of alkyl aryl condensate, has the following structural formula:

$$\bigcirc\!-\!C_{12}H_{25}$$

Variations of this material were mooted in the past but from the early 1950s to the mid-1960s the choice fell almost exclusively on the kind made from tetra propylene (C_{12}) condensed with benzene. The nature of the propylene polymerization is such that the tetramer (also the trimer when used) has a branched chain. When the problems with biodegradation arose the cause was quickly pinpointed to this branched chain structure and it was proved that linear alkyl benzene sulphonates (LAS) are biodegradable. The propylene tetramer type is still manufactured for use in countries which have not as yet legislated against it and for specialized purposes such as agricultural emulsifiers, where the problems of re-formulating, both administrative and technical, are great.

The chemistry of the alkylation is represented by the equations:

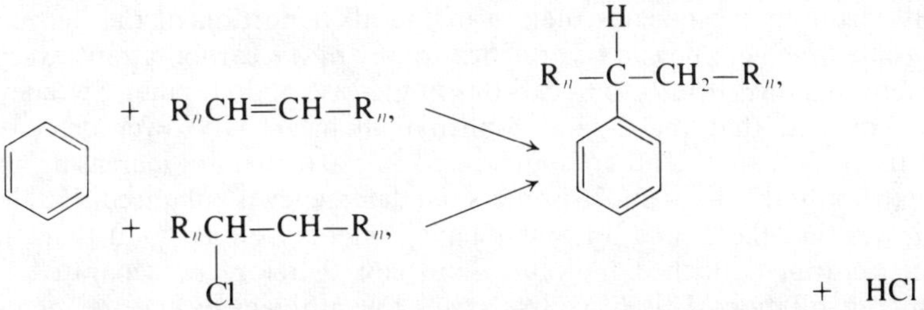

$$\text{C}_6\text{H}_6 + R_{II}CH=CH-R_{III} \longrightarrow \text{C}_6\text{H}_5-\underset{H}{\overset{H}{\underset{|}{C}}}-CH_2-R_{III}$$

$$\text{C}_6\text{H}_6 + R_{II}\underset{Cl}{\underset{|}{CH}}-CH-R_{III} \longrightarrow + HCl$$

Certain side reactions can and do take place; dimerization of the olefin, alkylation of this dimer, di-alkylation of the benzene, possibly di-phenylation of the olefin or chlorparaffin and cyclo-alkylation of the benzene. Thus in the reactor together with the alkylbenzene, we can find di-phenyl alkane, di-alkylbenzene, di(alkylbenzene), all in the ortho- or para-positions:

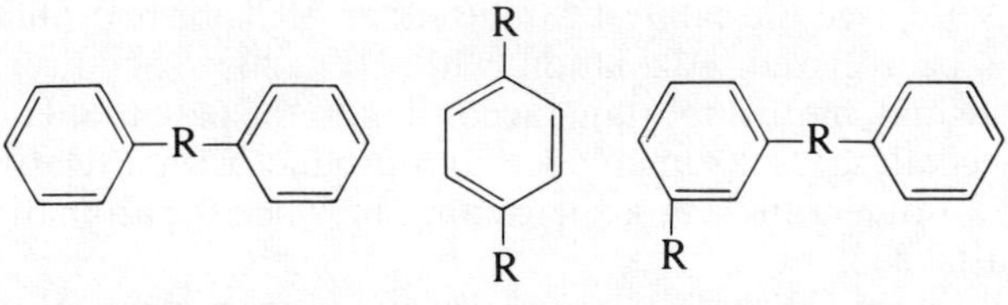

and from cyclo-alkylation, 1,3-dialkyl-indanes and 1,4-dialkyl-tetralins:

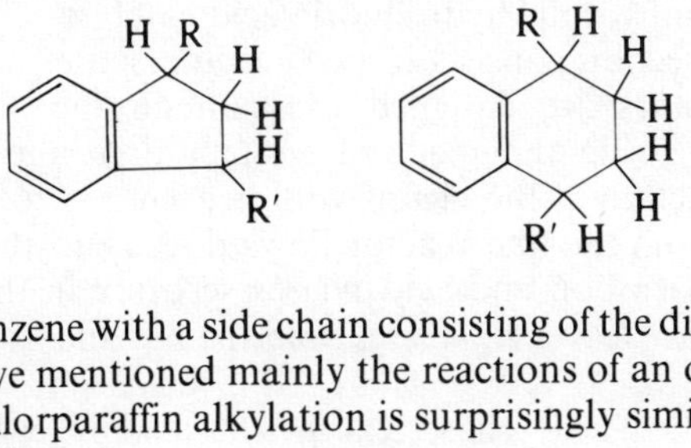

and also alkylbenzene with a side chain consisting of the dimer of the olefin. Although we have mentioned mainly the reactions of an olefin, the 'heavy alkylate' from chlorparaffin alkylation is surprisingly similar to that of the olefin based product.

When propylene tetramer is used as the alkylating medium a further by-product is obtained, light alkylate, this from alkylation of fractions lower than C_{12}, naturally present in the tetramer.

Both these extraneous materials are separated from the main alkylate by fractionation, the light alkylate is normally disposed of as a solvent or plasticizer. We shall discuss heavy alkylates in a later section.

Much work has been done on alkylation processes and nowadays the finished product can be removed from the plant with almost no after-treatment and able to comply with the stringent specifications imposed by the industry.

By virtue of the linear structure of the alkyl portion of the benzene molecule, the benzene can be attached to any of the carbon atoms except the terminal ones (the 2- to 6- carbons in the case of dodecane). Sweeney[7] has indicated that the 2- and 3-phenyl alkanes have lower detersive effectiveness than the other isomers and in particular by increasing the proportion if the 5-, 6- and 7-isomers the detergency is enhanced. He also indicates that the 2- and 3-isomers have a higher boiling point than the more centrally attached molecules and can therefore be separated by simple distillation. This of course poses the problem of end use of the separated terminal isomers. Processes have also been developed for isomerism of the olefin to minimize the 2-isomer.[8]

As stated above, alkylation is performed by the Friedel-Crafts reaction. Paraffins do not enter into this reaction, the reactants need to be either chlorinated paraffins or olefins.

If chlorparaffins are used, the catalyst needs to be aluminium chloride, and the catalyst sludge has to be disposed of as it becomes deactivated.

If olefins are used the catalyst can be either $AlCl_3$ or hydrofluoric acid, which can be recycled. Alkylation with HF allows of lower working temperature and, more important, according to the patent of Huang,[9] the use of this catalyst results in an alkyl aryl product with a relatively small portion of 2-isomers and a large percentage in which the aromatic group is centrally attached.

We shall describe the UOP process, Pacol, using an olefin (*Hydrocarbon Processing*, **63**(11), 6–8 (1985)).

The mixture of n-olefins and unreacted n-paraffins coming from the dehydrogenation unit, together with benzene previously dehydrated, is combined with anhydrous hydrofluoric acid catalyst.

The reaction section consists of two reactor–settler units operating in series. The reactors are designed to maintain the acid–hydrocarbon emulsion and provide the required contact time and temperature to complete the reaction of the olefins with benzene.

The effluent from the first reactor flows directly to the first stage acid settler; the hydrocarbon and acid phases separate in the vessel and the hydrocarbon phase leaves the vessel from the top and flows to the second stage reactor.

All of the settled acid phase, except the acid regenerator charge, is recycled back to the first stage reactor. The first stage acid inventory is held constant by acid spillback from the second stage settler from the discharge of second stage reactor circulating pump.

Hydrocarbon feed to the second stage unit combines with HF acid from the HF settler and is combined with pumped acid from the second stage settler. The alkylation reactions are completed in the second stage reactor. The effluent from the second stage reactor goes to the second stage settler; the hydrocarbon phase flows to the HF stripper.

Most of the settled acid from the settler is returned to the second stage reactor; acid inventory is maintained in the second stage settler by balancing the second stage acid spillback to the first reactor system with the returned acid from the HF stripper overhead receiver.

Bottom liquids from the HF stripper are sent to the benzene column; HF stripper overhead vapours are condensed along with HF refrigerator overhead vapours in a common condenser. The collected liquid is pumped to the liquid HF stripper regenerator overhead receiver where hydrocarbon and acid phase are separated; hydrocarbons are returned to the first stage reactor feed while acid phase is recycled to the second stage reactor.

The HF regenerator serves the purpose of eliminating the acid tars.

The acid tar remaining in the bottom of the regenerator needs to be drained periodically. Reaction products are then fractionated.

The bottom of HF stripper column is fed to the benzene column: benzene is taken overhead as a side cut and the alkylate plus the unreacted normal paraffins are taken from the bottom.

If an accumulation of impurities takes place in the benzene recycle, a drag system of benzene is foreseen. The bottom of the benzene column is pumped to the paraffin column which operates under vacuum like the alkylate rerun column.

Paraffins are separated from the overhead of the paraffin column while the alkylate product is taken from the bottom. Crude alkylate is rerun in the next column to be fractionated into linear alkyl benzene and heavy alkylate fractions.

A flow sheet illustrating the process is given in Fig. 5.2. Table 5.7 indicates the properties of a linear alkylate produced by the Pacol process.

At the Seventh International Congress on Surface Activity, held in Moscow in September 1976, a process was reported by Volkner *et al.* wherein the heavy alkylate mixed with spent catalyst and fresh catalyst is fed back into the alkylating system. As a result, the heavy alkylate becomes depolymerized and further linear alkyl benzene is formed. This process is already on stream at VEB DWH, Rodleben, East Germany (DDR).

An interesting alkylation process was employed by the German firm of Rheinpreussen. In this case, monochlorinated straight-chain petroleum fraction is also used in mixture with olefins in an alkylation process using metallic aluminium as the catalyst. The aluminium is transformed *in situ* into the active aluminium chloride. The reaction is probably triggered by the addition of spent sludge or hydrochloric acid.

Whereas in alkylation (with olefins derived from monochlorinated

SYNTHESIS OF DETERGENTS

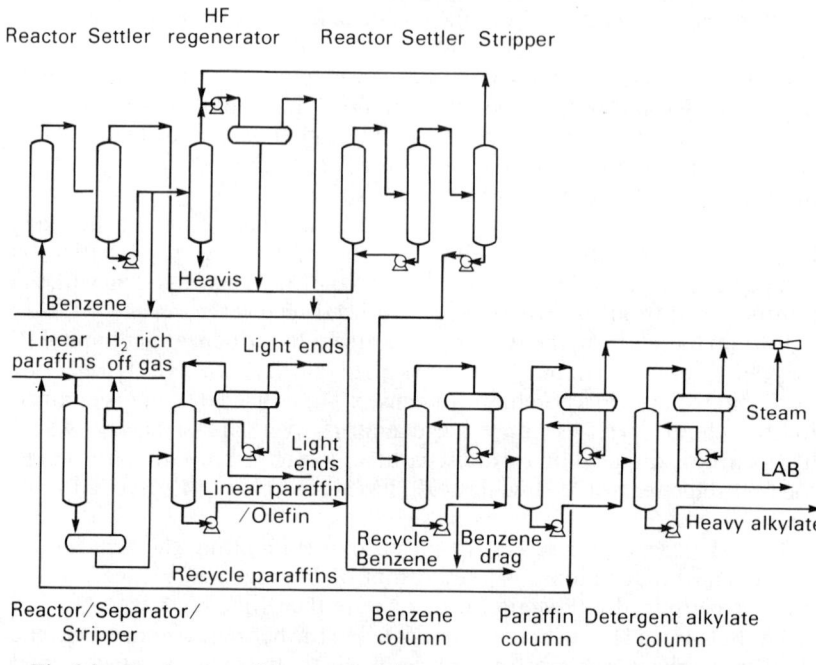

Fig 5.2 Production of linear alkyl benzene schematic flow (UOP process)

TABLE 5.7

Typical Properties of Linear Alkyl Benzene, Produced by the Pacol Process

Bromine no.	0·02
Saybolt colour	+30
Alkyl benzene content	97·4%
Doctor test	Negative
Unsulphonatable matter	1·0%
Water	0·1%
Specific gravity 15°C	0·8612
Refractive index n_D^{20}	1·4837
Flash point	138°C
Average molecular weight	240
Distillation range °C	
IBP	281
50 vol %	290
EP	309
2-Phenyl isomer	20%
Biodegradability (ASTM D-2667)	>95%

paraffinic C_{12-14} petroleum fraction) the HCl is recovered during the dehydrochlorination process, in the Rheinpreussen process HCl from the monochlorinated paraffinic petroleum fraction (or a synthetic paraffin fraction from the Fischer-Tropsch process) is recovered during alkylation. The idea of using pure aluminium rather than the hygroscopic and corrosive aluminium trichloride is rather attractive and has become fairly universal in the industry when chlorparaffins are used.

Figure 5.3 shows the flow sheet of the Rheinpreussen process (published in *British Chemical Engineering*).

The Arco Technology Inc. system of alkylation (described in *Hydrocarbon Processing*, November 1985) also uses pure aluminium in its process and the aluminium chloride produced is recycled with a top-up, under stringent control, of fresh (pure) aluminium.

The material balance for the conditions of reaction is given as:

Raw materials to produce 1000 kg linear alkyl benzene:

	kg
n-Paraffin	76
Benzene	359
Chlorine gas	368

By-products produced:

Heavy alkylate	81
HCl	365

Chemicals used:

Aluminium powder	1
Flake caustic soda	5
Caustic soda solution (50 per cent)	74

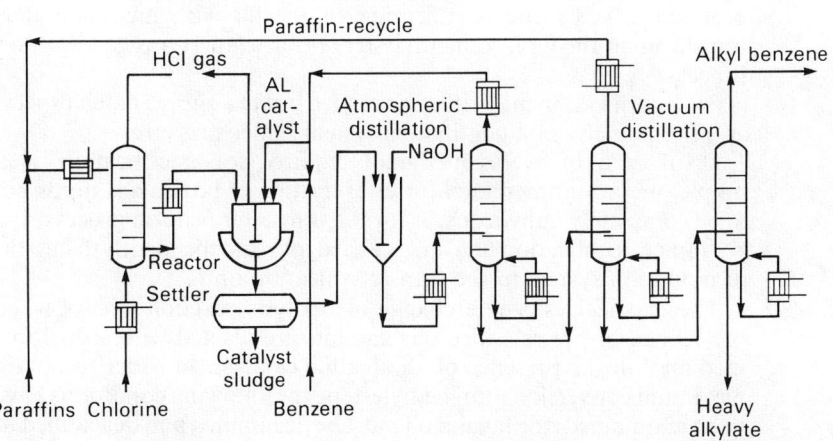

Fig 5.3 Alkylation process using aluminium metal as catalyst, forming aluminium chloride *in situ*—first introduced by Rheinpreussen

Methyl Esters

Methyl esters are used in the industry as an intermediate for three purposes, high-pressure hydrogenation of fats to produce fatty alcohols, production of 'superamides' and as the raw material for sulphonation to produce the sulphonated methyl esters. Each of these has its own requirements as to quality, etc. In addition, with the present low price of methanol, methyl esters are being mooted as the raw material for soap production.

For converting a fat or oil into a methyl ester, certain pretreatments are necessary.

No matter what the end use, the fat needs to be de-gummed to remove both gums and other extraneous matter. This is usually achieved by contacting the fat with slightly diluted sulphuric acid. The water, acid, coagulated gums and dirt are then separated by settling. If the colour of the fat is poor it is then treated with an absorbent earth and filtered.

For sulphonation the fat needs to be hydrogenated or to have an intrinsic low iodine value. The hydrogenation is of the normal fat hardening kind, not as described in the production of alcohols, ie, using a nickel catalyst at moderate pressure. For the preparation of alkanolamides the fat is not normally hydrogenated unless special properties are sought.

Thirdly, the free fatty acid in the oil must be reduced to less than 0·4 acid value as the presence of the fatty acids will react with and impair the performance of the catalyst.

The free fatty acid reduction can be achieved by one of the conventional methods used in the production of edible oils, using a caustic wash or steam distillation under vacuum. An alternative method has been developed by one of the authors (A.D.) and his collaborators[10] and we quote from their paper delivered at the International Symposium on Natural Based Cleaning Agents, Marseilles, April 1985:

> An alternative method of reducing free fatty acid for methylester production is by pre-esterification of the ffa with methanol thus transforming the ffa into methylesters prior to the transesterification process.
>
> Pre-esterification may either be carried out in a simple batch process or, alternatively, in a continuous system under pressure.
>
> Distillation of excess methanol required for esterification, is a simple operation; methanol for esterification of fatty acids needs not to be completely anhydrous, as for the transesterification process, and a simpler rectification may be carried out independently from the recirculation system in the transesterification unit.
>
> The classical ester-interchange alcoholysis reaction occurs when pre-refined or pre-esterified oils and fats are reacted with anhydrous methanol in the presence of an alkaline catalyst. In order to obtain maximum conversion into methylesters, the following conditions have to be maintained: the fat and oil must be degummed and deacidified to reduce the acid value below 0·4. Methanol must be anhydrous (max. 0·3 water).

The catalyst of choice is sodium methylate. The catalyst originally used by Bradshaw and Meuly[11] was dry NaOH or KOH, dissolved in methanol.

During extensive laboratory and pilot work, we could however not confirm a conversion rate as good as with sodium methylate, when a conversion rate or at least 95 per cent could easily be obtained, provided the sodium methylate was fresh and the methanol practically anhydrous.

Sodium methylate powder should however immediately be dissolved in methanol as a stock solution, otherwise it very quickly absorbs water and even CO_2 from the air, thereby reducing its efficiency as a catalyst.

Two alternatives exist to separate the glycerine layer set free after completion of the interchange reaction. One is the classical method, namely to separate it from the reaction mixture in the form of crude glycerine containing the soap formed by the catalyst, some methylesters and methanol.

This glycerine layer then undergoes a prerefining step: distilling off methanol, splitting the soap with acid and separating the fatty acids and some methylester on top of a glycerine solution slightly diluted with water to give a prerefined glycerine of about 70 per cent concentration, which can of course easily be brought up to a higher concentration.

The alternative method to separate glycerine is as follows: after the transesterification reaction is terminated, excess anhydrous methanol is distilled, and recovered for recirculation from the total reaction mixture containing methylester, glycerine, soap formed by the catalyst and some small percentage of unconverted triglyceride, at about 70°C. A low temperature is recommended in order to prevent any reverse reaction of methylester with glycerine. The last traces of methanol are then stripped off preferably under reduced pressure.

When all the methanol has been removed from the reaction mixture, about 3 per cent water containing sufficient acid to neutralize any free sodium methylate and to split the soap formed, is added.

Glycerine separates rapidly from the methylester in the form of an about 70 per cent glycerine solution. The sodium salt content of the glycerine solution, either sodium sulphate or chloride, is less than the salt content of crude glycerine obtained from soap spent-lye.

Organic matter, too, is less. The methylester may be washed free from last traces of glycerine and salt by again adding about 3 per cent water. This rinse is again made up with acid and used for the next process batch.

Obviously, when using this system for glycerine recovery, the methylester contains a small percentage of free fatty acid (ffa) from the split soap. However, when the methylester is used for alkanolamide production or for soap, the presence of ffa does not represent a disadvantage. Only when the methylester has to undergo distillation and the distilled methylester is marketed as a commodity, the acid value has to be reduced to a low level below 1.

This can easily be accomplished by conventional alkali or carbonate neutralization, a process which is not much more difficult, than washing the methyl ester free from glycerine and soap in the conventional process.

Alternatively, it was found that, on neutralizing the ffa with monoethanolamine prior to distillation, no foam problems occur and no highly viscous sticky bottom fraction is obtained.

The modified method of separating glycerine is especially attractive, when transesterification is carried out continuously.

Figure 5.4 shows a batch system for transesterification.

The components, fats and oils, sodium methylate stock solution in methanol, and methanol are introduced, by automatic batching, into the reactor. The reaction mixture is kept boiling under reflux at about 70°C for about two hours, when conversion of triglyceride into methyl ester reaches 95 per cent for tallow and even 97 per cent for coconut or palmkernel oil.

Subsequent processing steps have already been described.

In the continuous process, Fig 5.5, which can also be carried out under pressure, residence time is considerably reduced; the components are dosed by triple-head dosing pump, the fat or oil passes through a preheater, so as to give a temperature of the total mixture, catalyst stock solution, methanol and fat of about 70°C in the static mixer.

From there the reaction mixture enters the reactor. The temperature is above 70°C when the alcoholysis is carried out at atmospheric pressure, and higher when carried out under pressure.

Correlation between temperature and pressure is, eg, 3·4 bar at 100°C, 10·4 bar at 140°C. From a practical constructional and operational point of view, operation at 5–7 bar at a temperature around 120°C would be preferable.

From the reaction column, the reaction mixture enters the distillation column for distilling off the excess anhydrous methanol for recirculation via the surge tank.

Splitting off the soap, separation of glycerine and washing are generally performed discontinuously. It may be mentioned that the batch and the continuous transesterification processes should be able to use the two alternate methods of separating the glycerine.

The flow sheets shown in Figs 5.4 and 5.5 describe integrated units manufactured by the Ballestra Group, Milan.

Sulphonation of Detergent Raw Materials

Sulphonation and sulphation are the most important processes for the production of anionic detergents, which, as we have seen, are by far the

SULPHONATION OF DETERGENT RAW MATERIALS

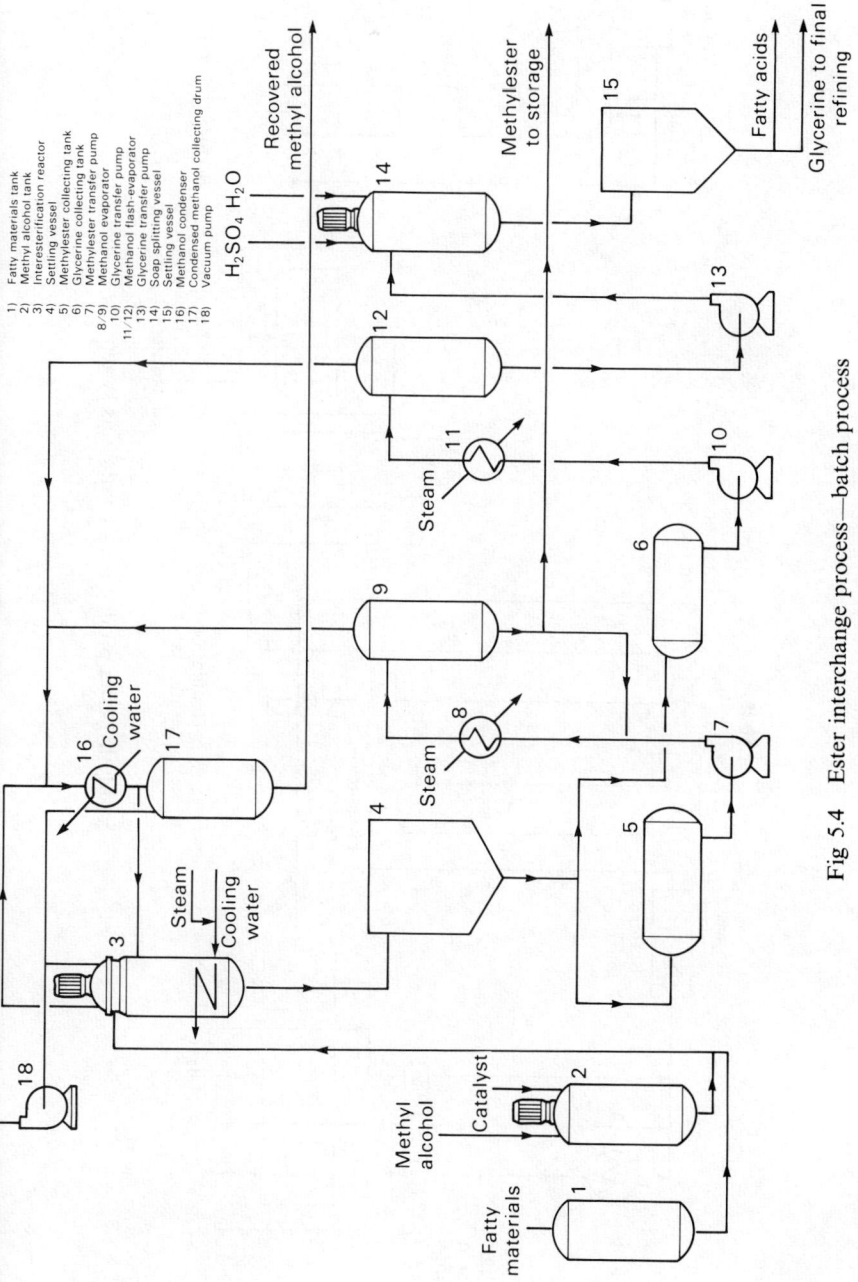

1) Fatty materials tank
2) Methyl alcohol tank
3) Interesterification reactor
4) Settling vessel
5) Methylester collecting tank
6) Glycerine collecting tank
7) Methylester transfer pump
8/9) Methanol evaporator
10) Glycerine transfer pump
11/12) Methanol flash-evaporator
13) Glycerine transfer pump
14) Soap splitting vessel
15) Settling vessel
16) Methanol condenser
17) Condensed methanol collecting drum
18) Vacuum pump

Fig 5.4 Ester interchange process—batch process

SYNTHESIS OF DETERGENTS

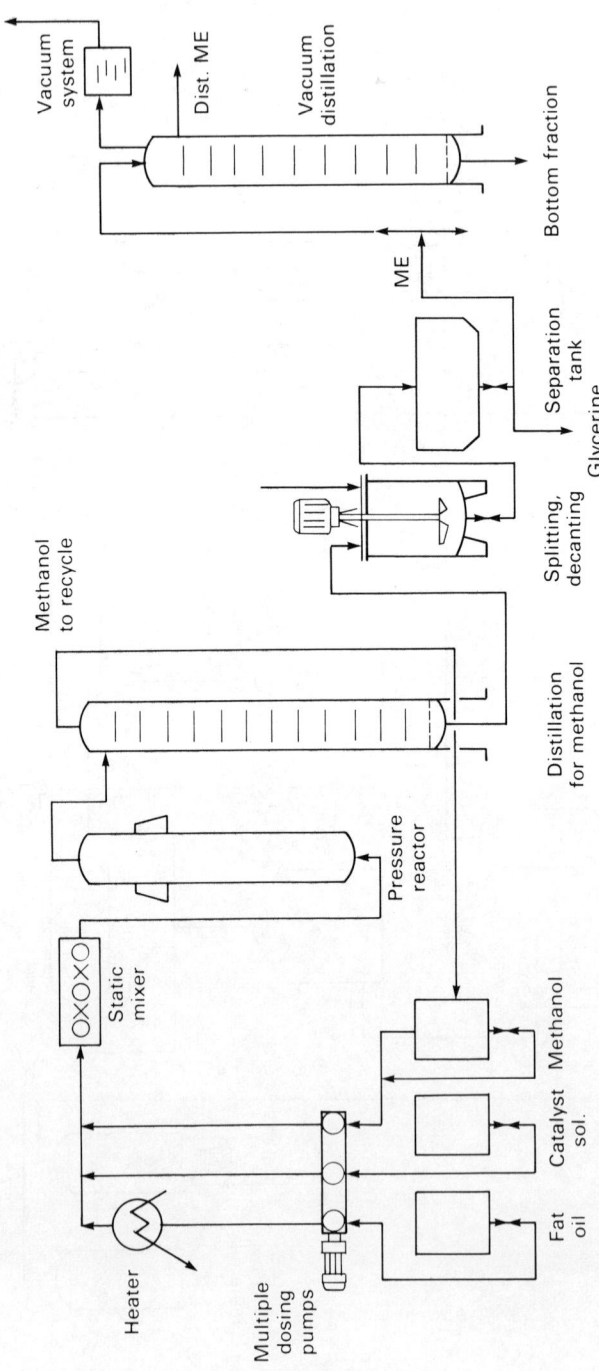

Fig 5.5 Methyl ester production—continuous system

most important synthetic detergents. In recent years sulphonation processes have been developed which have revolutionized the detergent industry. Sulphonation and sulphation with acids, such as sulphuric acid, oleum of various concentrations, and chlorosulphonic acid, is now being replaced by methods which use gaseous SO_3, obtained either by vaporization of liquid stabilized SO_3 (sulphuric acid anhydride) or directly as converter gas from a contact sulphuric acid plant. However, sulphonation with acids is still widely practised, so that a description of the more conventional acid-sulphonation processes will also be given in this chapter.

The Inorganic Raw Materials for Sulphonation

A precondition for properly carrying out sulphonation is the knowledge of the inorganic raw materials. We will therefore give a short description of the most important inorganic acid raw materials used in detergent sulphonation processes. (In this context we consider liquid SO_3—ie, sulphuric acid anhydride—as an acid, too. We will also give the properties of sulphur since this is the starting material for the most modern sulphonation process.)

(A) *Sulphuric Acid and Oleum (Fuming Sulphuric Acid)*

The oldest, and still widely used, acid in the detergent industry is sulphuric acid and oleum of various strengths. The chemical formula of sulphuric acid is H_2SO_4, molecular weight 98·08. For sulphonation processes either the so-called monohydrate of 98/99 per cent acid strength is used, or oleum, ie, sulphuric acid with various percentages of free SO_3 dissolved. In Tables 5.8 and 5.9, the properties of sulphuric acid and oleum of various strengths are given. The physical properties of more dilute sulphuric acid are of interest for the practical man in the detergent industry, as he might well encounter dilute sulphuric acid in the form of so-called 'spent acid' and in acid cleaners.

The tables show that, especially at a percentage between 97 and 100 per cent sulphuric acid, there is a curious irregularity in density. The specific gravity of an acid of 97 and 98 per cent is identical, and the density of an acid of 99 and 100 per cent strength is practically the same as that of an acid of 94–95 per cent strength. This fact is of great practical importance, because it makes determination of acid strength by simple hydrometer reading impossible within this range of concentration, which unfortunately is the most important concentration range for acid sulphonation processes. There are other physical control methods for the determination of acid strength in the range of 93–100 per cent, such as the measurement of conductivity and 'heat of dilution'. Incidentally, these physical methods have been worked out for the control of SO_2 absorption and air-drying in the classical contact sulphuric acid production processes. It is, however, beyond the scope of this book to go into details about these test methods.

SYNTHESIS OF DETERGENTS

TABLE 5.8

Properties of Sulphuric Acid

Degrees Baumé	Specific gravity 60°F 60°	% sulphuric acid H_2SO_4	Weight of 1 Imp gal in lb av	Normal boiling-point °C	Freezing (melting)-point °C
0	1·0000	0·00	10·011	100·0	0·00
1	1·0069	1·02	10·079	100·1	— 0·39
2	1·0140	2·08	10·150	100·3	— 0·83
3	1·0211	3·13	10·222	100·4	— 1·28
4	1·0284	4·21	10·295	100·6	— 1·72
5	1·0357	5·28	10·369	100·8	— 2·22
6	1·0432	6·37	10·443	100·9	— 2·77
7	1·0507	7·45	10·517	101·1	— 3·44
8	1·0584	8·55	10·595	101·3	— 4·11
9	1·0662	9·66	10·673	101·5	— 4·72
10	1·0741	10·77	10·752	101·7	— 5·39
11	1·0821	11·89	10·832	101·9	— 6·11
12	1·0902	13·01	10·915	102·2	— 7·00
13	1·0985	14·13	10·996	102·4	— 7·94
14	1·1069	15·25	11·082	102·7	— 8·72
15	1·1154	16·38	11·167	102·9	—10·00
16	1·1240	17·53	11·251	103·2	—11·11
17	1·1328	18·71	11·340	103·6	—12·28
18	1·1417	19·89	11·430	103·9	—13·61
19	1·1508	21·07	11·521	104·3	—15·33
20	1·1600	22·25	11·612	104·7	—17·22
21	1·1694	23·43	11·707	105·1	—19·17
22	1·1789	24·61	11·802	105·4	—21·39
23	1·1885	25·81	11·898	105·8	—23·94
24	1·1983	27·03	11·996	106·3	—26·67
25	1·2083	28·28	12·096	106·8	—29·94
26	1·2185	29·53	12·198	107·3	—33·33
27	1·2288	30·79	12·301	107·9	—37·56
28	1·2393	32·05	12·407	108·6	—42·39
29	1·2500	33·33	12·513	109·2	—48·50
30	1·2609	34·63	12·622	110·0	—56·39
31	1·2719	35·93	12·732	110·9	—61·50
32	1·2832	37·26	12·845	111·8	—59·56
33	1·2946	38·58	12·959	112·7	—57·78
34	1·3063	39·92	13·077	113·6	—56·27
35	1·3182	41·27	13·197	114·6	—54·72
36	1·3303	42·63	13·317	115·7	—42·50
37	1·3426	43·99	13·441	116·9	—49·44
38	1·3551	45·35	13·565	118·2	—45·88
39	1·3679	46·72	13·694	119·7	—42·22
40	1·3810	48·10	13·819	121·2	—39·00
41	1·3942	49·47	13·952	158·3	—36·10
42	1·4078	50·87	14·087	124·6	—33·70

SULPHONATION OF DETERGENT RAW MATERIALS

TABLE 5.8 (*continued*)

Properties of Sulphuric Acid

Degrees Baumé	Specific gravity 60°F 60°	% sulphuric acid H_2SO_4	Weight of 1 Imp gal in lb av	Normal boiling-point °C	Freezing (melting)-point °C
43	1·4216	52·26	14·226	126·4	−31·90
44	1·4356	53·66	14·365	128·6	−30·40
45	1·4500	55·07	14·510	130·8	−29·40
46	1·4646	56·48	14·656	133·1	−28·50
47	1·4796	57·90	14·806	135·8	−28·40
48	1·4948	59·32	14·958	138·4	−28·80
49	1·5104	60·75	15·114	141·4	−30·00
50	1·5263	62·18	15·274	144·4	−31·90
51	1·5426	63·66	15·437	147·8	−34·20
52	1·5591	65·13	15·601	151·1	−36·80
53	1·5761	66·63	15·772	155·0	−38·20
54	1·5934	68·13	15·944	159·2	−39·70
55	1·6111	69·65	16·121	163·6	−42·80
56	1·6292	71·17	16·302	168·3	−40·40
57	1·6477	72·75	16·488	174·1	−39·70
58	1·6667	74·36	16·678	179·7	−33·60
59	1·6860	75·99	16·872	186·3	−22·80
60	1·7059	77·67	17·071	193·1	−11·40
61	1·7262	79·43	17·273	200·6	− 1·50
62	1·7470	81·30	17·482	208·6	+ 4·20
63	1·7683	83·34	17·695	218·8	7·60
64	1·7901	85·66	17·914	230·7	7·10
64¼	1·7957	86·33	17·969	234·2	6·10
64½	1·8012	87·04	18·024	238·1	4·40
64¾	1·8068	87·81	18·080	243·4	− 2·20
65	1·8125	88·65	18·138	247·2	− 0·40
65¼	1·8182	89·55	18·194	252·2	− 4·20
65½	1·8239	90·60	18·252	258·9	− 9·40
65¾	1·8297	91·80	18·310	268·3	−16·40
66	1·8354	93·19	18·366	279·4	−29·40
	1·8373	93·77	18·386	285·0	−34·90
	1·8381	94·00	18·392	286·7	−33·30
	1·8407	95·00	18·419	296·7	−22·20
	1·8427	96·00	18·439	307·8	−13·90
	1·8437	97·00	18·449	318·9	− 6·90
	1·8437	98·00	18·449	327·2	− 1·10
	1·8424	99·00	18·436	310·0	+ 5·00
	1·8391	100·00	18·402	274·4	10·90

Note: Data shown in tables compiled from various sources including:
— Manufacturing Chemists Association, Manual Sheets T-7 (1904) and T-7A (1938).
— Gable, Betz and Maron, *JACS*, BB, 1445–8 (1950).
— Engineering Department, General Chemical Division, Allied Chemical Corporation.

Allowance for Temperature
At 10° Be, 0·029° Be or 0·00023 sp gr = 1°F.
At 20° Be, 0·036° Be or 0·00034 sp gr = 1°F.
At 30° Be, 0·035° Be or 0·00039 s0 gr = 1°F.
At 40° Be, 0·031° Be or 0·00041 sp gr = 1°F.
At 50° Be, 0·028° Be or 0·00045 sp gr = 1°F.
At 60° Be, 0·026° Be or 0·00053 sp gr = 1°F.
At 63° Be, 0·026° Be or 0·00057 sp gr = 1°F.
At 66° Be, 0·0235° Be or 0·00054 sp gr = 1°F.

Note: Since sulphuric acid above 93 per cent may have same specific gravity at different concentrations, a direct analysis of acid strength is recommended.
At 94%   0·00054 sp gr = 1°F.
At 96%   0·00053 sp gr = 1°F.
At 97·5% 0·00052 sp gr = 1°F.
At 100% 0·00052 sp gr = 1°F.

TABLE 5.9

Properties of Oleum
Free SO_3 total SO_3 and equivalent H_2SO_4

% free sulphur trioxide SO_3	% actual sulphuric acid H_2SO_4	% total sulphur trioxide SO_3	% equivalent sulphuric acid H_2SO_4	% free sulphur trioxide SO_3	% actual sulphuric acid H_2SO_4	% total sulphur trioxide SO_3	% equivalent sulphuric acid H_2SO_4
0	100	81·63	100·00	21	79	85·49	104·73
1	99	81·82	100·23	22	78	85·67	104·95
2	98	82·00	100·45	23	77	85·86	105·18
3	97	82·18	100·68	24	76	86·04	105·40
4	96	82·37	100·90	25	75	86·22	105·63
5	95	82·55	101·13	26	74	86·41	105·85
6	94	82·73	101·35	27	73	86·59	106·08
7	93	82·92	101·58	28	72	86·78	106·30
8	92	83·10	101·80	29	71	86·96	106·53
9	91	83·29	102·03	30	70	87·14	106·75
10	90	83·47	102·25	31	69	87·33	106·98
11	89	83·65	102·48	32	68	87·51	107·20
12	88	83·84	102·70	33	67	87·69	107·43
13	87	84·02	102·93	34	66	87·88	107·65
14	86	84·20	103·15	35	65	88·06	107·88
15	85	84·39	103·38	36	64	88·24	108·10
16	84	84·57	103·60	37	63	88·43	108·33
17	83	84·75	103·83	38	62	88·61	108·55
18	82	84·94	104·05	39	61	88·80	108·78
19	81	85·12	104·28	40	60	88·98	109·00
20	80	85·31	104·50	41	59	89·16	109·23

TABLE 5.9 (*continued*)

% free sulphur trioxide SO_3	% actual sulphuric acid H_2SO_4	% total sulphur trioxide SO_3	% equivalent sulphuric acid H_2SO_4	% free sulphur trioxide SO_3	% actual sulphuric acid H_2SO_4	% total sulphur trioxide SO_3	% equivalent sulphuric acid H_2SO_4
42	58	89·35	109·45	72	28	94·86	116·20
43	57	89·53	109·68	73	27	95·04	116·43
44	56	89·71	109·90	74	26	95·22	116·65
45	55	89·90	110·13	75	25	95·41	116·88
46	54	90·08	110·35	76	24	95·59	117·10
47	53	90·27	110·58	77	23	95·78	117·33
48	52	90·45	110·80	78	22	95·96	117·55
49	51	90·63	111·03	79	21	96·14	117·78
50	50	90·82	111·25	80	20	96·33	118·00
51	49	91·00	111·48	81	19	96·51	118·23
52	48	91·18	111·70	82	18	96·69	118·45
53	47	91·37	111·93	83	17	96·88	118·68
54	46	91·55	112·15	84	16	97·06	118·90
55	45	91·73	112·38	85	15	97·25	119·13
56	44	91·92	112·60	86	14	97·43	119·35
57	43	92·10	112·83	87	13	97·61	119·58
58	42	92·29	113·05	88	12	97·79	119·80
59	41	92·47	113·28	89	11	97·98	120·03
60	40	92·65	113·50	90	10	98·16	120·25
61	39	92·84	113·73	91	9	98·35	120·48
62	38	93·02	113·95	92	8	98·53	120·70
63	37	93·20	114·18	93	7	98·71	120·93
64	36	93·39	114·40	94	6	98·90	121·15
65	35	93·57	114·63	95	5	99·08	121·38
66	34	93·76	114·85	96	4	99·27	121·60
67	33	93·94	115·08	97	3	99·45	121·83
68	32	94·12	115·30	98	2	99·63	122·05
69	31	94·31	115·53	99	1	99·82	122·28
70	30	94·49	115·75	100	0	100·00	122·50
71	29	94·67	115·98				

Figure 5.6, showing the freezing point of concentrated sulphuric acid and oleum, again demonstrates another irregularity of great practical importance in handling oleum at various strengths, especially during the colder months. For sulphonation of detergent raw materials, mainly alkyl benzene, grades of oleum with 10–22 per cent free SO_2 are used. It is thus of interest to note that it is within this range of free SO_3 that the greatest irregularity in the freezing temperature occurs. 100 per cent H_2SO_4 has a relatively high freezing point, whereas an oleum of 15–20 per cent has a relatively low freezing point. Oleum with 60 per cent free SO_3 and a rather low freezing point serves as a source of SO_3 in some detergent sulphonation processes.

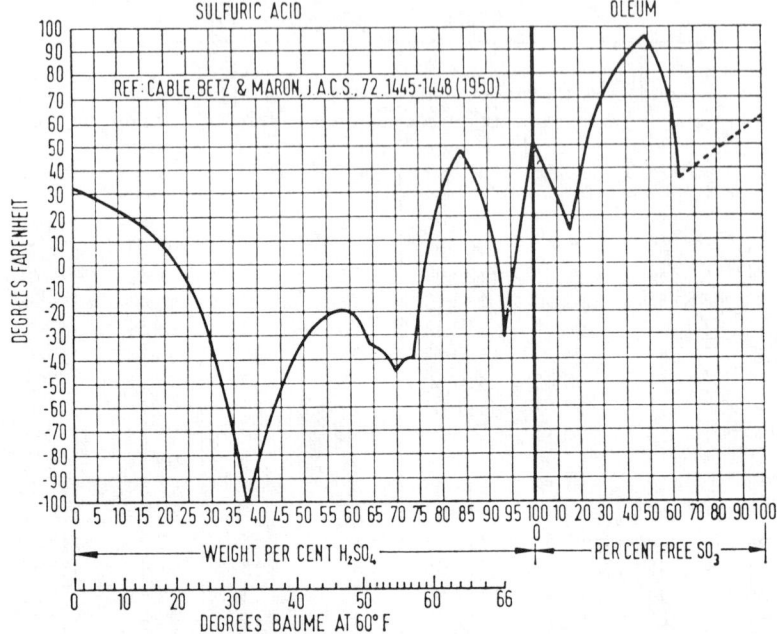

Fig 5.6 Freezing points of sulphuric acid and oleum

Handling and Storage of Sulphuric Acid and Oleum: Excellent descriptions of handling and storage procedures as well as of safety precautions and materials of construction are published by the largest US sulphuric acid and oleum producer Allied Chemical, in their booklet on sulphuric acid, from which we quote some information with their kind permission. We emphasize, however, that in many countries explicit regulations exist for the safe handling, storage, etc., of sulphuric acid and oleum. These regulations should be carefully studied and followed by anybody starting a sulphonation plant.

Storage areas should have facilities for drainage and for the washing down of spills with water. Major spills of acid should not be discharged into a sewer, or washed into a sewer or river until they have been neutralized. Where convenient, the use of crushed limestone as a foundation under storage tank areas is considered good practice since it both provides good natural drainage and neutralizes the spills before they are discharged into the sewer.

Storage tanks should be equipped with vents of such size as to maintain the tanks at atmospheric pressure. Metal catwalks should be provided for working on top of tanks.

Drums should be stored with the plugs up. The plugs should be loosened upon receipt and every week thereafter, or more frequently depending upon the weather, to release the internal pressure created by the evolution of hydrogen.

Although sulphuric acid is not inflammable, it should not be stored near organic materials, nitrates, carbides, chlorates and metal powders. Contact of high concentrations of sulphuric acid with these materials may cause ignition.

Sulphuric acid in drums, tank cars and metal storage tanks evolves hydrogen. Therefore open lights, flames and spark-producing tools should not be permitted near such containers.

Heating of storage tanks or storage areas is necessary for certain strengths of acid which freeze at common winter temperatures.

Sulphuric acid, like any other corrosive liquid, is dangerous when improperly handled. However, if suitable precautionary steps are taken, proper handling procedures followed, and adequate protective equipment worn, there is relatively little danger in working with this chemical.

The storage of sulphuric acid in bulk storage tanks, and piping it directly to process, is by far the easiest and safest way to handle the acid. This not only results in a considerable saving of space, but eliminates the hazards involved in lifting, transporting and discharging of drums and carboys. With properly designed and maintained equipment, handling hazards are reduced to a minimum.

Before being moved, drums should be inspected for loose plugs, for leakage and for bulging. To remove drum plugs, a pipe wrench or a plug wrench with a long handle should be used. The plug should be turned very slowly for one full turn, and accumulated pressure should then be permitted to vent itself. After the hydrogen is released, the plug may be further loosened and removed.

Drums of sulphuric acid should be emptied by gravity, using a faucet or safety siphon. A pump may also be used. Air pressure should never be used. Since the drums may contain hydrogen gas, they should never be struck with a spark-producing tool.

Employees should wear face shields and rubber gloves when opening drums.

For all practical purposes the dilution of sulphuric acid and/or oleum with water should not be carried out in a detergent plant. There is a strong exothermic reaction, and as a general rule one should never add water to acid. Dilution of acid is a process step during sulphonation, and will be described later in this chapter. (The elimination of spillages has already been mentioned.)

The handling of sulphuric acid and/or oleum as well as of 'spent acid' (see p 150) presents a corrosion problem, too. Table 5.10 gives some basic information about the materials for constructable tanks, pumps, gaskets, etc.

Personal protection methods for handling sulphuric acid will not be given here. As a general rule the user of sulphuric acid should ask the supplying firm for explicit safety data, and should always take into ac-

SYNTHESIS OF DETERGENTS

TABLE 5.10

Storage and Handling Equipment for Various Strengths of Sulphuric Acid and Oleum

Products	Storage tank materials	Pipe and fittings	Valves	Gaskets	Pumps	Pump packings
Sulphuric acid 54–60° Be 66° Be—100%	Heavy steel	Steel, Sch 80 pipe, welding fittings and 150 lb welding flanges Cast-iron pipe, with cast-iron flanged 250 lb fittings	FA-20 Alloy‡ with TFE plastic packing and 150 lb flanges Plug type, FA-20 Alloy, TFE plastic sleeve and 150 lb flanges	TFE plastic* CFE plastic† Compressed asbestos	Centrifugal type FA-20 Alloy All iron with FA-20 Alloy impeller High-silicon cast iron	Packless or mechanical seals TFE plastic CFE plastic
Sulphuric acid electrolyte and other grades up to 60° Be	Steel or wood with chemical lead lining	Chemical lead pipe, ¼″ minimum wall thickness, with 6% antimony lead 125 lb flanges or lap joint flanges High-silicon cast-iron pipe and fittings Rigid, unplasticized, normal impact polyvinyl chloride (PVC) Sch 80 pipe and fittings	FA-20 Alloy or high-silicon cast iron with TFE plastic packing and 150 lb flanges 6% antimony lead with TFE plastic packing and 125 lb flanges Plug type, FA-20 Alloy, TFE plastic sleeve and 160 lb flanges	TFE plastic CFE plastic Compressed asbestos	Centrifugal type FA-20 Alloy High-silicon cast iron	Packless or mechanical seals TFE plastic CFE plastic
Sulphuric acid electrolyte 66° Be	Steel or wood with chemical lead lining Steel with baked phenolic lining such as Heresite P-403	Chemical lead pipe, ¼″ minimum wall thickness, with 6% antimony lead 125 lb flanges or lap joint flanges High-silicon cast-iron pipe and fittings	FA-20 Alloy or high-silicon cast iron with TFE plastic packing and 150 lb flanges 6% antimony lead with TFE plastic packing and 125 lb flanges Plug type, FA-20 Alloy, TFE plastic sleeve and 150 lb flanges	TFE plastic CFE plastic Compressed asbestos	Centrifugal type FA-20 Alloy High-silicon cast iron	Packless or mechanical seals TFE plastic CFE plastic

TABLE 5.10 (*continued*)

Products	Storage tank materials	Pipe and fittings	Valves	Gaskets	Pumps	Pump packings
Sulphuric acid chemically pure (reagent ACS) 95.5–96.5%	Glass-lined steel	Pyrex glass TFE plastic-lined hose	Porcelain 'Y' valves with TFE plastic discs and 150 lb flanges	TFE plastic envelope type	Centrifugal type, glass lined Diaphragm type with TFE or CFE plastic diaphragm	TFE plastic CFE plastic
Oleum all strengths	Heavy steel	Steel, Sch 80 pipe, welding fittings and 150 lb welding flanges§	FA-20 Alloy with TFE plastic packing and 150 lb flanges Plug type, FA-20 Alloy, TFE plastic sleeve and 150 lb flanges	TFE plastic CFE plastic Compressed asbestos	Centrifugal type FA-20 Alloy Mechanite CB-3	Packless or mechanical seals TFE plastic CFE plastic
Sulfan (stabilized sulphuric anhydride)	Heavy steel	Steel, Sch 80 pipe, welding fittings and 150 lb welding flanges§	FA-20 Alloy with TFE plastic packing and 150 lb flanges Plug type, FA-20 Alloy, TFE plastic sleeve and 150 lb flanges	TFE plastic CFE plastic	Centrifugal type FA-20 Alloy Mechanite CB-3	Packless TFE plastic ‡ CFE plastic

* TFE—Polytetrafluoroethylene—Representative trade name—Teflon.
† CFE—Polychlorotrifluoroethylene—Representative trade names—Halon (trademark of Allied Chemical Corporation), Kel-F, and Fluorothene.
‡ FA-20 Alloy—Representative trade names—Durimet 20, Aloyco 20, Carpenter 20.
§ Cast iron and high-silicon cast iron must not be used for oleum.

count the prevailing government regulations and other official or semi-official safety rules for handling acid. For a detailed description of handling procedure, precautions, etc, we refer to the book *Manufacture of Sulfuric Acid* by Duecker & West: Reinhold Publ Co, 1959.

(B) *Liquid SO_3*

Appearance and colour:	Clear to slightly turbid liquid. 30–50 APHA
Assay:	99·5 per cent SO_3
Boiling point:	44·8°C
Freezing point:	16·8°C
Density at 20°C:	1·922 g/ml
Heat of reaction with water:	1·096 kJ/g
Vapour pressure (mm of mercury)	25°C 265
	50°C 950
	75°C 3000

Sulphur trioxide in the vapour form always exists as the monomer, but when cooled below its boiling point, in the presence of minute traces of moisture or even on standing for any length of time, it can polymerize to liquid and solid polymers. Thus liquid SO_3 can only be transported or used in its stabilized form.

Stabilized liquid sulphur trioxide is chemically the gamma form of SO_3. The alpha and beta forms are solids having a higher melting point. Since transformation between these three forms of sulphur trioxide can result in formation of the solid varieties, stabilized liquid SO_3 (gamma) which is protected against the formation of high melting-point solids by the addition of stabilizers such as borates or sulphonic acids,[12] is the preferred sulph(on)ating agent.

Gamma form of sulphur trioxide (Trimer)

In the detergent industry, liquid SO_3 is not used directly in its liquid form as a sulphonation agent, but mainly in its vaporized form strongly diluted with dry inert air (see p 152). It must be mentioned that liquid SO_3 is also used as a liquid sulphonation agent diluted with undercooled liquid SO_2.

This, however, is a highly specialized process, employed, for example, by the US firm Pilot. Another method dispenses with a carrier and injects vaporized liquid SO_3 directly into the reactor(s) under vacuum[13] (see also p 153).

The actual vaporization and dilution processes will be described later in this chapter.

(C) SO_3 from Oleum Stripping

In areas where liquid SO_3 is not readily available (there are stringent restrictions on the marine transport of this type of material), and where for some reason or another a sulphur burning plant is not to be established, SO_3 can be obtained by the stripping of oleum—60 per cent oleum is preferred although 25 per cent can be used.

The method is to distil the oleum, condense the evolved SO_3 gas into a storage vessel and add a stabilizer. Continuous evolution of SO_3 to feed into a reactor cannot be employed because as the SO_3 is removed from the oleum, the rate of evolution will fall, and a constant supply of SO_3 can never be maintained.

The liquid SO_3 thus formed needs to be treated as described above, ie, to be vaporized prior to feeding into the reactor(s). When the oleum concentration falls to 20 per cent, distillation is discontinued. This 20 per cent oleum has to be disposed of, usually by returning to the sulphuric acid manufacturer for enrichment.

Another possibility is to blow dried, heated air through the liquid, possibly in a heated column, thus obtaining directly, diluted SO_3 gas for sulphonation. To maintain an accurate and constant gas flow, the concentration of SO_3 in the evolved gas should be monitored with a feedback to control the air supply.

As can be seen, problems of corrosion and handling will be enormous, and the plant is in essence two plants, one for vaporization and the other for treatment of the liquid SO_3. These factors should influence the potential manufacturer to install a sulphur burning plant.

(D) Sulphur

As already mentioned, sulphur is an important raw material for the most modern sulphonation method, which applies converter gas for direct sulphonation of detergent raw materials.[14,15] It is this process which will be described in much detail later in this chapter. For this process elementary, technically sulphur of high purity (atom wt 32·06, specific gravity 2·03) is required. Three types of sulphur are most commonly used. Crude, run-of-mine 'bright' 99·5 per cent pure sulphur, free of arsenic, selenium and tellurium (eg 'Texas sulphur'). Refined sulphur, which is an elemental sulphur produced by distilling crude sulphur, of not less than 99·8 per cent purity. Free-burning; available in lumps or cast sticks. A more recently available variety is the so-called 'gas-sulphur' which is obtained from natural gas or petroleum-refining processes (eg, Lacqsulphur from France). These types of sulphur solidify at about 114·5°C.

For practical purposes, data on the viscosity of liquid sulphur are important. The graph (Fig 5.7) shows the change in viscosity of sulphur. As will be seen later, an exact pumping of liquid sulphur by dosing pump is the basis of the modern sulphonation process with converter gas. The pumping is best carried out at the lowest viscosity, ie 150–155°C (302–311°F). In this connection it must be mentioned that carbonaceous and hydrocarbon impurities tend to cause 'sticking valves' in the dosing pump.

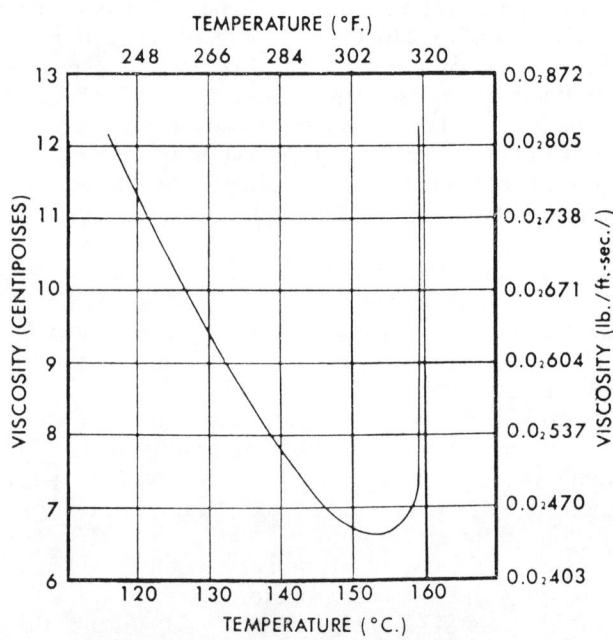

Fig 5.7 Viscosity of molten sulphur

The advantage of sulphur as basic inorganic raw material for sulphonation is obvious indeed. It is safe to handle, practically non-toxic and non-corrosive to the skin. (Irritation of ocular conjunctivae and the mucous membranes of the respiratory organs may occur.) What has to be taken into account, however, is its inflammability. Even so, for all practical purposes of sulphonation, solid sulphur would not have to be ground in a mill, which would represent a certain ignition hazard. Storage of solid sulphur should exclude humidity and ignition hazards. Solid sulphur has to be molten in steam-jacketed vessels (see p 155).

A very detailed description of the handling of sulphur in the molten stage with data on material of construction is to be found in the *Sulphur Manual* published by the Texas Gulf Sulphur Co. However, for all practical purposes of the sulphonation process starting with sulphur burning, these problems are taken care of by the supplier of sulphonation plants based on this patented sulphonation process (see p 158).

(E) Chlorosulphonic Acid

This acid is still widely used in the detergent industry for the sulphonation of fatty alcohols, ethylene oxide condensation products, etc (see p 170). For alkyl benzene it cannot be used, as in the presence of an aromatic ring structure sulpho-chlorination would occur, whereas in all other sulphation procedures using chlorosulphonic acid free hydrochloric acid (HCl) in gaseous form is evolved during the sulphation reaction and has to be absorbed in water, to give liquid HCl of various strengths (see p 171).

Physical and chemical properties

Chemical formula:	$ClSO_2OH$
Molecular weight:	116·53
Colour:	Colourless
MP:	$-80·0°C$
BP:	151·5°C at 101·75 kPa
Solubility:	Decomposes in water, decomposes in alcohol, insoluble in CS_2
Index of refraction nD:	1·437 at 14°C
Specific gravity:	1·787 at 25°C
Specific heat:	1·18 J/g, °C, 15°C to 80°C
Heat of vaporization:	406 kJ/kg
Heat of formation (from elements):	597·1 kJ/mol
Viscosity:	2·5 mPa s at 30°C
Inflammability:	Non-flammable, but may cause ignition of combustible materials on contact. In contact with metals hydrogen gas is formed which can be very explosive in certain mixtures with air.

The physical data listed summarize the bulk of information available on chlorosulphonic acid and are adequate for many engineering calculations. Additional data on a good commercial grade acid (eg, Monsanto) are:

Appearance:	Pale yellow liquid
Assay $ClSO_3H$:	99·4 per cent
Iron—(Fe):	$< 5·0$ ppm
Free SO_3:	0·40 per cent

For handling chlorosulphonic acid it is advisable again to ask the supplier for detailed safety data, and also about materials of construction, etc. We quote, however, some safety data from the Monsanto booklet on chlorosulphonic acid:

Chlorosulphonic acid is a strong acid capable of exerting severe local corrosive action. Either the liquid or vapour is dangerous when in contact with the eyes or skin, or when taken by mouth. The vapour has such a sharp penetrating odour that the inhalation of toxic quantities is unlikely unless the victim is trapped in such a location that escape is not possible. No systemic effects of liquid or vapour are noted; however, any contact of liquid with the body results in immediate severe burns. CS reacts violently with water, and spattering or explosion is likely to occur. For this reason, potential sources of contact of CS with water should be eliminated.

Cleaning Up Spilled Acid: Depending on the location of spilled acid (either outdoors or within an enclosed space) the method for cleaning up the spills differs. CS spilled outdoors may be flushed to the nearest sewer by gradually diluting the spilled acid with a large stream of water. A solid stream of water is preferred to a spray to minimize the possibility of spattering acid. It is desired to have the wind to the back of the workman to avoid inhalation of fumes of HCl which are evolved upon dilution. Very thorough washing is desirable to prevent corrosion damage to equipment. Personnel should wear goggles, rubber shoes, rubber gloves and a gas mask, if there is any danger of inhalation of HCl.

Because of fumes given off when chlorosulphonic acid is diluted with water, this method is not suitable for cleaning up spills in enclosed areas where ventilation is not good. The recommended method consists in absorbing the spilled acid in an inert material such as dry sand, vermiculite, or powdered clay. Materials containing moisture or organic materials should not be used because of the possibility of spattering the acid or charring the organic material sufficiently to cause a fire. Sufficient absorbing material should be used to remove the bulk of the spill, and then the traces remaining should be treated with a dilute (5 per cent) solution of soda ash or caustic followed by a thorough wash with plain water. The absorbent mass may be disposed of by burying it and covering the area with lime, or by washing with water as previously outlined.'

(F) Sulphamic Acid

HSO_3NH_2 (also called amido-sulphonic acid) is occasionally used commercially for specialized sulphonations. It is non-corrosive and the plant required is relatively simple. Alcohol and phenol ethoxylates have been sulphated by sulphamic acid, although results with linear alcohols have not been successful.[16]

Compared to other methods of sulphation, this process is rather expensive (see also pp 172, 297).

Materials sulphated with sulphamic acid are neutralized naturally to the ammonium salt:

$R-O-CH_2-CH_2-OH + HSO_3NH_2 \rightarrow R-O-CH_2-CH_2-O-SO_3NH_4$

In general, the fact that the material is neutralized with ammonia is usually no detriment, but if the final formulation has a pH of over 8, a smell of ammonia will develop. The ammonium cation can then be replaced by adding the stoichiometric amount of caustic soda solution and heating to 100°C, when the ammonia will be distilled off.

Sulphonation with Sulphuric Acid and/or Oleum

Sulphonation with sulphuric acid and/or oleum of various free SO_3-strengths is still widely carried out in the detergent industry. The sulphonation with acid is mainly used for the sulphonation of alkylate, either ABS or LABS. The procedure of sulphonation is described in detail in publications issued by the producers of alkylate.

The chemical reaction is as follows:

R—⟨benzene⟩ + H_2SO_4 → R—⟨benzene⟩—SO_3H + H_2O

alkylate

As is to be seen, water is set free by the reaction. The presence of water retards the sulphonation reaction, or even prevents it altogether. This is the reason why either a very large surplus of H_2SO_4 or a lesser surplus of oleum ($H_2SO_4 + SO_3$) is required to bring the reaction to completion.

Because alkyl benzene (see p 124) is produced to precise specifications, the detergent manufacturer can work out sulphonation procedures for detergent alkylates from various sources, without changing process conditions which, once established, have become standard procedures for an established sulphonation set-up. This does not mean, however, that there are no differences in quality among detergent alkylates of various provenances. Another important point must be mentioned: that sulphonation procedures vary to a certain extent depending on whether straight-chain alkylate or PT alkylate is used. In the following we will give a description of a batch sulphonation process of alkylate.

For sulphonation, a glass-lined (or stainless) jacketed vessel with an anchor type agitator of c 120 rpm is suitable. During sulphonation, optimum temperature is maintained by cooling with water through the jacket. In hot climates it may be necessary to augment the cooling effect by a refrigeration system and/or a side-arm heat exchanger through which the reaction mass is recirculated by means of a gear pump. The pipes through which the reaction mass is circulated should be of stainless (type 316) steel. It is our opinion that in all cases where the batch size is above 1000 kg a side-arm heat exchanger should be installed.

For acid sulphonation the following process variables are decisive for the quality of the finished product:

1. Acid strength (and, of course, purity of acid, colour, etc).
2. Ratio alkylate/acid.
3. Sulphonation temperature and time of acid addition.
4. Digestion time and temperature.

After the actual sulphonation step the separation of spent acid (ie, the greater part of surplus acid) has to be accomplished. Sulphonation without subsequent spent-acid separation is rarely carried out, because in this case all the surplus acid has to be neutralized with caustic soda, transforming it to sodium sulphate. This is rather uneconomical and, owing to the high reaction temperature for neutralizing considerable amounts of free sulphuric acid, special precautions are required to keep neutralization temperatures under control. We will therefore describe a sulphonation process with subsequent spent-acid separation.

Table 5.11 shows the variables in sulphonation for various acid strengths.

This table should serve as a guide. In practice, some variations will be necessary, eg, when sulphonating with 98 per cent acid, it might be advisable to warm the alkylate to c 40°C, in order to start the reaction smoothly. Otherwise, when starting at too low a temperature, the reaction might not begin until a very large amount of acid is added, and then it might be quite sudden, with an excessive amount of heat which might cause the reaction to get out of control. Anyway, it is a good rule to follow. Reaction should start immediately with the addition of acid, so that the time of acid addition and cooling together keep the reaction well within the range of optimum conditions. Naturally the reaction temperature also has an effect on the ratio alkylate/acid. Saving on acid generally

TABLE 5.11

Acid Sulphonation Conditions			
Acid strength	Ratio alkylate/acid	Reaction temperature (°C)	Digestion time (°C)
20–22% SO_3 oleum	1:1·1 (min. 1:1)	30° until half of acid addition 45° at the end	2 h at 40–45°
12–14% SO_3 oleum	1:1·2 (min. 1:1·1)	45° half, 55° end	2 h at 50–55°
100% H_2SO_4	1:1·5	50° half, 55° end	2 h at 55–60°
98% H_2SO_4	1:1·65 (min. 1:1·55)	45° half, 55° end	2 h at 50–55°

means a sulphonation temperature at the upper limit. Furthermore it may often be preferable not to aim at too complete a degree of sulphonation, and to be satisfied with, say 1·2–1·5 per cent of unsulphonated oil calculated on 100 per cent active matter, rather than 0·8 per cent, which would require more acid, higher temperature, longer ageing time, etc. This would result in a product of poorer colour and odour properties—on the whole a disadvantage that might not be warranted by a decrease of unsulphonated content from 1·5 to 0·8 per cent. With straight-chain alkylate, the reaction conditions are very much the same as for PT alkylate; however, the reaction temperatures should generally be about 5°C lower at the end of the reaction than those given for PT alkylate.

Now to the separation of spent acid. As a rule of thumb, 10 per cent of water calculated on the total of alkylate plus acid are added to obtain a good spent-acid separation. The water is added as it is, or in the form of crushed ice (not too large lumps of ice, in order to prevent damage to the sulphonation equipment). Under no circumstances should the temperature rise above 60°C, when all the water has been added. Separation takes about six to eight hours, and the resulting spent acid has a strength of 70–80 per cent. In case the sulphonic acid is to be used for the production of liquid detergents, where freedom of electrolytes is of great importance, the authors recommend a modified 'double-wash' spent-acid separation.

As previously mentioned, sulphuric acid and/or oleum sulphonation is still used in the detergent industry, mainly for the sulphonation of alkyl benzenes, PT benzene, or straight-chain alkylates. It is not suitable for the production of fatty alcohol sulphates, or for the sulphation of fatty alcohol and/or alkyl phenol ethoxylates.

Sulphonation with SO_3

The greatest advance in sulphonation technique in recent years is the use of SO_3 as sulphonating agent, made possible by the commercial availability of this material (see p 145). Whereas sulphuric acid and/or oleum is mainly used for the sulphonation of alkyl benzene, and chlorosulphonic acid is only suitable for detergent raw materials where an OH group is present, SO_3 is suitable for practically all types of detergent raw materials.

The reaction with SO_3 takes place without formation of water (see above). The following equation clearly shows this:

$$R\text{-}C_6H_5 + SO_3 \rightarrow R\text{-}C_6H_4\text{-}SO_3H$$

alkylate

One hydrogen wanders from the aromatic ring to form SO_3H, sulphonic acid.

As SO_3 is a vigorous sulphonating and dehydrating medium, on occasion certain side reactions take place:

(a) Formation of anhydride

R–C₆H₄–S(=O)₂–O–S(=O)₂–C₆H₄–R

(b) Formation of sulphone (not to be confused with sultone)

R–C₆H₄–S(=O)₂–C₆H₄–R

In practice, either or both of these materials can be formed in small quantities, usually less than 1 per cent of the total mass.

Various methods of hydrolysing these two products have been patented, all consisting of adding small quantities of water, dilute alkali, aromatic solvents or alcohols to destroy either the anhydride or the sulphone.

It can be seen that the anhydride hydrolyses to two molecules of sulphonic acid, whereas the sulphone produces one molecule of sulphonic acid and one of alkyl benzene on hydrolysis.

For fatty alcohols the reaction is as follows:

$$RCH_2OH + SO_3 \rightarrow RCH_2OSO_3H$$

Again, a hydrogen wanders to the SO_3 to form the acid sulphate which in the case of fatty alchohols has to be neutralized immediately to prevent hydrolysis.

The most widely accepted method of using liquid SO_3 as sulphonating agent is to employ it in its vaporized form diluted with inert carrier gas, mostly dry air, to give a dilute SO_3 gas stream of 7–12 per cent (vol) SO_3. (A special sulphonation process with liquid SO_3 diluted in liquid undercooled SO_2 is used by the US firm 'Pilot'.) Naturally the carrier gas must be as dry as possible, otherwise the SO_3 would react to give H_2SO_4 fumes which would cause poor quality of the finished product. Vaporization of SO_3 and air-drying is an essential part of process equipment in sulphonation plants using liquid SO_3. (Air-drying units are also essential components in SO_3-sulphonation plants starting with elementary sulphur (see below).) Detailed descriptions of vaporizers for liquid SO_3 are somewhat beyond the scope of this book. Producers of liquid SO_3 generally give advice on the proper type of vaporizer to be used. Usually, the liquid SO_3 is fed by dosing pump (eg, Teflon-lined pumps) into the electrically heated vaporizer, so that not much liquid SO_3 enters at a time, and vaporization is aided by the stream of dry air passing through the vaporizer. Often the equipment is designed as a film vaporizer, so that at the lower end, the 'sump', the stabilizer collects as a liquid, which is often re-shipped to the SO_3 supplying firm. A very detailed description of the sulphonation process is given by Carlson *et al*, *Ind Eng Chem* **50**, 277–84 (1958).

For the air-dilution of the vaporized liquid SO_3 two principal methods may be used: (*a*) a 'once-through' system, where the dry air used for dilution of the SO_3 passes from the sulphonation unit to vent; and (*b*) a 'closed' system, where the air leaving the sulphonation unit is recircled to the SO_3 vaporizer. The advantage of the first method is that no droplets and fumes recycle to the vaporizer, which otherwise might cause serious colour deterioration. The disadvantage is that the entire dilution air must be dried all the time. The obvious advantages of the second method are the elimination of continuous air drying and the absence of exhaust fumes from the vent. But, in order to be efficient and to prevent colour deterioration, an extremely efficient mist-collector must be inserted between the vent from the sulphonation unit and the vaporizer. In a closed system, nitrogen may be used as the carrier gas, which otherwise would be too expensive, but has the advantage of being completely inert.

Mention must be made of the 'Jergens' process, marketed by the US Metrochem Corporation. This process eliminated the use of dry inert carrier gas for gaseous SO_3 by reacting the alkylate, fatty alcohol, or other raw material with SO_3 vaporized in a partial vacuum (about 0·66 kPa or even less). The reduction in SO_3 concentration is affected by the partial vacuum rather than by dilution with dry air or nitrogen.

The actual sulphonation technique with vaporized liquid SO_3 is the same as with converter gas from a sulphur-burning unit. However, sulphonation with liquid SO_3 has the advantage that such a plant may be operated on a daily shift basis, whereas a plant based on sulphur burning and conversion to SO_3 is only economical when operated continuously, or at least on a weekly basis. Some sulphonation plants using dilute SO_3 vapour obtain the SO_3 by stripping it from 60 per cent SO_3 oleum.

This method of obtaining vaporized sulphur trioxide diluted to about 7–10 per cent in dry air may seem to be unduly complicated in view of the fact that the converter gas which enters the sulphur trioxide absorber in any contact sulphuric acid plant has the same percentage of gas as one wishes to obtain by vaporization of the liquid. However, up to now, there have been good reasons why converter gas has not been so widely used as a direct sulphonating agent. For instance, it may happen that the industry wanting to use converter gas from a sulphuric acid plant is not situated near a contact plant. Furthermore, it is not always economically advisable for the industry producing sulphonated materials to become dependent on a sulphuric acid plant. Technically, it is not a simple matter to use 'split-off' converter gas from a sulphuric acid plant, mainly because of the difficulty of measuring exactly the amount of sulphur trioxide taken from the cooler to be used in the sulphonation process.

To control the actual amount of sulphur trioxide, one must know the volume of converter gas at constant temperature and the concentration within a very narrow range. This does not appear to be too difficult, but in practice it is not a simple matter. For many sulphonation processes the amount of sulphur trioxide used must be accurately measured to within ± 1 per cent, as any greater variations will lead to either over- or undersulphonated material. In the case of detergent alkylate, this would mean a product with too much unsulphonated matter or with an unacceptable dark colour. With fatty alcohols it is even more dangerous, as inaccurate addition of sulphur trioxide might even lead to a reversal of the sulphonation reaction.

However, methods for controlling the degree of sulphonation have been worked out, which automatically record the amount of unsulphonated material, and by feedback adjust the flow of alkylate to the sulphonation plant. This method of control might overcome the difficulties in working with 'split-off' converter gas.

It is only with those sulphonation processes where accuracy is less important that the use of 'split-off' converter gas is a relatively easy matter. Thus, white oil and highly refined kerosene can be produced by treating the appropriate petroleum fraction with converter gas either in a continuous or discontinuous process.

A Davidsohn, in 1960, investigated the possibility of designing a plant to produce sulphur trioxide on a small scale, where all the sulphur trioxide from the converter is cooled and used for the sulphonation of detergent raw material. To do this, certain conditions had to be fulfilled. First of all, to ensure a constant dosing of sulphur into the system, a metering pump was used. Then, in order to have the sulphur transformed into a constant amount of dry sulphur trioxide, the air used for burning the sulphur in the sulphur-burner had to be completely dry, as otherwise sulphuric acid would be formed by the reaction with the water present. Drying of air was therefore carried out in a silica-gel drying unit, which brought the humidity down to about 0.5 g/m^3 air. If a still higher degree of drying is required, it is possible to carry out a kind of 'threshold drying' by having the silica-gel-dried air pass through a molecular sieve system, which brings the final humidity down to as little as 0.05 g/m^3 air.

A more efficient method to reach a very low dew point is by pre-cooling the air to about $5°C$, condensing and then removing the bulk of the moisture present. The silica gel, operating at a lower temperature works better and the final dew point of the air can be $-30°C$ or even $-60°C$.

A simple method of controlling the efficiency of the drying unit is to use the Draeger system of gas detection. The method is to draw by means of a hand pump, a constant volume of air through the appropriate Draeger tube and the concentration of moisture is read off. For more exact determinations special instrumental methods are available.

The same principle can be applied to the measurement of the SO_2 content of the exhaust gas leaving the sulphonators. Determination of SO_2 gives an indication of the original degree of conversion and also is important from the point of view of air pollution.

As already mentioned, the sulphur fed into the sulphur-burner in a converter-gas plant must also be measured very exactly. Furthermore, it was found in practice that the sulphur-burner used should preferentially be so designed that the sulphur is not sprayed under pressure through a nozzle into the burner, but flows through a relatively large orifice. Such a design is less liable to irregularities because of the clogging of the nozzle, and there is also the advantage that the pump is not working against a high pressure.

The sulphur used in small converter plants should be of high quality. Sulphur from Texas is suitable, as is also sulphur recovered from natural gas (see p 145). Even pure sulphur of the 'Texas'—or 'Lacq'—type may contain some impurities, and small amounts may lead to trouble. Bituminous impurities, for example, tend to react at certain temperatures with sulphur to give highly polymerized compounds, which are of a very 'sticky' nature, and may impair the action of the valves of the metering pump. The following precautions are thus recommended: (1) the temperature of the molten sulphur in the melting-pot must not be allowed to rise above 150°C; (2) the sulphur pipes leading to the metering pump must be situated in such a manner that as much as possible of the impurities of the sulphur settle on top or on the bottom of the melting-pot; (3) filters should be used either inside or outside the melting-pot, to retain impurities; (4) no local overheating should occur; and (5) the temperature must not be higher than 152°C, as above this temperature there is a sudden rise in viscosity (from 6·7 cP at 152°C to 12·8 cP at 159°C) (see p 146). The melting-tanks have to be cleaned from time to time, and it is therefore customary to use two of them. The sulphur filters should be of such a type that they can either be cleaned *in situ* or speedily dismantled for this purpose.

If these precautions are taken, the delivery of sulphur can be kept within the range of ± 0·5 per cent of the desired amount. To do this, it is advisable to lead the sulphur from the melting-pot to the pump and from the pump to the burner through steam-jacketed pipes with steam at 500 kPa.

The sulphur is burnt in a stream of dried air to give an optimum concentration of 7–8 per cent sulphur dioxide in the gas reaching the converter. The approximate consumption of dry air required to obtain burner gas of various concentrations is given in Table 5.12.

The burner-gas leaving the sulphur-burner has to be cooled, either by simple radiation cooling or by an air heat exchanger, so as to bring the burner-gas down to a temperature of 415–420°C at the entrance to the converter. If an air heat exchanger is used, the cooling air is used for regeneration of the silica-gel (generally the heat from the sulphur trioxide gas cooler is sufficient, but an arrangement using the air from the sulphur dioxide cooler or converter may serve to augment the supply). Alternatively, the heat content of the burner-gas may be used to produce the steam for the sulphur melting-pot and the steam-lagged pipes leading from the sulphur melting-pot to the sulphur pump.

The burner-gas then enters the converter, where it is transformed into sulphur trioxide over a vanadium catalyst. The gas leaving the converter

SYNTHESIS OF DETERGENTS

TABLE 5.12

Air Requirements for Sulphur Burning

Volume-per cent of sulphur dioxide	Air, m^3/h (for 10 kg/h sulphur)
6·0	117·0
6·5	108·0
7·0	100·0
7·5	93·0
8·0	87·5
8·5	82·5
9·0	77·5

has a temperature of c 420–450°C, and has to be cooled to c 70–80°C in a heat exchanger.

The air used for converting the sulphur to sulphur dioxide must be forced through the pre-cooling system, the silica-gel drying unit, the sulphur burner, and the converter by a 'master-blower', which must be capable of overcoming not only the flow resistance in the sulphur-trioxide production unit, but also that of the sulphonation unit. Generally a pressure of about 1400–1800 mm water is sufficient, provided that no special high-pressure sulphonation is to be carried out.

For a small sulphur-trioxide production unit, one converter using a vanadium catalyst is sufficient, and generally four layers of catalyst are applied. The conversion of sulphur dioxide to sulphur trioxide being an exothermic process, the temperature in the converter must be controlled by air cooling, the air stream passing around the catalyst chamber. As a rule, the cooling of the converter is controlled manually but thermostatically controlled cooling can be installed.

Some of the characteristics of a typical plant are set out below:

Working pressure of air from the 'master-blower'	1500 mm water
Pressure loss in silica-gel drying unit	30 mm water
Pressure loss in the sulphur-trioxide production section	30 mm water
Temperature in sulphur burner	850–900°C
Temperature at the entrance of converter	500°C
Temperature at the first catalyst layer	530°C
Temperature at the second layer	540°C
Temperature at the third layer	430°C
Temperature at the fourth layer	450°C
Temperature of converter gas after air cooling	65°C
at the entrance to the sulphonation vessels	50°C
Temperature of air used to regenerate silica gel (from converter cooling)	200°C
Temperature during silica-gel regeneration	145°C

For optimum conversion of SO_2 to SO_3 (above 98 per cent), the burner gas should have 7–7·5 volume per cent SO_2.

Sulphonation Processes

In addition to the advantages previously mentioned, the use of sulphur trioxide for sulphonation, generated as described, has the additional asset that its oxygen content is low compared with vaporized liquid sulphur trioxide, where air acts as the carrier gas. In the production of the trioxide direct from sulphur, about half of the oxygen in the air is used for conversion. This reduction in the amount of oxygen in the sulphur trioxide gas stream eliminates side effects often encountered with vaporized gas. Furthermore, the traces of sulphur dioxide in converter gas are also beneficial in this respect.

In the first plant employing this process, in Dalia, Israel, sulphonation was carried out semi-continuously in two small sulphonating vessels of about 250 litres capacity. The unit consumes 1000 kg sulphur/24 h, and this produces 2400 kg of sulphur trioxide (as 7–8 volume per cent converter gas), ie, enough for about 7500 kg alkyl benzene/24 h.

Some 175 kg (200 litres) of alkyl benzene are charged into one of two sulphonation vessels and sulphur trioxide admitted. Sulphonation requires about 60 min and, while the reaction is proceeding in one vessel, ageing is taking place in the other. After ageing, which takes about 15 min, a small amount of water or NaOH solution is added, and the alkyl benzene sulphonic acid is either run to storage or immediately converted to the neutralized AB-sulphonate. The small amount of water (about 2–3 per cent in the sulphonate mixture) helps to prevent 'anhydride-sulphonate' formation and keeps the colour of the AB-sulphonic acid stable, even on prolonged storage in mild steel drums or tanks. By introducing a time control and automatic-operated valves, etc, the semi-continuous sulphonation system can be converted to a continuous process.

The Sulphurex process of the Ballestra Company of Milan (see Fig 5.8) represents a further development. Here both the sulphur and the detergent raw material are metered into the system in the requisite proportions.

After being cooled, the SO_3 gas stream passes into the sulphonators at about 50°C. These consist of two or more reactors—different in number and size depending on the capacity of the plant. The SO_3 gas stream enters the reactors in parallel, with the major portion entering the first reactor and the minimum portion entering the last reactor. This proportioning of the SO_3 gas into the various reactors is done automatically and requires no attention from the operator.

The alkyl benzene (or any other material to be sulphonated or sulphated) essentially passes through a series of reactors in cascade, but because of the 'free oil' controller, a small amount is proportioned into the last reactor, since it is this quantity which is varied according to the completion of reaction.

SYNTHESIS OF DETERGENTS

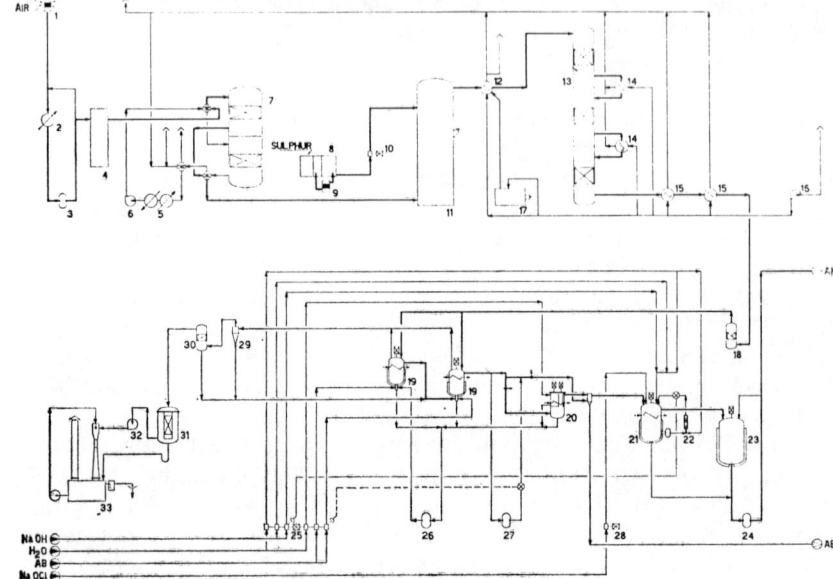

Fig 5,8   Flow diagram of Ballestra Sulphurex Process

1. Air filter
2. Pre-cooling unit
3. Blower
4. Air cooling unit
5. Silica gel regeneration unit
6. Fan
7. Silica gel air driers
8. Sulphur melter
9. Filter
10. Proportioning pump
11. Sulphur combustion furnace
12. SO_2 cooler
13. Converter
14. Intermediate coolers
15. SO_3 coolers
16. Cooling fan
17. Pre-heating furnace
18. Filter
19. Sulphonators
20. Digestor
21. Neutralizer
22. pH control group
23. Final homogenizing unit
24. Active matter transfer pump
25. Raw material proportioning pumps
26. Start-up pump
27. Free oil control group
28. Bleaching proportioning pump
29. Cyclone
30. Filter
31. SO_3 filter
32. Exhaust gas fan
33. Gas scrubber

Special high-speed turbine mixers disperse the SO_3 gas and the alkyl benzene with a high degree of efficiency, which prevents any local overheating of the reactants. Inside the reactor are precisely located cooling coils and baffles which immediately remove the large amount of heat released during the reaction.

The temperatures normally used for various raw materials are:

(a) for the usual types of alkyl benzene (branched or straight-chained):
  Sulphonators: 50°C
  Digestion: 55°C
  Neutralization: 45°C

(b) for lauryl alcohol:
    Sulphonators: 35°C
    Neutralization: 45°C
(c) for tallow alcohol:
    Sulphonators: 55°C
    Neutralization: 50°C
(d) for C_{14}—C_{18} Ziegler alcohols:
    Sulphonators: 50°C
    Neutralization: 45°C

After the sulphonators, the main stream of sulphonic acid passes on to the digestion vessel, where it is held, to allow the reaction to go to completion. A side portion of this stream is directed to a patented device which automatically analyses the sulphonic acid and determines its 'free oil' content. If the free oil is increasing, this control unit automatically cuts down on the quantity of alkyl benzene being fed to the last reactor, and conversely, if the free oil is going too low, the unit automatically increases the flow of alkyl benzene to the last reactor.

The gases from the reactors are essentially free of SO_3. However, the plant is equipped with a scrubbing system for removing any SO_3 gas that may be in the exit stream, and for scrubbing out the SO_2, so that the exhaust gas is free of these undesirable components as it passes to the atmosphere.

The scrubbing solution containing mainly sodium sulphite may be used directly as a slurry component in an adjacent spray-drying unit (sulphite has no detrimental effect on optical brighteners) or better still, it could be oxidized by a special air injection system to sodium sulphate, which again may safely be used as a slurry component, or in case no spray-drying unit is near the sulphonation unit, be discharged to the sewer. The scrubbing system is efficient enough to absorb the SO_2 so that the exhaust gas exits with as little as 5 ppm SO_2.

After the digestion vessel the sulphonic acid stream can be passed into another small reactor where water or NaOH solution can be added if desired. This unit is to take care of any anhydride formation that may occur.

Of course, no water or NaOH solution is used during the sulphation of the fatty alcohols. Because of the combination of reactors it is very easy to perform co-sulphonation–sulphation reactions and produce a high-quality product.

The sulphonic acid then passes on to the neutralizer, where caustic soda and water are continuously added to give a white paste containing the desired percentage of active ingredient. An automatic pH control unit controls the addition of caustic by sending a signal to the controller which varies the stroke of the proportioning pump of the neutralizing agent by means of a servomotor.

The neutralized paste overflows into a storage tank, where it is continuously homogenized. From here it is pumped to any further processing operations, such as spray-drying, liquid production, etc.

It is important to note that after the sulphonators no transfer pumps are required. The reactants and acid flow from vessel to vessel by gravity. Operation and control of this plant are extremely precise and yet simple. One operator runs the entire plant—sulphur-burning, conversion, and sulphonation.

Neutralization of sulph(on)ated products can be an important operation, particularly for materials other than alkyl benzene sulphonates made by SO_3 sulphonation. If oleum sulphonation is used, the sulphonic acid contains both free sulphuric acid and water, rendering it corrosive. Unless special storage conditions are observed the acid material will both corrode iron and darken on standing. Neither of these restrictions applies to SO_3 sulphonated alkyl benzenes, they can be stored and transported in mild steel with no deleterious effects. For all that even the SO_3 sulphonated variety is often neutralized immediately after sulphonation.

Alcohol sulphates and alcohol ether sulphates need to be neutralized immediately, otherwise, in acid medium, the sulphuric ester will split to its original components (the reverse of the sulphation reaction). Olefin sulphonates need to be hydrolysed in addition to neutralization (see p 167). Sulphonated methyl esters need special treatment to prevent (or minimize) ester saponification in an alkaline medium.

For all the above reasons, the Ballestra Group has developed its Neutrex system of double step neutralization, which eliminates the local concentrations of acid or water (or alkali), preventing hydrolysis of the ester group, or the formation of highly viscous hydrated lumps which dissolve slowly, causing pH drift. If hydrolysis were to occur, free sulphuric acid and free oil are formed. Free oil increases the unsulphonated portion of the finished paste, free sulphuric acid requires considerably more neutralizing agent which can increase both the costs and the amount of inorganic salts present. Increase of free acid (even for a short period) can cause corrosion problems to the equipment and this is enhanced if, as is sometimes the case, bleaching agents are added during the neutralization step. The plant is also so designed that the temperature of the paste is kept within reasonable limits.

To overcome all these problems, the Neutrex system neutralizes 90 per cent of the incoming material in the first of two reactors, equipped with specially designed low speed impellers, which mix and homogenize. The heat of neutralization is removed in this vessel. The remaining 10 per cent is fed into the second (larger) mixer and this mixes with the incoming neutralized paste from the first. The neutralizing base which can be any of the alkaline materials commonly used is accurately dosed with linked pH control. The mass in the second neutralizer is kept at a temperature such that it can be easily transported further.

This unit can be coupled to any SO_3 sulph(on)ation system, and by using this system the hitherto difficult to produce 70 per cent alcohol ether sulphates can easily be made.

Comparison of Oleum and Sulphurex Processes

	Oleum process	Sulphurex process
Sulphonating agent	Oleum	Sulphur
—characteristics	Corrosive liquid	Inert powder
—supply	Tanks, drums	In bulk (powder)
—storage	Tanks, drums	In bulk (powder)
Sulphonic acid purification	Washing and separation	No purification required
By-products	Dark 70% spent sulphuric acid	No by-products
Products storage	Lead-lined tanks	Mild steel tanks

For the production of 100 kg of 100 per cent active material from 72·5 kg alkyl benzene, the acid sulphonation process requires 75 kg of oleum, while the Sulphurex process consumes 10·5 kg of sulphur.

Only one man per shift is needed to operate the 'switch-over' and the Sulphurex processes. A sulphur-trioxide absorption tower is not essential for warming up and cooling down the unit. While the unit is being brought into equilibrium, the converter gas of irregular composition may be absorbed, for example, in alkyl benzene.

To use the converter-gas plant for the sulphonation of other compounds, such as toluene, xylene, etc, some modifications are necessary. Thus, for sulphonating low-boiling aromatic compounds, special cooling condensers have to be provided, to prevent undue losses by carrying over of unsulphonated product. By special modification of the sulphonation process on the basis of the 'dominant bath' system for low-boiling aromatics, a less expensive condensing system is obtained.

It is also possible to co-sulphonate toluene or xylene in conjunction with detergent alkylate, to give the proper proportion of hydrotrope or anticaking agent in the finished detergent.

Production of toluene sulphonates with oleum or SO_3 is rather costly, so the price of separately bought toluene sulphonate is higher than the price of AB-S. Since only a small percentage of toluene sulphonate is used (5–10 per cent calculated on AB-S used), it would be worth while to co-sulphonate toluene with the alkylate. The main difficulty encountered here is the low boiling point and high volatility of toluene. There is danger of 'carry-over' of the toluene with the inert gas stream leaving the system through the exhaust, before the toluene is sulphonated completely in the system. This leads to losses, unpleasant exhaust problems, and danger of ignition; problems which can be overcome in the laboratory by adding condensers and scrubbers to the sulphonation unit, but which are costly and difficult to set up on a plant scale. A different approach is used. The toluene is fed into the sulphonation system (in this case a 'switch-over' system (see p 157)) via the side-arm heat exchanger. Once the alkylate is sulphonated, toluene is fed into the suction line of the circulation pump for the side-arm heat exchanger in a ratio corresponding to the feed of SO_3 entering the system. The toluene is thus pre-mixed with the AB-sulphonic acid in the circulation pump and heat exchanger, and then further mixed into the 'dominant bath' of the AB-S in the sulphonation system. No losses occur, and the toluene can be properly co-sulphonated. (The same principle can, with modifications, be applied to the fully continuous system.) If an automatic control system is not available, or if it is desired to control the system from time to time, the unsulphonated toluene can be controlled by a quick and simple technique: a sample of the AB-sulphonic acid is withdrawn and heated under an infra-red drying balance. Weight losses during five minutes give the amount of unsulphonated toluene, since neither AB-sulphonic acid nor toluene-sulphonic acid is volatile enough to disturb this quick test. Of course the same technique applies to co-sulphonation of cumene or xylene.

Practical test runs on a large scale have been made in a Sulphurex plant in which the standard Sulphurex set-up was used. Sulphonation in the first vessels was carried to about 85–90 per cent, and toluene was fed by the second dosing pump into the last sulphonation vessel, where the practically fully sulphonated AB-S (or for that matter fatty alcohol sulphate) was used as the 'dominant bath' for the toluene to be co-sulphonated. No special mixing pump was required in this case. Adjustment of the automatic instrument for sulphonation control can easily be made.

Recently there has been an increased interest in α-sulpho-fatty-acid methyl ester. They form useful components for many types of powder detergents, and are also used as components for syndet bars and syndet/soap combinations. They can be produced by sulphonating methyl esters of coco fatty acids, stearic and palmitic fatty acids. It is important for the quality and colour of the sulphonated end product, that the methyl esters should have a low iodine value, ie, below 0·5.

With the introduction of large-scale SO_3 sulphonation which, as we have stated, revolutionized the industry, considerable work has been done on improving the process both to produce higher quality materials and to adapt the procedure to allow of the sulphonation of the hitherto hard to sulphonate hydrophobes.

Sulphonation units for detergent production based on the falling film reactor are now being almost universally supplied to the exclusion of all other types, as these have proved to be able to cope with practically all the base materials except of course paraffin sulphonates.

Basically these units use a film of the material to be sulphated/sulphonated, injected on to the wall(s) of a vertical tube or series of tubes (bundle type). The SO_3 is fed concurrently into the centre of the space. The reaction takes place on the surface of the tube(s) and the heat of reaction is dissipated through the walls of the conduit.

As the reaction is almost instantaneous, and as the film is flowing downwards continuously, and as 70–80 per cent of the SO_3 is consumed in the first third of the reactor, the sulph(on)ated material does not remain in contact with a large concentration of SO_3 for any length of time.

The advantages of the falling film reactor are:

1. Contact time of already sulphonated material with SO_3 is minimal.
2. Low residence time, thus acid sensitive materials (alcohol, alcohol–ether sulphates) can be transported rapidly to the neutralizing section.
3. Relative flow rates and dilution of the SO_3 can be adjusted so that olefins, which react very vigorously, are contacted with dilute SO_3 for a relatively short time, on the one hand, and slow reacting materials, such as methyl esters, can have their residence time increased, on the other hand.

The problems in designing the reactor are:

1. Maintaining a constant uniform film of the organic feed on the walls.
2. Metering almost stoichiometric quantities of organic feed and accurately diluted SO_3.
3. Dissipating the heat of reaction but not at such a rate that the already sulph(on)ated material becomes a viscous mass and ceases to flow.
4. Controlling the velocity of the gas stream so that the organic feed is not carried over as a mist or the lower boiling components are not volatilized.
5. Having and maintaining a uniform small diameter (or aperture) as this increases the contact between the SO_3 and the feed.

There are several types of film reactors available and designers are constantly improving existing designs or adding new concepts. We shall describe the basic features of some commercially available plants:

Allied-Mecchaniche Moderne uses two very long concentric water jacketed cylinders, the reacting surfaces being between the two and these surfaces are polished to a mirror finish. The SO_3//air mixture is fed into the annular space and at the bottom of the reactor, the acid mix is separated from the air dispersed in it.

Chemithon also supplies a reactor consisting of two concentric cylinders and the organic feed is again pumped to the two walls. They have a shorter column and induce turbulence by a cage rotor rotating in the upper portion of the annular space.

Both Mazzoni and Ballestra supply a bundle type tubular reactor. The problem in these multitube reactors is maintaining an equilibrium in each individual tube. We shall describe the Ballestra reactor to show how this has been overcome.[17]

SYNTHESIS OF DETERGENTS

The Ballestra Multitube Film Reactor consists of a vertical bundle of long, small-diameter tubes.

The liquid film, which forms inside the tubes, comes into contact with the gaseous reactant. The length and diameter of each tube is so dimensioned as to provide the optimum product quality performance.

The most interesting feature of the Ballestra reactor is the simple way in which the exact reactants ratio is maintained in all the tubes.

The automatic adjustment of the reactants in each of the tubes is the result of the geometric design of the reactor itself and is completely independent of the actual number of tubes. It is a simple matter, therefore, to scale up the reactor to meet any production capacity requirement. All that is required is to increase the number of reaction tubes.

Figure 5.9 illustrates the principle of the reactor. The SO_3 gas enters at the top of the reactor and then flows, without restriction, into each of the parallel-mounted reactor tubes. The liquid is uniformly distributed, by means of a simple concentric slot, over the inner walls of each tube.

The concentric slot permits the same amount of liquid to be fed to each tube. A difference of about 20 per cent between each tube is permissible,

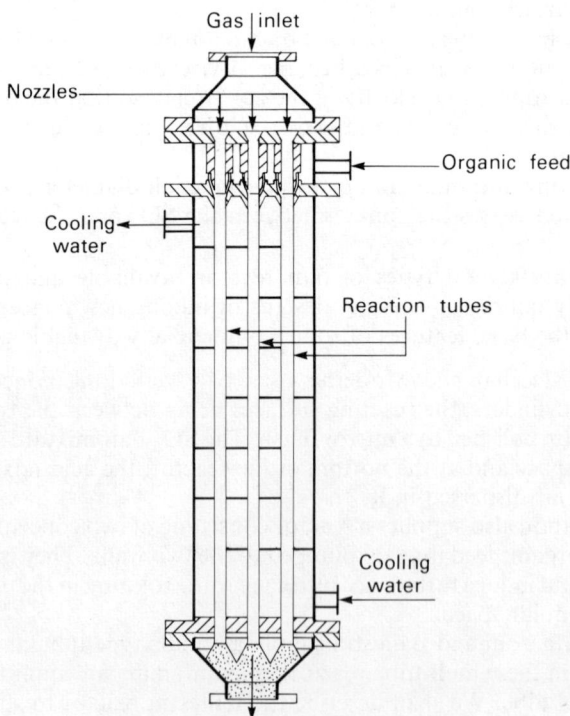

Fig 5.9 Ballestra multitube film reactor

and the correct amount of gas is automatically fed to each tube, due to the aforementioned geometric design of the reactor.

The special patented feature of the reactor is that the required molar ratio between the reactants in each tube is self-adjusting and, furthermore, is close to the preset ratio of the total liquid to the total gas flow rates in each of the tubes. The total gaseous reactant is fed to the reactor at the proper rate and at the proper concentration. The total liquid reactant is pumped to the reactor by means of the proportioning pump.

The flow of gas and liquid inside the reactor is parallel, with the liquid phase forming the film on the inner surface of each tube and the gas flowing through the tubes at very high speed. The outsides of the tubes, the shell side, are provided with cooling to ensure correct reaction temperature. The reaction takes place at a very high rate, with about 70 to 80 per cent of conversion occurring in the first third of the reactor.

Each individual tube, being fed at a particular average rate of liquid flow which is different from the next, has a correspondingly different film thickness and viscosity profile. This contributes to the self-adjustment of the gas-to-liquid mol ratios to the optimum average values, since the flow of gas in each tube then is directly proportional to the flow of liquid itself. The sulphonated liquid and exhaust gas leaving the reactor enter the gas separation vessel.

The separated exhaust gas is sent to the cleaning, electrostatic precipitation and scrubbing section and then released to the atmosphere. The liquid remaining in the separation vessel is transferred by pump to the neutralization section.

Should the flow of liquid to the reactor be interrupted for any reason, an independently powered stand-by emergency system takes over and purges the reactor with fresh raw material. In this way charring of sections of the reactor due to stagnant liquid in the presence of SO_3 is avoided.

Figure 5.10 shows the flow sheet for the complete sulphonation process.

All data are shown on a control panel detailing the complete process. Operation can be by push-button or computer.

Different feedstocks require differing concentrations of SO_3 in the gas stream. These can range from 2 to 7 volume per cent. The air-drying section is therefore dimensioned to supply the requisite dilution of dry air.

A combination of the film reactor and sulphonation under vacuum (p 153) principles has been patented by TrinLo Corporation, New Jersey, USA (USP 4,335,079). Liquid SO_3 is vaporized under vacuum and this vacuum is maintained throughout the whole system till the sulphonated material is discharged. The organic feed is fed into a specially designed egg-shaped reactor by a rotor to form a thin film on the internal wall of the egg.

These patentees claim that all the feedstocks normally used as detergent raw material intermediates can be sulph(on)ated by this unit. Advantages are that no air-drying or gas–liquid separation is necessary, considerably reducing both capital costs and space required.

Film reactor plants are being used to sulphonate all types of conventional detergent materials. They are, however, not suitable for co-sul-

SYNTHESIS OF DETERGENTS

phonating volatile materials, such as toluene, xylene, or cumene. They have found great success in the sulphonation of α-olefines which are playing an increasing part in detergent formulations and which hitherto had required very special methods of sulphonation.

ALPHA OLEFIN SULPHONATES (AOS)

The sulphonation of α-olefins produces in approximately equimolar proportions:

(a)
$$R-\underset{\underset{H}{|}}{\overset{\overset{H}{|}}{C}}=\underset{\underset{H}{|}}{\overset{\overset{H}{|}}{C}}-SO_3H$$

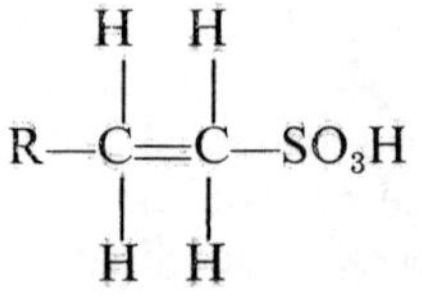

Alkene sulphonate

(b) Alkyl sultone

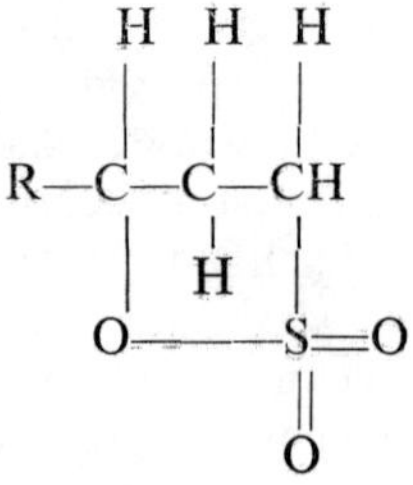

with minute amounts of

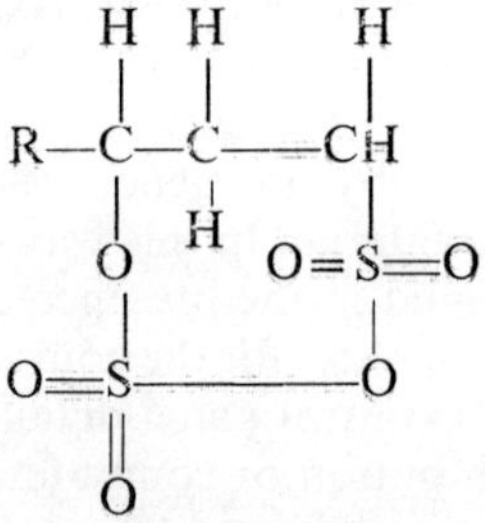

Alkyl disultone

The sultone shown is the 1,2-alkyl sultone but this is not the only one that can be formed. More often than not equal quantities of the 1,2- and 1,3-sultones:

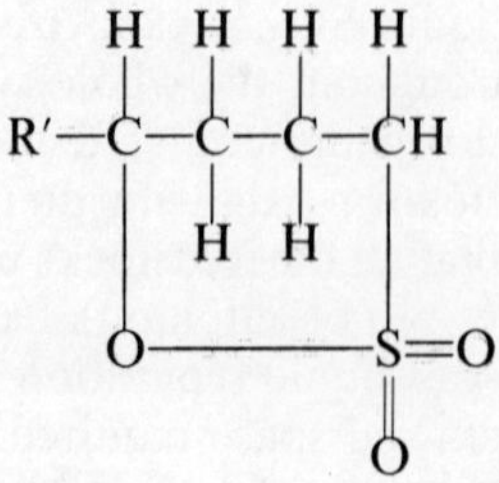

are formed.

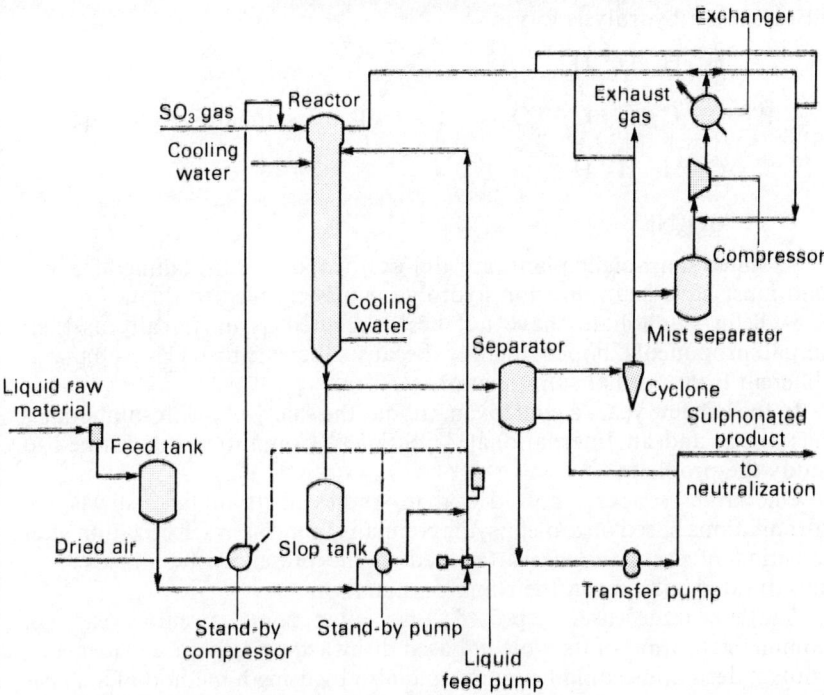

Fig 5.10 Ballestra's film sulphonation system

These sultones need to be hydrolysed to convert them to detergent active materials. This hydrolysis is done in an alkaline medium, when the ring opens to form the sodium hydroxy sulphonate:

$$R'-\underset{\underset{OH}{|}}{\underset{|}{C}}\underset{H}{\overset{H}{|}}-\underset{H}{\overset{H}{\underset{|}{C}}}-\underset{H}{\overset{H}{\underset{|}{C}}}-\underset{H}{\overset{H}{\underset{|}{C}}}-SO_3Na$$

β-Hydroxy sulphonate is an undesirable product of this hydrolysis as it is not soluble in water, giving a hazy solution. Fortunately the 1,2-sultone from which it is formed is thermodynamically unstable, and will isomerize to the 1,3- or 1,4-sultone on standing. This isomerization takes only a matter of minutes so that sulphonation/neutralization/hydrolysis conditions need to be arranged so that there is a 'hold' period between the sulphonation on the one hand and the neutralization and hydrolysis on the other.

Formation of disultones is dependent on sulphonating conditions, excess of SO_3 and temperature, and should be kept as low as possible. The disultone on hydrolysis forms

$$R'-\underset{\underset{SO_3Na}{\overset{|}{O}}}{\overset{H}{\underset{|}{C}}}-\overset{H}{\underset{H}{\overset{|}{C}}}-\overset{H}{\underset{H}{\overset{|}{C}}}-\overset{H}{\underset{H}{\overset{|}{C}}}-SO_3Na \qquad \text{Alkane sulphato-sulphonate}$$

As stated the α-olefin plants are almost invariably of the falling film type and must have neutralization/hydrolysis units coupled to them.

α-Olefin sulphonates have not been accepted as universally as their initial proponents hoped, mainly because the chemistry is somewhat different from normal sulphonation.

In the last few years a controversy about the safety of olefin sulphonates has arisen and an International AOS Safety Committee was formed to study the problem.

One larger soaper reported that in studies of its liquid dishwashing formulations based on α-olefins, the company found skin sensitization after a period of use. This was attributed to the presence of ppm levels of unsaturated sultones in their final formulation.

Another large user reported that after some twenty years of commercialization of its α-olefin based dishwashing liquid no evidence of contact dermatitis could be found, and also by using a method of analysis able to detect sultones down to 0·2 ppb quantities no significant levels of sultones were found.

The difference between the two was attributed to the fact that in the first instance, where sensitization had been encountered, the α-olefin sulphonate came out of the reactor with a relatively dark colour and needed to be bleached by the addition of hypochlorite in the hydrolysis step.

The second firm indicated that the olefin sulphonate produced by it was of light colour and needed no bleaching. From this the conclusion can be drawn that it is not advisable to bleach the sulphonate, reaction conditions need to be so arranged that a light-coloured material is formed *ab initio*.

Table 5.13 shows, among other items, the colour of unbleached detergent materials sulphonated in a Ballestra Sulphurex F film reactor. It can be seen that the colour of all the materials, including the olefin sulphonate, is such that no bleaching is required.

METHYL ESTER SULPHONATES

The basic chemistry of the sulphonation of methyl esters of fatty acids has been touched on (p 22), as has their method of preparation (p 130). The chemistry is not completely elucidated but appears to be the rapid formation of the complex, which can be termed a 'sulpho-anhydride' and this rearranges (relatively) slowly to the sulphonic acid.

Table 5.13

Product Characteristics of Materials Sulphonated in a Ballestra Sulphurex F Film Reactor

Material	Neutralized material			On 100% material		Klett colour*
	AM %	Uns %	Na_2SO_4	Uns %	Na_2SO_4	
LABS	—	—	—	0·8	0·9†	25
Lauryl alcohol sulphate sodium salt natural or synthetic	35	0·3	0·5	0·9	1·4	7
Lauryl ether sulphate 3 mol EO- sodium salt	30	0·3	0·4	1·0	1·3	15
Alcohol C_{16-18} sodium salt	25	0·65	0·4	2·6	1·6	60
α-Olefin sulphonate sodium salt	40	0·72	0·56	1·8	1·4	70

* On 5 per cent AM solution, 40 mm cell, #42 filter, UNBLEACHED PRODUCTS.
† Free sulphuric acid.

From work done by Davidsohn and his co-workers[10] it is evident that sulphonation is best done in a film reactor, requires an excess of SO_3, a relatively high temperature and a holding period to allow the intermediate complex to rearrange to the true sulphonic acid. Colour of the sulphonate will be high but this is easily bleached, either on the acid material using peroxide, or on the neutralizate using hypochlorite.

Neutralization needs to be done with care, the pH, even localized, should not rise above 8·5, otherwise saponification of the ester might occur. The Neutrex double step neutralization unit is therefore an ideal solution to the after-treatment of the sulphonic acid.

Heavy Alkylate Sulphonates

The heavy alkylate produced as the 'bottoms' in the alkylation of benzene is sulphonated by either oleum or SO_3 to produce synthetic petroleum sulphonates, not to be confused with paraffin sulphonates (see p 175). Natural petroleum sulphonates are produced as by-products in the production of white oils, and the 'synthetic' variety is almost identical functionally, although not chemically.

As mentioned on p 125, the heavy alkylates are complex mixtures but all the constituents of the mixture contain an aromatic ring, coupled to alkyl,

alkyl–aromatic, other aromatic or cyclic nuclei in the ortho- or para-positions. Sulphonation of alkyl benzene is almost invariably at the para-position, so if the para- position is blocked, sulphonation in the ortho-position is not feasible due to steric hindrance.

Heavy alkylates can therefore be only partially sulphonated and commercial products normally contain 60–70 per cent sulphonated matter.

Sulphonation is best achieved by SO_3 plus a non-reactive paraffinic solvent (boiling range c 210–230°C) to facilitate the separation of any oil-insoluble acid sludge. The unsulphonated and unsulphonateable portion acts as a solvent for the neutralized material allowing neutralization to be done with concentrated caustic soda or alkaline earth hydroxide slurries. The small amount of water introduced and produced in the neutralization process can either be left in the material or distilled off if the specifications require.

The sodium salts of the heavy alkylate sulphonates are used as emulsifiers, wetting agents and dry-cleaning additives.

The alkaline earth salts (calcium, barium and magnesium) are used in lubricants and greases and their 'over-based' varieties as rust preventatives. Over-basing is achieved by adding surplus hydroxide or carbonate to the neutralize in such a way that the base remains in colloidal suspension.

Sulphonation with Chlorosulphonic Acid

Chlorosulphonic acid (see p 147) is widely used for the sulphonation of fatty alcohols, fatty alcohol-ethoxylates, alkylphenol-ethoxylates, and related detergent raw materials with OH groups available for the attachment of an SO_3H group. The reaction of lauryl alcohol with chlorosulphonic acid illustrates the chemistry involved:

$$C_{12}H_{25}OH + ClSO_3H = C_{12}H_{25}O\text{---}SO_3H + HCl$$

The hydrogen chloride must be absorbed to give a 30 per cent HCl solution as a by-product. One can well consider the HCl in the formula of chlorosulphonic acid as a kind of 'carrier gas' present as liquid bound to SO_3, and set free during the sulphation reaction. Whereas sulphonation with SO_3 within an inert carrier gas consists of a gas/liquid reaction, sulphonation with chlorosulphonic acid is a liquid/liquid reaction, with liberation of HCl.

Sulphonation with chlorosulphonic acid requires special corrosion-proof equipment, either glass-lined steel or all-glass, the latter especially suitable for small-batch-size sulphonation. The HCl-absorber, too, is built either of glass-lined steel or is all-glass. Some firms use glass-lined steel reactors and all-glass HCl-absorbers. A typical batch sulphonation process of lauryl alcohol is carried out as follows:

250 kg lauryl alcohol are charged into the glass-lined reaction vessel (see p 148). 156 kg chlorosulphonic acid (the calculated stoichiometric

amount of the lauryl alcohol used) are added slowly with cooling. The reaction temperature is kept at an optimum of 25°C (not exceeding 30°C). 47 kg HCl gas are set free and have to be absorbed to 30 per cent HCl acid solution. (Some small amounts of HCl remain dissolved in the acid sulphonate and form a neutral chloride with the neutralization agent.) If NaOH is used as the neutralization agent, c 85 kg NaOH are required. Generally, the resulting 359 kg of acid lauryl-alcohol-sulphate are run into a crushed ice/water mixture c 1000 kg, while the NaOH is run into the neutralization vessel (as 255 kg 38° Be NaOH), keeping a small excess of NaOH, so that the mixture is always at a slightly alkaline pH. This prevents splitting off of the sulphuric acid by hydrolysis of the $O-SO_3H$ group, which would rapidly occur at an acid pH in the presence of water. Neutralization should be carried out as quickly as possible after the sulphonation process step; otherwise hydrolysis will occur, even in the absence of water.

The slurry (c 35 per cent active matter) is adjusted to a pH of 7·7–8·0, and is then stable on storage. Good mixing during neutralization is very important to prevent agglomeration of acid sulphonate, which again would lead to hydrolysis and a high unsulphonated alcohol content in the finished paste. Any bleaching required is carried out with sodium hypochlorite at a near neutral pH. Neutralization with ethanolamines is also often practised, and this method, and neutralization with ammonia, is, in fact, less critical: the mixture remains liquid and agglomeration of acid sulphonate does not occur so frequently. The heat of neutralization is more easily dissipated than in the case of caustic soda neutralization.

In sulphating ethers (and to a lesser extent alcohols) with chlorosulphonic acid, the HCl liberated in the initial stages of the reaction dissolves exothermically in the reaction mass, and in the second stage is displaced from solution together with the fresh portion of HCl being formed. The evolution of this double quantity of gas in the second stage of the reaction causes troublesome foam.

This foam formation, in continuous process particularly, can be overcome by arranging the flow of the liberated HCl on the one hand, and the incoming alcohol or ether on the other hand in such a manner that the evolved HCl is allowed to flow upwards through an absorption tower, down which the incoming alcohol or ether is allowed to trickle. This saturates the unsulphated material with HCl and does not allow any further HCl to dissolve during the reaction process. To a large extent this eliminates foaming in the case of alcohols but when ethers are being sulphated, in addition to the above arrangement it is necessary to dissolve in the unsulphated ether about 40 ppm of an anhydrous, oil-soluble, silicone antifoam. With these precautions, although a certain amount of foaming takes place, the foam formation remains within reasonable limits.

It is interesting to note that in the sulphation of ethers with SO_3, the same troublesome foam is encountered due to the large excess of air or other inert gas. Addition of silicone will prevent undue foam formation. In SO_3 film sulphonation systems this trouble does not occur.

SYNTHESIS OF DETERGENTS

Commercial specifications of various anionic detergents produced by the several sulphonation methods are given below, where the first comparison made is between commercial sodium alkyl benzene* sulphonate pastes produced by either oleum or sulphuric acid sulphonation with spent-acid separation and those produced by SO_3 sulphonation (either with vaporized SO_3 or directly with converter gas).

	By acid sulphonation	By SO_3 sulphonation
	%	%
Sodium alkyl benzene sulphonate	50	50
Unsulphonated free oil	0·5	0·5
Sodium sulphate	7·5	0·8
Water to make	100	100
Colour	White	White

Typical specifications for converter-gas-derived AB-sulphonic acid are as follows:

	Dodecyl benzene sulphonic acid	Tridecyl alkylate sulphonic acid†
Active detergent matter calculated as the sodium salt	100% ± 1%	100% ± 1%
—calculated as sulphonic acid with molecular weight 325	94% ± 1%	94% ± 1%
Unsulphonated free oil (calculated on 100% am)	1·7% max	1·7% max
Acid number	182–7 mg KOH/g	174–81 mg KOH/g

† Specifications for sulphonic acid produced with converter gas from 'straight-chain' alkylates have principally the same data, depending on the molecular weight of the alkylate, generally either in the range of PT alkylate or of tridecyl benzene. Heat of neutralization with NaOH is about 45 cal/g.

Not all sulph(on)ated materials are made using either oleum or SO_3. Sulphamic acid can be used for special applications, sulphosuccinates and sulphosuccinamates are made using sulphite addition and alkanes are 'sulphonated' using SO_2 and oxygen.

SULPHATION BY MEANS OF SULPHAMIC ACID

A method of sulphating alcohols and ethoxylates is by the use of sulphamic acid. The acid is quite expensive so the final cost of the finished product will

* The alkylate may be either of the PT benzene, tridecyl-benzene, or 'straight-chain' alkylate type.

be high compared to conventional methods of sulphation, but against this must be balanced the fact that no large investment in plant is needed.

The plant for sulphamation is relatively simple; all that is required is a closed stainless steel reactor with stirring, heating and cooling arrangements and a coil for the introduction of a nitrogen purge.

The reaction of sulphamic acid with an —OH group is:

$$-CH_2OH + H_2NSO_3H \longrightarrow -CH_2-O-SO_3H + NH_3$$
$$-CH_2-O-SO_3H + NH_3 \longrightarrow -CH_2-O-SO_3NH_4$$

Thus the ammonia liberated immediately neutralizes the sulphuric ester produced and an ammonium salt is formed. Sulphamic acid is particularly useful in sulphonating alkyl phenol ethers as in this case the aromatic ring is not attacked, whereas with SO_3 and chlorosulphonic acid up to 20 per cent ring sulphonation takes place.

It is not considered that ring sulphation detracts from the quality of the finished product, but the two types of sulphates (the ring has a sulphonate group) will be different. If 20 per cent ring sulphonation occurs, an equivalent amount of the terminal —OH will remain unsulphated (assuming stoichiometric quantities are used). This molecule will therefore be somewhat more hydrophilic than the non-ring sulphonated material.

Despite the remarks by Gilbert[16] that sulphation of alcohols with sulphamic acid has not been successful, General Aniline has patented a process whereby all types of alcohols and alcohol ethoxylates can be sulphated using this acid.[18]

The procedure is relatively simple but varies for different materials. In principle the material being sulphated is charged into the reactor, the stirrer started and 1·05 mol of sulphamic acid, which should be finely divided, is fed in rapidly. The rate of stirring and rate of feed of the sulphamic acid should be such that the powder is well dispersed throughout the mass of the material. When all the sulphamic acid has been added, the lid of the reactor is closed and nitrogen passed through the vessel to purge all the air. Simultaneously the reactor is heated to 110°C rapidly and kept at 110–115°C for $1\frac{1}{2}$ hours. After this period the reaction mass is cooled rapidly to 70°C and dropped into water or a mixture of water and alcohol, the amounts depending on the final concentration required and the viscosity of the finished sulphate. pH is then adjusted with ammonia to 6·5 to 7.

SULPHOSUCCINATES AND SULPHOSUCCINAMATES

Synthesis of these materials can be done with relatively simple equipment, a glass-lined or stainless steel reactor equipped with a dual purpose heating/cooling jacket and a stainless steel dissolving tank.

The chemistry of the reaction was stated on p 20. The choice of hydrophobe is wide, it can be any of the long-chain alcohols (natural or synthetic), long-chain primary amines (if a succinamate is to be made) or an alkanolamide or a mixture of any of these.

For this reason hard and fast rules as to quantities cannot be given but it is essential for the manufacture that the (average) molecular weight be known.

For alcohols and alkanolamides, the molecular weight can be obtained from suppliers' specifications or it can be calculated from the hydroxyl number by the formula:

$$\text{Molecular weight of mono-alcohol} = \frac{56,100}{\text{Hydroxyl number}}$$

The hydroxyl number if not available can be determined by standard procedures on the mixture. If an alkanolamide is being analysed for hydroxyl number, the natural alkalinity present will affect the result but as this alkalinity can also be a factor in the coupling to the maleic anhydride, no correction need be made.

For amines the hydroxyl procedure will also give a result for molecular (or equivalent) weight but a simpler method is to determine the base number by titrating in alcoholic solution with acid using bromcresol green indicator.

The hydrophobe, which needs to be practically anhydrous, is charged into the reactor and heated to 60–70°C. Heating is now stopped and maleic anhydride in the proportion of 1·04 moles of the anhydride calculated on the hydrophobe, is added slowly. Maleic anhydride is normally supplied as lenses or flakes and the heat required to melt roughly balances the heat of reaction of the esterification. The rate of addition of the anhydride is adjusted so that the temperature remains within the range 60–70°C.

Concurrently with this the sodium sulphite or bisulphite is dissolved in water to make an approximately 25 per cent solution. The sulphite should be freshly prepared and free of thiosulphite. If monoethanolamine sulphite is to be used (p 22), prepare a solution of 8 per cent monoethanolamine in water and bubble in approximately half its weight of SO_2, for the sulphite, and an equal weight for the bisulphite. Completeness of the reaction can be checked by alkalinity and iodine titrations.

In every case the amount of sulphite salt should be the molar equivalent of the amount of maleic anhydride used. For determining these figures we give the molecular weights of the reactants:

Maleic anhydride	91·1
Maleic acid	134·1
Na_2SO_3	126·1
$Na_2SO_3 \cdot 7H_2O$	252·2
$Na_2S_2O_5$ (the anhydrous form of sodium bisulphite)	190·1 (equivalent weight 95·05)
Monoethanolamine sulphite	187·0
Monoethanolamine bisulphite	126·6

When all the maleic anhydride has been added, maintain a temperature of 60–70°C for 1 hour and check acidity. Figures for acidity should be set as a standard based on laboratory work, because if the half-ester is being produced this also gives an acid reaction in addition to the unreacted maleic anhydride. For the di-ester only maleic acid will react if no half-ester is formed concurrently.

Once the reaction is found to be complete, the maleic ester (or amide) is transferred by gravity or pump to the solution tank containing the sulphite solution. Stirring is continued for 1 hour without further heating till the residual sulphite content has fallen to below 0·5 per cent (checked by iodimetric titration). Surplus sulphite is then destroyed by adding peroxide.

This procedure produces a paste of 40–45 per cent active matter. Colgate has patented a process whereby anhydrous sulphosuccinates can be made without drying, using a 'plasticizer' as a diluent.[19]

ALKANE (PARAFFIN) SULPHONATES

This group of anionic detergents, initially obtained by sulpho-chlorination under ultra-violet irradiation, was first produced on an industrial scale in Germany before and during the Second World War. Essentially, the reaction involved is as follows:

$$R_1.CH_2.R_2 + SO_2 + Cl_2 \rightarrow R_1.\underset{\underset{SO_2Cl}{|}}{CH}.R_2 + HCl$$

The source of the paraffins being the Fischer/Tropsch coal hydrogenation process. This process is carried out in special reactors under the influence of ultra-violet rays. The next step is saponification of the chlorosulphonate by treatment with NaOH:

$$R_1.\underset{\underset{SO_2Cl}{|}}{CH}.R_2 + 2NaOH \rightarrow R_1.\underset{\underset{SO_3Na}{|}}{CH}.R_2 + NaCl + H_2O$$

In brief, water and n-paraffin (which needs to be specially purified, see p 124) are fed into a photochemical reactor equipped with lamps emitting ultra-violet radiation of wavelength between 3300 and 3600 Å. Sulphur dioxide and oxygen are introduced with vigorous agitation.

After the war Hoechst in Germany continued with this process using SO_2 and oxygen and ultra-violet radiation. In the 1960s Esso in the USA developed a sulphoxidation process using cobalt-60 radiation as the initiator instead of the ultra-violet light. This process, however, has not been commercialized.

ATO Chemie in association with Ballestra has developed the process further whereby sulphur is burnt in oxygen to produce the correct ratio of SO_2 to O_2 for sulphoxidation. Separation of the sulphoxidized paraffin from by-product dilute sulphuric acid is performed by extraction with hexanol.

The paraffin for the ATO (and also the Hoechst) process is n-paraffin from the appropriate petroleum fraction separated from the iso-paraffins by molecular sieve (p 123). It has to be practically free of aromatics the presence of which would severely inhibit the sulphoxidation reaction.

Pressure and temperature in the reactor are close to ambient.

As the reaction proceeds, a portion of the reaction mass is withdrawn to be replaced by fresh or recycled reactants.

The outflow is passed to a decanter and consists of sulphonic acid, n-paraffins, dissolved SO_2, water and sulphuric acid. This is separated into two phases, n-paraffins which are recycled to the top of the reactor, and a sulphonic/sulphuric acid mixture.

The aqueous phase is degassed and passed to the solvent extraction unit, where it is mixed with an alcohol of at least five carbon atoms. This again separates into two phases, sulphonic acid/solvent and a sulphuric acid solution in water of about 25 per cent acid.

The sulphuric acid solution can be either neutralized to sodium sulphate (for captive use) or after concentration sent to a sulphuric acid/fertilizer plant.

The organic phase is neutralized with caustic soda and contains sodium alkane sulphonate, solvent, water and unreacted paraffin. The water and solvent are stripped off under vacuum and finally the unreacted paraffin is distilled off and recycled.

A flow sheet for the complete process is shown in Fig 5.11.

To produce 1000 kg active matter the plant uses the following raw materials:

n-Paraffins	675 kg
SO_2	410 kg
Oxygen	110 kg
Caustic sode (100%)	145 kg
Process water	1865 kg

and utilities:

Cooling water	310 m^3
Steam	2000 kg
Power	0·65 kWH
Fuel	215,000 kcal

As far as the detergent processor is concerned it would be better if future development would lead to a change in the *raw material* for sulphonation (either straight-chain alkylates or synthetic fatty alcohols) rather than in the production of ready-sulphonated material supplied by giant concerns, which leaves nothing to the detergent industry but the last stages of processing: spray-drying and compounding. Producers of primary raw materials for anionic detergents have recently begun to give special attention to this requirement, and are now supplying manufacturers with new raw materials rather than finished products.

The authors venture to suggest that it is quite conceivable that in the future purified n-paraffins may form the basic raw material for the de-

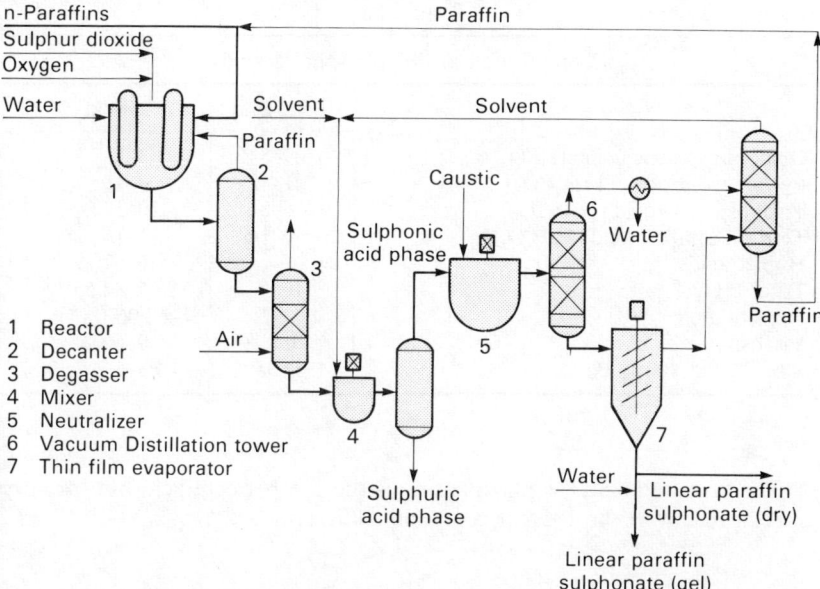

Fig 5.11 Process for making linear paraffin sulphonate detergent

tergent manufacturer as packaged sulphoxidation units using ultra-violet irradiation have now been developed. These could eventually be set up by a modern detergent manufacturing concern; the limiting factor of course would be the capacity which would make the plant economical.

Phosphate Esters

Details of the method of manufacture are both patented and the subject of trade secrets. The reactions involved are outlined on p 25.

The phosphating media are either P_2O_5, a well-defined inorganic chemical, or polyphosphoric acid. This acid is available in two commercial grades, 105 per cent and 115 per cent, calculated as orthophosphoric acid. They are made either by molecular dehydration of orthophosphoric acid under vacuum, or by dissolving P_2O_5 in orthophosphoric acid or a dearth of water to give a mixture of complex acids. Typical compositions of the acids are given in Table 5.14.

The variations in the composition and the viscosity of the 115 per cent acid are dependent on the P_2O_5 content which differs for different manufacturers. Handling of this polyphosphoric acid at ambient temperatures is difficult and the containers should be heated over 50°C for efficient flowing or pumping. It is also to be noted that these acids are hygroscopic. Absorption of water tends to hydrolyse the complex acids to orthophosphoric acid rapidly.

TABLE 5.14

Composition of Polyphosphoric Acids

	105% acid	115% acid
Orthophosphoric acid (H_3PO_4)	49–54	3–5
Pyrophosphoric acid (H_4PO_7)	41–42	9–16
Triphosphoric acid ($H_5P_3O_{10}$)	5–8	10–17
Tetraphosphoric acid ($H_6P_4O_{13}$)	0–1	11–16
Higher acids	—	46–67
Total P_2O_5%	76	83·5–84·5
Density g/cm³	1·92	2·06
Viscosity at 25°C Pa s	0·8	30–60
at 100°C Pa s	0·035	0·5–1·7

The favoured acid for phosphating is the 115 per cent acid but for very special reactions the 105 per cent acid suffices.

Non-ionic Detergents

Ethylene and Propylene Derivatives

The most important non-ionic detergents are those obtained by reacting various hydrophobic groups with ethylene oxide or propylene oxide. It is essential that any compounds to be condensed, with either ethylene oxide or propylene oxide, should contain a suitably reactive group, and Table 5.15 shows the principal types of reaction involved.

Before describing the various groups of compounds for condensing, it will be useful to outline the processes for making ethylene oxide and propylene oxide. Ethylene oxide is produced, either by reacting ethylene with hypochlorite followed by hydrolysis with calcium hydroxide or by direct oxidation of ethylene with oxygen. The method first employed commercially in the period from 1925 to 1930 is based on the classical Wurtz synthesis (1859). It is essentially a two-stage process, in which ethylene is first chlorohydrinated and then hydrolysed with calcium hydroxide. The reactions may be represented by the following equations:

$$CH_2 : CH_2 + HOCl \longrightarrow CH_2ClCH_2OH$$

$$2CH_2ClCH_2OH + Ca(OH)_2 \longrightarrow 2O \cdot CH_2 \cdot CH_2 + CaCl_2 + 2H_2O$$

Ethylene dichloride, CH_2ClCH_2Cl and beta-beta'-dichloroethyl ether, $O(C_2H_4Cl)_2$ are produced by side reactions.

TABLE 5.15

Preparation of Non-ionic Surfactants from Ethylene Oxide

Hydrophobic portion	Product
$R\text{-}C_6H_4\text{-}OH + n\,CH_2\underset{O}{-}CH_2 =$	$R\text{-}C_6H_4\text{-}(OCH_2CH_2)_n OH$
$R\text{---}OH + n\,CH_2\underset{O}{-}CH_2 =$	$R\text{---}(OCH_2CH_2)_n OH$
$R\text{---}\overset{O}{\overset{\|}{C}}\text{---}OH + n\,CH_2\underset{O}{-}CH_2 =$	$R\text{---}\overset{O}{\overset{\|}{C}}\text{---}(OCH_2CH_2)_n OH$
$R\text{---}SH + n\,CH_2\underset{O}{-}CH_2 =$	$R\text{---}S(CH_2CH_2O)_n H$
$R\text{---}\overset{O}{\overset{\|}{C}}\text{---}NH_2 + n\,CH_2\underset{O}{-}CH_2 =$	$RC\text{---}N \begin{smallmatrix}(CH_2CH_2O)_n H \\ (CH_2CH_2O)_{n'} H\end{smallmatrix}$ with C=O
$R\text{---}NH_2 + n\,CH_2\underset{O}{-}CH_2 =$	$R\text{---}N \begin{smallmatrix}(CH_2CH_2O)_n H \\ (CH_2CH_2O)_{n'} H\end{smallmatrix}$

n and n' = the number of ethylene molecules

This process is presently of historical interest only.

The newer process, invented in 1931 by Lefort, reacts ethylene with oxygen or air at temperatures between 220 and 300°C over a silver catalyst at elevated pressure.

The reaction involved is as follows:

$$H_2C=CH_2 + \tfrac{1}{2}O_2 \rightarrow H_2C\underset{O}{-}CH_2$$

A competing reaction which can take place is:

$$CH_2=CH_2 + 3O_2 \rightarrow 2CO_2 + 2H_2O$$

This obviously reduces the yield but modern technology has settled on reaction parameters to minimize this reaction. In modern plants the yield is

SYNTHESIS OF DETERGENTS

70–72 per cent, thus 100 kg ethylene oxide will produce approximately 100 kg ethylene oxide.

Initially the conditions were that a mixture of 3–5 volume per cent ethylene in air was fed into a tubular reactor with a fixed bed supported silver catalyst at a pressure of 10–20 bar with a temperature of 220–280°C. Selectivity is highly sensitive to temperature, dropping as temperature rises. The reaction is highly exothermic and temperature control is maintained by heat exchangers with a circulating heat transfer agent.

The gas leaving the reactor is cooled by another heat exchanger, and scrubbed by water to retain the ethylene oxide produced. The waste gas, containing unreacted ethylene is then split, part is fed back as recycle to the reactor and part is passed through a second reactor for further reaction.

Ethylene oxide is then stripped from the water solution and condensed.

Many inhibitors have been suggested to lower the competing carbon dioxide reaction, the one in most common use is ethylene dichloride.[20]

Modern plants, instead of using air as the source of oxygen use pure oxygen, which eliminates the necessity of purging the system of inert and possibly reactive components.

Propylene oxide is obtained from propylene generally via its chlorhydrin by reacting it with hypochlorite similar to the Wurtz reaction described for ethylene oxide. A recent process for the direct oxidation of propylene to propylene oxide has now appeared (*Hydrocarbon Processing*, November 1976).

Further developments are the production by oxidation of propylene by hydrogen peroxide via perpropionic acid and still another by the epoxidation of propylene with peracetic acid (*Hydrocarbon Processing*, November 1983).

Tables 5.16 and 5.17 give selected physical properties of pure ethylene oxide and propylene oxide. The commercially produced compounds are very pure and have virtually the same properties.

TABLE 5.16

Physical Properties of Pure Ethylene Oxide	
Auto-ignition temperature	571°C
Boiling-point (101·1 kPa)	10·7°C
Explosive limits in air by vol %	3–100%
Flash-point (open cup)	$< -17\cdot8°C$
Heat of vaporization (101·1 kPa)	569·4 J/g
Molecular weight	44·05
Refractive index (8·4°C)	1·3599
Solubility in water	Complete
Specific gravity 20/20°C	0·8711
Specific heat, liquid (0°C)	0·44
Vapour pressure (20°C)	146·0 kPa
Viscosity (0°C)	0·32 mPa s
Density (20°C)	870 kg/m^2

TABLE 5.17

Physical Properties of Pure Propylene Oxide	
Boiling-point (101·1 kPa)	34·1°C
Explosive limits in air by vol %	2·1–38·5%
Flash-point (open cup)	− 37·2°C
Freezing-point	− 104°C
Heat of vaporization (101·1 kPa)	372·6 kJ/kg
Molecular weight	58·08
Refractive index, n_D (20°C)	1·3657
Solubility in water, weight % (20°C)	12·5
Specific gravity 20/20°C	0·8305
Specific heat, kJ kg^{-1} K^{-1} (20°C)	2·1304
Vapour pressure (20°C)	60·5 kPa
Viscosity (20°C)	0·33 mPa s
Density (20°C)	829 kg/m^3

The handling and storage of ethylene oxide and propylene oxide, as well as their reactions, must be carried out with the utmost caution. When starting a plant where ethylene and/or propylene oxide is used it is advisable to contact the supplier for details of safety precautions, and to follow these instructions conscientiously. An outline of these precautions will be given later, when we describe ethylene oxide reactions.

We will now describe some of the compounds that constitute the hydrophobic portion of surface active ethylene oxide compounds. We will confine ourselves to three main groups: (*a*) fatty acids; (*b*) fatty alcohols; and (*c*) alkyl phenols.

Fatty acids react with ethylene oxide to form esters. The same product can theoretically be formed by the esterification of the fatty acid with the appropriate polyethylene glycol, but in practice some di-ester is formed. Table 5.18 shows the properties of fatty acids (representative commercial products) suitable for reaction with either ethylene oxide or propylene oxide, and also for reaction with alkylolamines (see p 191). The properties of fatty alcohols, natural and synthetic, have already been described (see pp 118–121).

Alkyl phenol ethylene oxide condensation products still comprise a substantial proportion of the total world production of non-ionic detergents. This in spite of the fact that the alkyl phenol ethoxylates, obtained from polymerized propylene or butylene, represent a similar sewage problem to that caused by anionic detergents derived from PT benzene (see pp 30, 45); less serious, however, owing to their low foam property.

The most important alkyl phenols are C_8 to C_{12} phenols. They are obtained by alkylation of phenol with the corresponding olefins. The olefins are the dimeric or trimeric polymers of propylene or butylene or their mixture.

TABLE 5.18

Properties of Fatty Acids

	Titre °C	Acid value	Saponification value	Iodine value	Unsaponifiable %
Lauric acid 94/96%	39–41	279–84	279–84	max 0·5	0·5
Lauric acid 98/100%	43–43·6	279–82	279–82	max 0·2	0·2
Dist. stripped coconut oil fatty acid	25–28	250–60	252–62	8–14	0·5
Dist. stripped palm-kernel oil fatty acid	23–27	250–60	250–60	16–22	1·0
Myristic acid 94/96%	50–53	244–54	245–55	max 1·0	0·5
Myristic acid 98/100%	51–54	244–54	245–55	max 1·0	0·5
Palmitic acid 94/96%	60–61	215–25	215–25	max 2·0	0·5
Palmitic acid 98/100%	61–62	219–24	219–24	max 2·0	0·5
Stearic acid 94/96%	64–66	195–8	196–9	1–3	1·5
Stearic acid 98/100%	66–68	195–9	196–200	max 2·0	1·5
Stearine single pressed	52–53	207–11	207–11	8–10	0·5
Stearine double pressed	53–54	208–11	208–11	5–8	0·5
Stearine triple pressed	54–56	207–11	207–11	1·5–5	max 0·5
Dist. tallow fatty acid	39–43	204–8	205–9	53–57	max 1·5
Oleine light coloured	7–10	186–204	188–206	85–92	2–6
Dist. palm oil fatty acid	42–46	207–13	209–15	44–54	max 1·5
Dist. refined tall oil light coloured	—	188–93	190–5	155–65	1·5–2·2
Tall oil refined	—	163–8	165–70	153–67	7–8

Table 5.19 gives the properties of three of the most important alkyl phenols for the production of surface active ethylene oxide condensates.

Before discussing laboratory and industrial ethoxylation processes, a general remark must be made on the hydrophobe/hydrophile balance of ethylene oxide and propylene oxide condensates: the higher the molecular weight of the hydrophobe component, the more moles of ethylene oxide (or propylene oxide) are needed to produce water-soluble condensates. Conversely, the lower the molecular weight, the smaller the number of moles required to obtain a water-soluble product. For the purpose of producing non-ionic detergents, it is generally desirable that the products be easily soluble in water. For the production of non-ionic emulsifying agents, on the other hand, a certain affinity to the oil phase is preferable. Thus, to condense the hydrophobe component, fewer moles of ethylene oxide will be needed.

As has been mentioned, the production of ethylene and propylene oxide condensates requires a great deal of experience and it is thus advisable to study the reaction of ethoxylation on a laboratory scale. We give details of

TABLE 5.19

Properties of Three Important Alkyl Phenols for the Production of Surface Active Ethylene Oxide Condensates

	Octylphenyl	Nonylphenol	Dodecylphenyl
Formula	C_8H_{17}—⟨⟩—OH	C_9H_{19}—⟨⟩—OH	$C_{12}H_{25}$—⟨⟩—OH
Molecular weight (theoretical)	206·2	220·2	262·2
Molecular weight of commercial products	200–7	221·4	262–7
Appearance	Yellowish crystalline mass	Colourless to yellowish oil	Yellowish oil
Density 30°/4°	—	0·93–0·95	0·92–0·94
Distillation range at 101·1 kPa	280–95°C	290–315°C	330–35°C

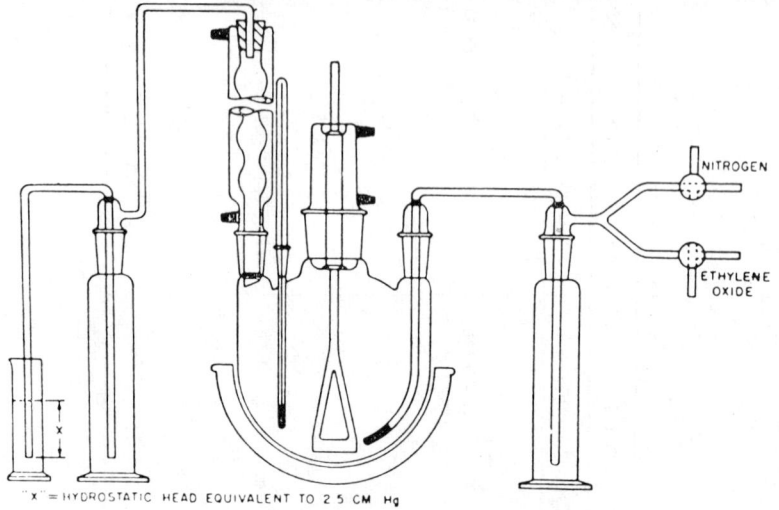

Fig 5.12 Laboratory ethoxylation plant

a laboratory synthesis of alcohol ethoxylates and alkyl phenol ethoxylates.*

Quantities of about a pound are readily made in the laboratory by the following procedure: A 1-litre four-necked flask, fitted with a water-cooled stirrer, condenser, thermometer, and gas-dispersion thimble is connected through two 500 ml safety bottles to a graduate containing white oil and a manifold connected to low-pressure nitrogen and a tank of ethylene oxide (Fig 5.12). 200 g (1·00 mol) of tridecyl alcohol is charged to the flask followed by 0·4 to 0·5 g of powdered sodium hydroxide. The flask, thermometer, gas disperser, and contents are weighed. A slow stream of nitrogen is started through the alcohol and the temperature raised to 150–160°C by means of a glass-insulated heating mantle. Ethylene oxide gas is admitted to the manifold and allowed to escape momentarily to the atmosphere through the three-way stopcock. The three-way stopcock is then turned to vent the nitrogen to the atmosphere and the ethylene oxide is simultaneously introduced into the flask. The operation should be carried out in a well-ventilated hood and away from naked flames.

Absorption takes place immediately, with a rise in temperature. The ethylene oxide rate is increased to the point where an occasional bubble of ethylene oxide escapes through the white oil. The temperature is permitted to rise to 180–220°C, and is maintained within this range until the required amount of ethylene oxide has been absorbed. For trial purposes, between 8·5 and 10 moles of ethylene oxide per mole of alcohol is suitable. This is

---

* *Higher OXO Alcohols*, pp 76–9, L. F. Hatch: New York (1957). Reprinted by kind permission of the publishers, John Wiley & Sons, Inc.

equivalent to an absorption of 374–440 g of ethylene oxide. Approximately three to six hours are required to absorb the above quantities.

In shutting down, the ethylene oxide is vented to the air, and the nitrogen is turned into the flask and a slow stream maintained until all of the ethylene oxide has been purged from the system. The flask and contents are weighed to determine the ethylene oxide absorption.

When the desired quantity of ethylene oxide has been absorbed, the free alkali is neutralized with either 30 per cent sulphuric acid or glacial acetic acid to a phenolphthalein end point (external indicator). Ten grams of decolorizing carbon are then added and the mixture stirred and heated at 100°C for 15 min. The mixture is filtered through a Büchner funnel using a small amount of a filter-aid on the paper as a precoat in order to ensure a bright filtrate. The small amount of water resulting from neutralization can be permitted to remain, or may be stripped from the product by heating under reduced pressure.

Another example is for the ethoxylation of 1 mole C_{12} lauryl alcohol (either natural or synthetic) with 4 moles of ethylene oxide:

Lauryl alcohol C_{12}	186 g	(1 mol)
Ethylene oxide	176 g	(4 mol)
NaOH powdered	0·6 g	

Proceed as described for tridecyl alcohol ethoxylate, including the same purification method.

Finally an example for the ethoxylation of nonylphenol with 9·5 moles ethylene oxide:

Nonylphenol	215 g	(1 mol)
Ethylene oxide	418 g	(9·5 mol)
NaOH powdered	0·7 g	

The same procedure is used.

This account of a laboratory ethoxylation synthesis makes it easier to understand the following description of an industrial batch ethoxylation plant.*

A typical ethylene oxide batch reactor is shown in Fig 5.13. The information contained in this diagram is presented to illustrate principle only.

An ethylene oxide kettle should have provisions for maintaining the desired reaction temperatures. Heating-cooling coils and a jacket are usually suitable. The kettle should be designed for predetermined pressures. A pressure-relief valve and rupture disc should be provided. A suitable vent should be supplied which will vent to a safe area through a flame arrester. The speed with which ethylene oxide reacts with other materials depends on intimate mixing; therefore, care should be exercised in the choice of the agitation system. Stainless steel or glass linings are recommended if the quality, especially colour, of the reaction product is important. Carbon steel is suitable for other purposes. Acetylide-forming

---

* The description of an industrial batch process set-up is taken from the technical brochure on ethylene oxide published by Jefferson Chemical Co Inc by kind permission.

SYNTHESIS OF DETERGENTS

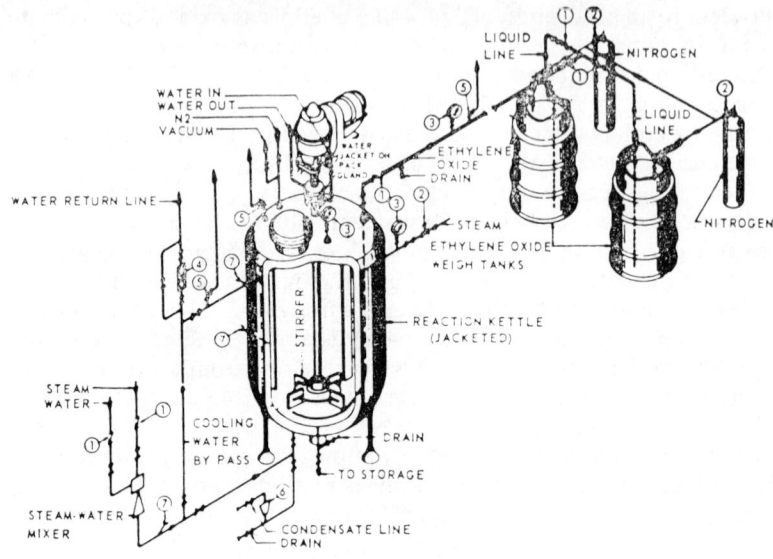

Fig 5.13   Ethylene oxide batch reactor

1. Check valve
2. Pressure controller
3. Pressure gauge
4. Rotameter
5. Safety valve (relief)
6. Steam trap
7. Thermowell

metals such as copper, silver, magnesium and their alloys should never be used in direct contact with ethylene oxide.

Heating of the charge can be accomplished with steam or some other heat-transfer medium. Cooling water should be available to remove heat of reaction, for emergency cooling, and for routine cooling of the finished product preparatory to removing it from the kettle. Precautions should be taken to avoid overcooling the reaction mass and thus allowing the accumulation of unreacted ethylene oxide in the kettle. With an excess of unreacted ethylene oxide in the vessel, the reaction could proceed with explosive violence. The inert gas, preferably nitrogen or methane, used to purge the system should be free of impurities such as oxygen, ammonia, hydrogen sulphide and acetylene. Prior to purging with an inert gas, vacuum equipment must be used to remove most of the air. Suitable facilities should be provided as required for introducing the charge materials, cooling water and inert gas. Automatic shut-off of ethylene oxide feed should be available for activation by: (1) drop in temperature below that at which the reaction rate is enough to consume the ethylene oxide being introduced; or (2) rise in pressure or temperature above that which has been determined as safe for the design of the equipment.

An intermediate ethylene oxide charge tank, located between the reaction kettle and ethylene oxide storage tank, should be available to serve as a work tank and to prevent the backflow of catalyst from the reactor to the ethylene oxide storage vessel. The work tank might also be used as a volumetric or weighing tank. Ethylene oxide transfer lines to process equipment should contain at least one check valve. Adequate storage facilities should be available for the other reactants and the product. Processing equipment and transfer lines should be thoroughly cleaned, leak tested and purged free of air before being placed in service. The equipment and transfer lines should be electrically bonded and grounded. Explosion-proof motors and spark-proof tools should be used.

The catalyst and oxide accepter are charged to the kettle. The charge is purged free of oxygen and brought to the temperature at which the reaction may be initiated. Ethylene oxide is cautiously added to the kettle. A temperature rise will indicate that the reaction has begun, at which point the ethylene oxide feed rate may be increased to the desired level, but never to a rate greater than that at which it is being consumed. The build-up of unreacted ethylene oxide in the vessel must be carefully avoided. Most ethylene oxide reactions take place in the temperature range of 120–200°C and in the pressure range of 205–410 kPa.

In general, ethylene oxide reactions are exothermic in nature, giving up about 80 kJ/mol of ethylene oxide reacted. This heat of reaction must be removed. Overcooling may result in stopping the reaction and allow accumulation of unreacted ethylene oxide in the vessel. The reaction may then resume and proceed with explosive violence.

After the predetermined amount of ethylene oxide has been added, flow to the kettle is stopped. The temperature of the kettle is held for a time to allow the last of the ethylene oxide to react, as indicated by a gradual drop in pressure. When the pressure is steady, the product is cooled, treated where necessary and removed from the kettle. Minimal moisture content of reactants is essential, otherwise side reaction with the formation of polyethylene glycol occurs (see p 31).

Completely enclosed explosion-proof magnetic mixers, without a stuffing-box, are very much to be recommended as agitators for the reactor. Furthermore, it is our experience that the safety of the set-up is greatly increased by a side-arm heat exchanger. Reaction time and reagent flow should be similar to the laboratory procedure, so that the process on a plant scale is, in fact, an upscaled laboratory process. Many ethylene oxide (and propylene oxide) condensation products and the processes for their production are subject to patents in various countries. Thus it is always advisable to study the patent situation carefully, before starting with the production of ethylene oxide and/or propylene oxide condensation products.

It must be mentioned that the condensation reaction may be carried out in a continuous 'run-through' pressure tube reactor. Such a system employs, eg, an 'in line' mixing system (without moving parts). The reactants are fed in proper proportion into the tube reactor by means of dosing pumps. Heat control—and thus control of reaction—is, in fact, easier than for a batch reactor system. The quantity of reaction mixture in the continuous reactor is relatively small; and the safety of such a system is to a

SYNTHESIS OF DETERGENTS

certain extent easier to guarantee than for a batch system. When using a continuous system, the NaOH catalyst is pre-mixed with the alcohol or alkyl phenol respectively, to be condensed with ethylene oxide or propylene oxide. The refining step with acid and active carbon can be carried out after the continuous condensation reaction is accomplished, either in a subsequent batch refining and filtering system or on a continuous basis.

In a commercial plant for the production of ethylene oxide adducts (Pressindustria S.p.a. Biassano, Milan) the production of 2500 kg of C_{16}—C_{18} alcohol natural or synthetic with 25 mol ethylene oxide condensed requires:

C_{16}—C_{18} alcohol	1250 kg
Ethylene oxide	1250 kg
Steam	150 kg
Nitrogen	5–6 m³
Cooling water at 15–20°C	16 m³
Electricity	45 kW

Operation of the plant is as follows (refer to Fig 5.14):

A preset amount of material to be reacted with ethylene oxide and/or propylene oxide by the RP programming system together with the catalyst is fed into the process unit through line 1. (The amount of catalyst is generally less than that usually introduced in a conventional ethoxylation plant, and this has a beneficial effect on the colour and purity of the end product.) After the starting material and catalyst have been introduced, the unit is set under vacuum, heating begins through F 1, and the recirculation through circulation pump PC 1 is set into operation. The mixture flows again into the upper part of SA 1, the gas–liquid contactor, and the desirable temperature is maintained by F 2. During this phase of operation the material is de-aerated and de-humidified. This operation is carried out in a very short period of time. Low residual moisture is essential to prevent any undesirable side reactions, eg, the formation of polyglycols and irregularity in the formation of the ethylene oxide and/or propylene oxide chain. Quick nitrogen injection and venting is then carried out; some nitrogen pressure is left in the system, and the actual condensation reaction starts by feeding the preset (RP programming system) amount of ethylene and/or propylene oxide into the pre-treated starting material and catalyst. Recirculation through circulation pump PC 1 into the gas–liquid SA 1 is carried out and the heat of reaction removed by the cooling heat exchanger F 2. In some cases it is desirable to pass the cooling medium, either water or a diathermic fluid, through a refrigeration system in a closed circuit.

An automatic control system regulates the flow of ethylene oxide and/or propylene oxide in relation to the temperature of the reaction mixture. This feedback system guarantees a constant reaction rate and constant temperature which accounts for uniform quality of the end product. The end product is cooled down by the F 2 cooling system and by the optional additional cooling vessel SA 2. No bleaching of the end product is required.

A small pilot plant for the continuous production of ethylene oxide

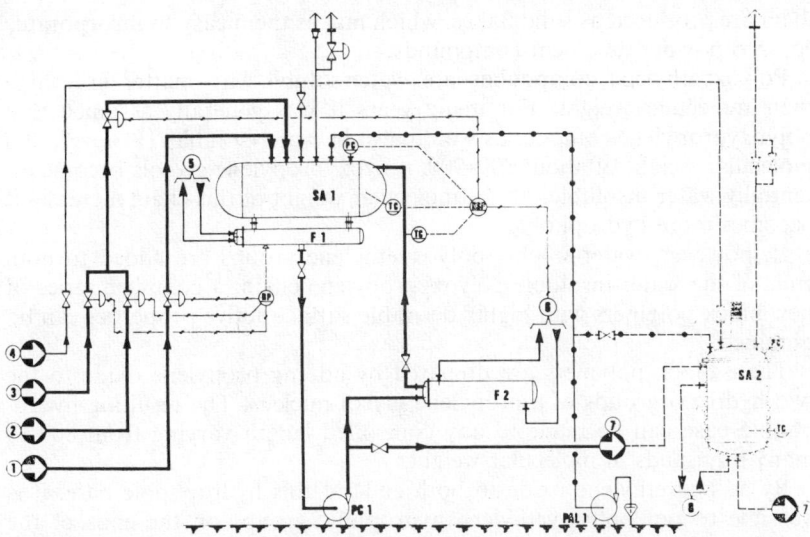

Fig 5.14 Pressindustria plant for the manufacture of ethylene oxide adducts

1. Basic raw material (eg alkylphenol or fatty alcohol)
2. Ethylene oxide from storage
3. Propylene oxide from storage
4. Nitrogen for purging and pressurizing
5. Heat transfer fluid for reaction start-up
6. Cooling circuit
7. End product discharge
8. Cooling-water for final cooling of product
SA 1. Special 'Pressindustria' gas/liquid contactor
RSP. Automatic reaction rate governing system
RP. Automatic programming system (formula preset system)
SA 2. Cooler (optional) for reaction product
PAL 1. Vacuum pump
PC 1. Process circulation pump
F 2. Heat-exchanger for reaction heat control
F 1. Heater for reaction start-up

condensate was described in an article by Ambach & Stein.[21] The reactor is in two parts, the first reactor working at a temperature between 170 and 240°C giving a condensate of molecular ratio 1:1 to 1:6, and the balance of the ethylene oxide being added in the second reactor at a temperature of 240–360°C. As a catalyst 0·1 to 1·5 per cent sodium as sodium methylate is used. Pressure is between 5 and 10 MPa and the residence time between 8 and 150 seconds. Use of continuous type reactors involving both small charges and contact time can obviously lessen considerably the danger inherent in this type of reaction.

In the last twenty years or so the block polymers (p 34) have assumed increasing importance, and all the varieties are now being produced by several firms under various trade names. Production of the original material was covered by US Patents 2,674,619 and 2,677,700 assigned to Wyandotte (inventors: Lundsted, Jackson *et al*). Their special qualities are the variable hydrophobe/hydrophile properties, and the fact that most of them are produced as solid flakes, which makes them easy to incorporate, eg, into powder detergent compounds.

Polyoxyethylene compounds are water-soluble—no matter how high their molecular weight. For many years it was generally assumed that polyoxypropylene compounds would also be water-soluble. However, at a molecular weight of about 800–900, polyoxypropylene glycols become essentially water-insoluble. As the molecular weight of this chain increases it becomes more hydrophobic.

If, however, water-soluble polyoxyethylene groups are added to both ends of the water-insoluble polyoxypropylene chain, a complete series of new block polymers with highly desirable surface active properties can be obtained.

These block polymers are prepared by adding propylene oxide to the two hydroxyl groups of a propylene glycol nucleus. The resulting hydrophobic base can be made to any controlled length varying from 800 to many thousands in molecular weight.

By adding ethylene oxide to both ends of this hydrophobic base, it is possible to put polyoxyethylene hydrophilic groups on the ends of the molecule. These hydrophilic groups are controlled in length to constitute anywhere from 10 to 80 per cent of the final molecule. The simplified structure can be represented as:

$$HO(CH_2-CH_2-O)_a(CH_2-CH_2-O)_b(CH_2-CH_2-O)_cH$$
$$\underset{CH_3}{|}$$

Other variations are described on p 35. If the hydrophobe portion of a normal non-ionic is calculated it will be found to be of the order of 30–40 per cent of the finished product. Block polymers are produced by (usually) propoxylating and then ethoxylating on to an initiator. This initiator is seldom 5 per cent, usually $\frac{1}{2}$–1 per cent of the final product. It can be seen that in batch process it will be extremely difficult to start the process with such a small amount of material in the reactor. It is common practice to produce a 'precursor', of the initiator with a definite portion of (say) propylene oxide. This precursor is then stored till required and the requisite amount(s) of ethylene and/or propylene oxides are added.

The newest development in the non-ionic detergent field is the extra-low-foam products. As has been mentioned, ethylene oxide condensates, especially with a short ethylene oxide chain (4–5 mol), produce poor foam but for all that the amount of foam generated in certain operations can still be a hindrance.

Modified non-ionics, usually made by coupling an alkyl group to the terminal—OH, are now available and are used in household powders for foam sensitive machines and more particularly for dishwashing machine powders which also contain active chlorine, as the blocking of the terminal —OH reduces the tendency for degradation of the material producing the active chlorine.

NON-IONIC DETERGENTS

Block polymers specially designed to give no (not low) foam are available and these are being used successfully in automatic dishwashing machine detergents, both liquid and powdered.

FATTY ACID ALKANOLAMIDES

This important class of non-ionic detergents is derived by condensing alkanolamine with fatty acids. These are the so-called fatty acid alkanolamides.*

Alkanolamines combine readily with long-chain fatty acids, such as oleic and stearic acids, to give neutral alkanolamine soaps.

$$HOCHRCH_2NH_2 + C_{17}H_{35}COOH \rightarrow HOCHRCH_2NH_2 \cdot HOOCC_{17}H_{35}$$

This neutralization reaction takes place at room temperature. The products are waxy, non-crystalline materials which have widespread commercial application as emulsifiers.

At elevated temperatures (140–160°C) n-alkylol amides are the chief products formed by reacting mono- and dialkanolamines with fatty acids in a 1:1 ratio.

$$HOCHRCH_2NH_2 + R'COOH \rightarrow HOCHRCH_2NH-\underset{\underset{O}{\|}}{C}-R' + H_2O$$

At the same time, significant quantities of amine esters and amide esters are formed by side reactions involving the hydroxyl moiety.

$$HOCHRCH_2NH_2 + R'COOH \rightarrow H_2NCH_2CHROOCR' + H_2O$$

$$HOCHRCH_2NH-\underset{\underset{O}{\|}}{C}-R' + R''COOH \rightarrow R''COOCHRCH_2NH-\underset{\underset{O}{\|}}{C}-R' + H_2O$$

In addition, when dialkanolamines are the starting materials, small amounts of amine diesters and amide diesters result.

$$(HOCHRCH_2)_2NH + 2R'COOH \rightarrow HN(CH_2CHROOCR')_2 + 2H_2O$$

$$(HOCHRCH_2)_2N-\underset{\underset{O}{\|}}{C}-R' + 2R''COOH \rightarrow (R''COOCHRCH_2)_2-N-\underset{\underset{O}{\|}}{C}-R' + 2H_2O$$

* *Alkanolamines Handbook*, Dow Chemical International, pp 43–9 (1979).

When diethanolamine is the starting dialkanolamine, some morpholine and some piperazine derivatives are also obtained.

$$\begin{array}{c} HO-C_2H_4 \\ \diagdown \\ NH \longrightarrow \\ \diagup \\ HO-C_2H_4 \end{array} \quad \begin{array}{c} C_2H_4 \diagdown \\ O NH + H_2O \\ C_2H_4 \diagup \end{array}$$

$$2 \begin{array}{c} HO-C_2H_4 \\ \diagdown \\ NH \\ \diagup \\ HO-C_2H_4 \end{array} \longrightarrow HO-C_2H_4-N \begin{array}{c} \diagup C_2H_4 \diagdown \\ \\ \diagdown C_2H_4 \diagup \end{array} N-C_2H_4OH + 2H_2O$$

Reaction of dialkanolamines with fatty acids in a 2:1 ratio at 140–160°C gives a second major type of alkanolamide. These products, in contrast to the 1:1 alkanolamides, display aqueous solubility. In composition, they are complex mixtures which comprise n-alkylol amides, amine esters and di-esters, plus a considerable percentage of unreacted dialkanolamine. This latter constituent chiefly accounts for the aqueous solubility of the 2:1 dialkanolamide.

Both the 1:1 and the 2:1 alkanolamides are of commercial importance as detergents and detergent additives.

The first patents for preparing alkanolamides were granted to Wolf Kritchevsky in 1937.[22,23] They covered the condensation of fatty acids, their triglycerides, esters, amides, anhydrides, and halides with not substantially less than two moles of an alkanolamine. The reaction was carried out at 100 to 300°C below the decomposition of the resulting product and at atmospheric pressure.

An improved process for making alkanolamides was revealed in a 1949 patent to Edwin M. Meades.[24] This process comprised mixing an ester of an aromatic or aliphatic carboxylic acid with an alkanolamine, adding an alkali metal alkoxide catalyst and heating the mixture to 100°C at atmospheric or above atmospheric pressures.

In 1958, Giuliana C. Tesoro patented a refinement of the Meade process.[25] It consisted of reacting a fatty acid ester or glyceride with a primary or secondary alkanolamine in the presence of a sodium methoxide catalyst at a temperature between 55 and 75°C and under a reduced pressure of 4–8 kPa.

A continuous process for making fatty alkanolamides in a thin-film reactor is covered in a 1958 patent issued to Jack W. Schurman.[26] The reaction involved condensation of a methyl ester of a fatty acid with a mono- or dialkanolamine, in the presence of an alkali metal, alkali metal alkoxide, or alkali metal amide catalyst. A short contact time in the reactor produces a high purity alkanolamide.

More recently (1962) Robert Ernst has patented another process for making high purity alkanolamides.[27] A fatty acid is first reacted with an excess of alkanolamine, forming amine and amide esters in addition to the intended unsubstituted alkanolamide. In a second step involving an alkali metal catalyst, the amine and amide esters are converted to the unsubstituted alkanolamide.

John W. Lohr was also granted a patent in 1962 for a process of making high-purity alkanolamides by condensing a dialkanolamine with a fatty triglyceride,[28] then adding phosphoric acid to remove the excess amine and most of the glycerine by-product.

There are two types of alkanolamide products. The first is the Kritchevsky-type liquid product, made by reacting an alkanolamine with a fatty acid or fatty acid derivative at elevated temperatures in a 2:1 ratio. Such a product contains 60–70 per cent alkanolamide, plus some amine esters and di-esters, amide esters and di-esters, and piperazine derivatives that are formed by side reactions. In addition there is significant unreacted alkanolamine. This excess alkanolamine renders the Kritchevsky-type alkanolamides water soluble.

The second type of alkanolamide is the so-called 'super' amide, prepared by reacting an alkanolamine and a fatty acid ester in a 1:1 ratio. These are generally solid products which have an alkanolamide content above 90 per cent. Some of the same by-products formed in preparing 2:1 alkanolamide are likewise formed in preparing super amides, but in much smaller quantities. For this reason, and because they contain only relatively small amounts of free alkanolamine, super amides have poor water-solubility. They are therefore always used in conjunction with a small amount of anionic or non-ionic surfactant which acts as a solubilizer, converting an aqueous alkanolamide dispersion into a viscous, clear solution. The isopropanolamine-based alkanolamides do however show an increased solubility in water over the ethanolamine types.

The starting material for the super amides are the methyl esters of fatty acids.

For the Kritchevsky type of fatty acid diethanolamide no catalyst is needed. The high surplus of diethanolamine (DEA) makes this unnecessary.

One mol lauric acid is reacted with 2 mol DEA in an electrically heated stainless steel reactor fitted with an agitator and a sparger-pipe for the introduction of nitrogen as purging gas. The reaction is carried out at 160°C for 5 h. (Purge nitrogen is introduced when the reaction mass has been heated to about 120°C.) The reaction may be controlled by having a water trap and cooler fitted to the reactor, so that the amount of water (1 mol) liberated during the reaction may be measured (see reaction equation, p 191). Of course, any water present in the DEA must be taken into account. In the same manner other fatty acid diethanolamides can be produced.

'Superamides' are generally prepared by reacting methyl esters of fatty acids with diethanolamine, using sodium methylate as a catalyst. From work done in the Dow laboratories[29] on their production, it appears that:

1. The optimum mol ratio of DEA to methyl ester is 1·1:1.
2. The optimum temperature is 105°C if the reaction is carried out at atmospheric pressure.
3. There is very little difference in the catalystic effect of sodium methylate between 0·15 and 0·25 per cent, calculated on the batch as a whole and under atmospheric pressure.
4. The undesirable amine and amide esters are not stable in sodium methylate catalysed reactions and if formed will revert to the amide.

All the above parameters yielded an amide content verging on 90 per cent. To produce over 90 per cent it was found that a reduced pressure (4 kPa) and temperature (60°C) with 0·75 per cent catalyst were necessary.

The temperatures are all above the boiling point of methanol which is liberated by the reaction. The plant needs to be equipped with condensing units to recover the methanol for recycle.

From infra-red studies of the superamide (again by Dow) it was found that on allowing the material to stand for a month or so at room temperature, there was a considerable drop in ester content with a corresponding rise in the amide, despite the fact that superamides are solid at room temperature.

To produce high quality superamides Dow suggests hastening this post-reaction by keeping the finished amide at 50°C for 3–4 hours when the bulk of the esters will have reverted to the amide.

We have found that in producing monoethanolamides, a surplus of monoethanolamine is useful. This reduces reaction time, and improves the quality and odour of the end product. The surplus amine is stripped at the end of the reaction under reduced pressure. In general, operating under reduced pressure, especially towards the end of the condensation, was found to be of advantage.

In Fig 5.15 we show a flow sheet for the commercial production of alkanolamides, using methyl esters. This process produces the 'superamides'. If the older types are to be produced, the methyl ester is simply replaced by the equivalent fatty acid or neutral fat.

The monoethanolamides and monoisopropanolamides of fatty acids are very easy to manufacture. The production process consists in simply heating stoichiometric amounts of fatty acids or neutral fats with monoalkanolamine at about 160°C for 2–3 hours.

If neutral fats (glycerides) are used, the glycerine liberated will remain in the final product, lowering its setting point. This will also add to the emolient effect of the finished formulation.

The monoethanolamides and monoisopropanolamides are waxlike substances, practically insoluble in water, but solubilized by another hydrophilic anionic or non-ionic detergent. Table 5.20 gives the setting points and main uses of various monoethanolamides and monoisopropanolamides. These types of detergents are generally marketed in the form of

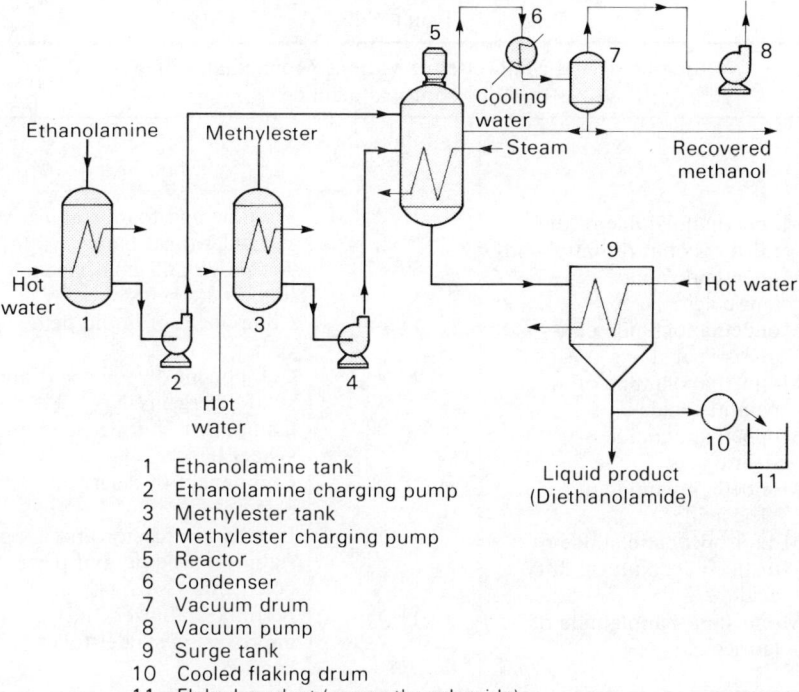

Fig 5.15 Plant for the production of normal and super alkylolamides

flakes produced by running the molten products over chilling rolls fitted with doctor blades for scraping off the flakes.

It will be noted from the table that the isopropanolamides show a lower setting point than the monoethanolamides of the same fatty acids.

Cationic Detergents

A very good description of the principle of manufacture and the use of these detergents is given by E. Kilner and D. M. Samuel.*

CATIONIC SURFACE ACTIVE AGENTS

Cationic surface active agents contain a long-chain cation which is responsible for their surface active properties. They are marketed in solid form or as pastes or in aqueous solution. Important examples include:

Amine acetates $[R.NH_3]O_2C.CH_3 (R = 8-18$ carbon atoms).

Alkyl trimethyl ammonium chlorides $[R.N(CH_3)_3]Cl$ $(R = 8-18$ carbon atoms

* *Applied Organic Chemistry*, E. Kilner and D. M. Samuel: Macdonald & Evans Ltd, London (1960), pp 185-6; quoted by permission.

TABLE 5.20

Setting-point and Main Uses of Various Monoethanolamides and Monoisopropanolamides

	Setting-point °C	Main use
Monoethanolamide of distilled coconut oil fatty acids	72–74	Perfume and foam stabilizer in syndet toilet bars
Monoethanolamide of lauric acid	80–82	As above and component of liquid detergents
Monoethanolamide of myristic acid	88–89	Component of liquid detergents
Monoethanolamide of palmitic acid	90–93	Component of powdered and solid detergents
Monoethanolamide of stearic acid	88–92	Component in toilet soap and syndet bars
Monoethanolamide of oleic acid	59–62	Component of detergent pastes
Monoisopropanolamide of distilled coconut oil fatty acid	46–50	Increases viscosity and foam stability of liquid and paste detergents
Monoisopropanolamide of lauric acid	54–58	Perfume stabilizer, 'feel improver' in syndet toilet bars

Dialkyl dimethyl ammonium chlorides $[R_2N(CH_3)_2]Cl$ (R = 8–18 carbon atoms)

Alkyl pyridinium chlorides and bromides, eg, $[C_5H_5N-R]Cl$ (R = 12–18 carbon atoms)

Lauryl dimethyl benzyl ammonium chloride

$[R.N(CH_3)_2-CH_2-C_6H_5]Cl$

Amine acetates are produced by neutralizing fatty amines with acetic acid, and are water-soluble. Quaternary ammonium compounds are normally prepared by one of four methods, viz, exhaustive alkylation of a primary or secondary fatty amine, alkylation of a low-molecular-weight tertiary amine with a fatty alkyl halide, alkylation of a tertiary fatty amine or by treating a tertiary amine or its salt with an epoxide.

In the detergent industry the main use of cationic detergents is for fabric softeners. For this purpose quaternary ammonium compounds derived from hydrogenated tallow are used; distearyl dimethyl ammonium chloride is preferred in Europe while in the USA distearyl dimethyl ammonium sulphate is also used. Although called 'distearyl' the tallow molecule contains an appreciable quantity of palmitic (C_{16}) acid and also, depending on the degree of hydrogenation, a proportion of oleyl amine (as the quaternary compound) might also be present.

Starting materials are either the fatty acid or alcohol, both of which are converted to the secondary (sometimes primary) amine.

If fatty acids are used they are reacted with ammonia successively to the ammonium salt of the fatty acid, dehydrated to the amide and again dehydrated to the nitrile:

$$R-COOH \rightarrow R-COONH_4 \rightarrow R-CONH_2 \rightarrow R-C\equiv N$$

This nitrile is then reduced (hydrogenated) to form as required the primary or secondary amine:

$$R-C\equiv N + 2H_2 \rightarrow R-CH_2NH_2$$

or

$$2R-C\equiv N + 4H_2 \rightarrow (R-CH_2)_2NH + NH_3$$

If an alcohol is to be the starting material, it is reacted with ammonia gas in the presence of a catalyst, when the nitrogen atom replaces the reactive hydrogen. The reaction for the preparation of a secondary amine is

$$2ROH + NH_3 \rightarrow R_2NH + 2H_2O$$

The primary or secondary amine can then be methylated with formaldehyde to the tertiary amine, an example of the reaction being

$$2R_2NH + 3HCHO \rightarrow 2R_2CH_3N + CO_2 + H_2O$$

or the primary or secondary amines undergo exhaustive methylation with methyl chloride according to the equations:

$$RH_2N + 3CH_3Cl \longrightarrow R(CH_3)_3N^+Cl^- + 2HCl$$
$$\text{or}$$
$$R_2HN + 2CH_3Cl \longrightarrow R_2(CH_3)_2N^+Cl^- + HCl$$

In both instances HCl is formed and as the reaction cannot go to completion in an acid medium, it is necessary to provide a scavenger for the acid produced. Industrially sodium or potassium carbonates are used in a polar medium such as water or alcohol and at a temperature between 60 and 95°C under slight pressure.[30]

Tertiary amines can also be methylated as above, but they are often reacted with dimethyl sulphate:

$$2R_3N + (CH_3O)_2SO_2 \rightarrow 2R_3CH_3N^+SO_4^=$$

forming the quaternary ammonium sulphate, fairly popular in the USA as a fabric softener.

The same reaction can be employed on secondary amines to add two methyl groups, producing for example a di-alkyl dimethyl ammonium sulphate.

Cationic surface active agents have wetting, foaming, and emulsifying properties. They are not, however, good detergents, being substantive to solids. Most of their applications are in fields where non-ionic detergents cannot be used and depend on this substantiveness to solid surfaces, whilst others arise from their excellent germicidal properties (all have high phenol coefficients). Amine acetates are used to eliminate the normal static electrical charges on resins and plastics, as collectors and frothers for a variety of minerals, eg, mica, phosphates, lead ores, etc., as emulsifying agents, as bactericides, and for other purposes. Alkyl trimethyl and dialkyl dimethyl ammonium chlorides are used as antistatic agents for plastics, as softening agents for textiles, as germicides, as flotation agents for the separation of certain minerals from low-grade ores, and for other purposes. The dialkyl dimethyl compounds are in addition used for emulsifying oils into water, as corrosion inhibitors, and as mould inhibitors. Cetyl trimethylammonium bromide is used as a skin-sterilizing agent, for cleansing wounds, and as a hospital disinfectant. Alkyl pyridinium chlorides and bromides and alkyl dimethyl benzyl ammonium chlorides are used in the textile industry for a variety of purposes (eg as antistatic and lubricating aids, in dyeing processes, and for stripping vat and azo dyes, in finishing processes, as germicidal and mildew-retarding agents); in the leather industry as bactericides and dye fixatives; as sterilizing agents for plant and equipment in food production and catering, brewing, and allied industries; in medicine; in the paper industry to control moulds and slime and as assistants in dyeing.

Amphoteric Detergents

Amphoteric detergents, described in the introductory chapter on classification of detergents, are specialty detergents, which find application mainly in the field of cosmetics. A more detailed description is somewhat beyond the scope of this book.[31]

References

1. *Ind. Eng. Chem.*, **46**, 1917–21 (1954).
2. *Fette-Seifen-Anstrichm.*, **64**, 326–9 (1962).
3. Wilson, G. R., *J Am Oil Chem Soc*, **31**, 564–8 (1954).
4. *Fatty Alcohols*, 2nd ed, Henkel KGaA, Dusseldorf (1982).
5. Hatch, L. F., *Higher OXO Alcohols*, Wiley, New York (1957).
6. Technical publication: AOS The Sulfonation Product of Ethyl Corporation's Alpha Olefins, Ethyl Corporation.
7. *US Patent* 3,349,141, W. A. Sweeney (24 Oct 1967).
8. *US Patent* 3,248,443, G. J. McEwan and S. G. Clark (1966).
9. *US Patent* 3,585,2534 S. K. Huang (15 June 1971).

REFERENCES

10. Davidsohn, A., Moretti, G. and Adami, I., Paper read at International Symposium on Natural Cleaning Agents, Marseilles (April 1985).
11. *US Patents* 2,271,619, G. B. Bradshaw and W. C. Meuly (to E. I. du Pont de Nemours & Co) (1942), and 2,360,844 (1944).
    Soap, **18,** 5, 23–4, 69–70, G. B. Bradshaw (1942).
12. C. F. P. Bevington and J. L. Pegler, *Chem Soc Spec Publ*, **12,** 283 (1958).
13. *US Patent* 3,535,339, H. H. Beyer and C. W. Motl (20 Oct 1970).
14. 3rd Int. Congress on Surface Activity, Cologne 1960, Vol. I, p 113, A. Davidsohn.
    Davidsohn, A., *Industrial Chemist*, 592–596 (Nov 1963).
    4th Int. Congress on Surface Activity, Bruxelles 1964, Vol. I, 281–91, A. Davidsohn.
15. 3rd Int. Congress on Surface Activity, Cologne 1960, Vol I, 104–6, M. Ballestra.
    Silvis, S. J. and Ballestra, M., *J Am Oil Chem Soc*, **40,** 618–20 (1963).
16. Gilbert, E. E., *Sulphonation and Related Reactions*, p 353, Interscience (1965).
17. Davidsohn, A. and Moretti, G., *Seifen-Ole-Fette-Wachse*, **111,** 9, 265 (1985).
18. *US Patent* 3,395,170, J. M. Walts and L. M. Schenck (30 July 1968).
19. *US Patent* 3,926,863, G. Perla and G. Mattielo (16 Dec 1975).
20. *Chemical and Process Technology Encyclopedia*, McGraw-Hill, New York.
21. Umbach, W. and Stein, W., *Fette-Seife,Anstrichstoffe* **71,** 938 (1969).
22. *US Patent* 2,089,212, W. Kritchevsky (10 Aug 1937).
23. *US Patent* 2,096,749, W. Kritchevsky (26 Oct 1937).
24. *US Patent* 2,464,094, E. M. Meade (to Lankro Chemicals Ltd) (9 Mar 1949).
25. *US Patent* 2,844,609, G. C. Tesoro (to Onyx Oil & Chemical Co) (22 July 1958).
26. *US Patent* 2,863,888, J. V. Schurman (to Colgate-Palmolive Co) (9 Dec 1958).
27. *US Patent* 3,024,260, R. Ernst (to Textilana Corp.) (6 Mar 1962).
28. *US Patent* 3,040,075, J. W. Lohr (to Andrew Jergens Co) (19 June 1962).
29. *The Alkanolamines Handbook*, The Dow Chemical Company (1979).
30. Jungerman, E., *Cationic Surfactants*, p 29, Marcel Dekker, New York (1970).
31. Bluestein, B. R. and Hilton, C. L., eds, *Amphoteric Surfactants*, Marcel Dekker Inc, New York, 1982.

6. Manufacture of Finished Detergents

Powders

The bulk of detergent materials are eventually converted into powders by one process or another. The problem in the production of powders is that all active detergent materials, with only a few exceptions, are not in themselves solids. It is, therefore, necessary to combine them with the builders and filling materials in such a way that the finished powder does not tend to cake or lump and remains dry to the touch. For each detergent raw material there is a limit of active material that can be incorporated into a powder, and the method of production of the powder also limits its active matter.

The principal methods of producing powders are:

(a) absorption of a liquid detergent on to inorganic salts
(b) simultaneous absorption and neutralization of a sulphonic acid by soda ash
(c) dry mixing of previously dried concentrated detergents with the other powder ingredients
(d) spray-drying
(e) a combination of spray-drying with one of the other above alternatives
(f) drum-drying

1. Simple Absorption

This method is the most limited in application, and the amount of active material that it is possible to incorporate into the finished powder depends on the physical form of the active ingredient. If the detergent is available in the form of a water solution, containing not more than 40 per cent active matter (ie, it is of relatively low viscosity), then 8 per cent active matter calculated on the final powder can easily be incorporated; and with special manipulation, up to 12 per cent can be achieved. If the detergent is in a more concentrated form, for example, a non-ionic detergent, which is usually available in 100 per cent concentration, 15 per cent active matter is usually the practical limit.

The plant required for this method is any one of the conventional bladed powder mixers available, such as ribbon mixers, plough mixers, incorporators, screw mixers. In recent years several sophisticated powder mixers have been developed and these are finding more and more applications in the absorption process of making detergent powders. Such mixers are manufactured by Apex, Lödige, Nauta and others, and to a large extent they have solved the problem of incorporating larger quantities of non-ionic detergents in powders.

Manufacture of powders by adsorption with a high percentage of active matter depends on the physical structure of the surfactant to be absorbed. A solid non-ionic is obviously no problem at all. In the other extreme, low ethylene oxide non-ionics can cause stickiness. We have found that a nonyl phenol with 12 EO gives better flow characteristics to the finished powder than does one with only 10 EO.

The physical characteristics or structure of the builders have a marked effect on adsorption. One has to distinguish between the water absorption capacity of the builder, ie, the water of crystallization and the adsorptive capacity for organic matter, the surfactant.

In these methods, anhydrous inorganic salts, which can be hydrated, are used, and on the addition of the detergent solution the salts are partially hydrated and the active matter is dispersed with this water of crystallization. It is essential that the detergent solution be of very low viscosity and that it contain more water than active matter. However, if low concentrations of active matter are to be incorporated, this condition need not necessarily be fulfilled. In this case, the detergent is merely adsorbed on to the surface of the inorganic material. To obtain the higher concentrations of active matter, it is necessary for the formula to contain a large proportion of soda ash, as it can theoretically absorb as water of crystallization 170 per cent of its weight of water. This figure is impossible to achieve in practice, as the organic detergent material does not allow the full 10 molecules of water of crystallization to be formed. The same effect can be achieved by the use of borax either in the anhydrous or pentahydrate forms. Borax is, of course, more expensive than soda ash but has two added advantages in that it is a better builder and can also to a certain extent eliminate stickiness.

As mentioned above, if a 40 per cent active matter solution is used, 8 per cent active matter can easily be incorporated. If, however, the vessel is steam-jacketed, or if the mixing blades are hollow and can be steam-heated, powders containing up to 11 per cent active matter can be produced.

The method of producing powders is to charge the mixer with the dry ingredients, start the mixing, and to pump (or pour) the solution through a jet into the powder slowly while mixing. The rate of addition of the liquid must be adjusted so that the liquid is absorbed as it reaches the surface of the powder; if not, hard crystalline lumps will be formed.

After all the liquid has been added, mixing is continued for at least 15 min, and the powder is discharged on to a concrete floor to age for 12–24 hours. If the total amount of liquid is small (not more than 5 per cent) this ageing can be dispensed with, but in general the ageing serves the purpose of allowing the crystals to form and cool. If one of the more sophisticated mixers, mentioned above (especially one with a disintegrating effect) is used, this ageing process and the grinding can often be dispensed with.

By these methods a free-flowing powder, free from lumps, should be obtained; but if any of the original dry ingredients contain an appreciable amount of lumps, it will be necessary to pass the powder through a hammer mill after ageing. This grinding can be obviated by charging the powders into the mixer through a 60-mesh sieve. In this case, the lumps are screened out and can by themselves be milled to recover for use in future batches.

Because this process permits only a limited amount of active matter to be incorporated into the powder, it yields materials with only limited application for household use. However, the process can be used for industrial cleaners where the active-matter requirement is low.

To illustrate this process, we give below one typical formulation which can be used for the washing of stone floors:

Formula 1   *Floor-washing Compound*
Soda ash	77
Sodium tripolyphosphate	5
Sodium metasilicate pentahydrate	5
ABS-Na (40 per cent active matter)	12
Pine oil	1

The soda ash, phosphate and metasilicate are charged into the mixer, the mixing started and the ABS-Na is added slowly with stirring. After all the ABS-Na has been added, the mixing is continued for 15 min and the pine oil added. Mixing is continued for a minute or two after the pine oil has been incorporated and the powder discharged on to the floor to age.

2. Combined Absorption and Neutralization[1]

The method of neutralization and absorption is more versatile than that of simple absorption. The process depends on the utilization of an unneutralized alkyl benzene sulphonic acid, and neutralizing it with soda ash which has previously been dry mixed with all the other dry ingredients of the formula. It is immaterial whether this sulphonic acid is of the 90 per cent (oleum or acid sulphonation) or 100 per cent (SO_3 sulphonation) type. If the manufacturer is doing his own sulphonation, it is obviously more economical to use his sulphonic acid without any intermediate manipulation, and if the manufacturer buys his raw detergent material, the use of a substantially anhydrous material will obviously save freight costs on water or other inert filling materials. It should be pointed out that if material other than 100 per cent sulphonic acid is being used, precautions against iron pick-up must be taken in storage and transport.

By this method there is obviously no lower limit to the amount of active material that can be incorporated into the powder, and a product such as Formula 1 can be manufactured with ease. The upper limit of active matter is generally 20 per cent, but it is possible to achieve as much as 24 per cent. Another advantage of this method is that powders incorporating solvents (solvent detergents) can be manufactured (in this case, however, milling of the powder should be dispensed with). If an attempt were made to manufacture solvent detergents by spray-drying, the solvent would be carried off in the air stream and could also constitute a fire and explosion hazard.

The plant required is the same as that mentioned in the section on absorption, such as plough mixers, ribbon mixers, incorporators and screw mixers, but in this case no heating arrangements are required. Although, theoretically, the powder produced by this process should not need to be ground, in practice a better appearance is obtained if the finished powder is passed through a hammer mill. The Lödige, Nauta, or Apex mixers are again eminently suitable for this process.

On p 227 we describe systems combining spray-drying with other forms of preparation. These ancillary units can also of course be used by themselves for powder production. With all of these specially designed mixers, ageing and grinding can usually be dispensed with. If perborate is to be added, the powder after manufacture can be discharged into a silo, fed on to a conveyor belt on to which perborate (and enzymes if required) is directly charged, and thence to a further simple drum-type mixer.

The method of manufacture is as simple as that for the absorption process. The dry materials are charged into the mixing vessel and the mixer started. The sulphonic acid is now poured in slowly while the mixer is running. If 90 per cent sulphonic acid is being used, the acid immediately reacts with the soda ash present to form the sodium salt, and the sulphonic acid should be run in at such a rate that no large unabsorbed excess of the acid is allowed to form. If 100 per cent sulphonic acid, which is more inert, is being used, it will not react so easily with the soda ash, and it can be added fairly fast and the mixer will merely disperse the acid. When the sulphonic acid has been well dispersed, ie, when there is a uniform brownish-blue discoloration of the powder, 2 per cent of water, based on the final weight of powder, is added. This water renders the sulphonic acid reactive and almost immediately the powder will assume a light-yellow colour without any dark lumps of sulphonic acid.* The sodium silicate can now be added, followed by the optical brightener, if the formulation calls for it.

In this reaction of alkyl benzene sulphonic acid with soda ash, no carbon dioxide is liberated, as in every case the formulation is arranged so that at least double the amount of soda ash is present over the stoichiometric amount and the reaction proceeds only as far as:

$$C_6H_4C_{12}H_{25}SO_3H + Na_2CO_3 = C_6H_4C_{12}H_{25}SO_3Na + NaHCO_3$$

Therefore, only one sodium atom of the two in the soda ash is utilized

* It is a curious fact that alkyl benzene sulphonic acids behave like titration indicators in that at a pH approaching that of neutrality they experience a colour change from dark brown to light yellow. To speed up the process sulphonic acid and water can be run in simultaneously (but not pre-mixed).

to form the sodium salt of the sulphonic acid and the soda ash is converted to sodium bicarbonate. If this were not the case, the evolution of carbon dioxide would cause the mixture to overflow the vessel. For this reason, when the 100 per cent type of sulphonic acid is used, it is necessary to disperse the acid well, so that no local concentrations are formed before the water is added.

Rhone-Poulenc has developed an alternative method for neutralizing sulphonic acid, by using anhydrous sodium metasilicate.[2] The reaction can be written:

$$C_6H_4C_{12}H_{25}SO_3H + 2Na_2SiO_3 \rightarrow$$
$$C_6H_4C_{12}H_{25}SO_3Na + Na_2O:2SiO_2 + H_2O$$

the metasilicate has neutralized the sulphonic acid and been converted to a colloidal type of silicate, obviating the use of energy intensive liquid or dry forms. The above equation converts the silicate to the 1:2 ratio. As mentioned on p 56, the preferred form can be the ratio 1:2·4. To obtain this form of the colloidal silicate, 3·2 parts of sulphonic acid are reacted with 1 part of anhydrous metasilicate to give a ratio of sodium sulphonate to colloidal silicate of 1:7.

Once the reaction is completed, which in practice should not take more than 30 min for a 300–500 kg batch, the powder is discharged on to a concrete floor and allowed to age overnight. After the ageing period, the powder is fed into a hammer mill to break up any lumps and is then ready for packing.

If one of the sophisticated mixers described later, is being used the ageing and grinding can of course be dispensed with.

Powders made by this process will have a cream to light-yellow colour, the exact shade depending on the colour of the sulphonic acid used. If a whiter powder is required, this can be achieved by bleaching it with sodium hypochlorite. When 100 per cent sulphonic acid is used, a 10 per cent solution of sodium hypochlorite is employed instead of the water, and when 90 per cent sulphonic is being used, the sodium hypochlorite is added thus:

The mixer is charged with all the powdered ingredients, except the optical brightening agent, and the mixing started. As soon as the powders are uniformly dispersed, the sulphonic acid (either 90 per cent or 100 per cent) is added. When the sulphonic acid has been well mixed in, the sodium hypochlorite solution, calculated at the rate of 2 per cent on the final weight of powder, is added. This will help to disperse the sulphonic acid, trigger the reaction with the soda ash and simultaneously bleach the powder. The silicate of soda is added after the bleaching is completed.

When sodium hypochlorite is used for bleaching of powders, the optical brightening agents should not be added to the mixing vessel immediately after the neutralization, as the hypochlorite will attack the dyestuff. If perborate is to be added to the powder (see below), the optical brightener can be incorporated simultaneously with the perborate. If the perborate is not to be added, after the reaction of the sodium hypochlorite has ended, the powder can be qualitatively checked for active chlorine.* If this test is negative, the optical brightener can then safely be added. If this test is positive, 50 g of an 'anti-chlor' (sodium thiosulphate, sodium sulphite or sodium bi-sulphite)/300 kg batch of powder are added, the mixing continued for 3–5 min and then the optical brightener can be added with safety. The qualitative test for chlorine could even be dispensed with and the anti-chlor added as a routine, but one must stress that the anti-chlor can only be added after the powder has been bleached to the required shade, as otherwise it will interrupt the bleaching reaction.

Sodium perborate, if called for in the formulation, cannot be added prior to the ageing or grinding (unless one of the modern powder mixers is used) whether sodium hypochlorite is used or not. If sodium hypochlorite is not used, the perborate will start decomposing when the powder is damp, prior to ageing, if it is introduced initially into the mixing vessel, or it will start decomposing in the hammer mill as a result of the heat of friction in the grinding process. If sodium hypochlorite is used, the hypochlorite and perborate (peroxide) ions are mutually incompatible (so much so that perborate can be used as a rather expensive form of anti-chlor if nothing else is available). Sodium perborate can only be added after the ageing and grinding processes by reintroducing the powder into the mixer and adding the requisite amount of perborate. It is preferable to make use of a separate simple powder mixer for this operation.

The active ingredient by this process need not necessarily be only alkyl benzene sulphonic acid, but this can be mixed with any other active material desired to give special properties. These other materials can be non-ionics of the ethylene oxide condensate type, alkylolamides (the preferred types being lauric acid monoethanolamide or monoisopropanolamide) and even fatty acids to form a soap *in situ*. If the material to be mixed with the sulphonic acid is a solid at room temperature, it should be warmed about 10°C above its melting-point, then added to the sulphonic acid immediately before the addition to the powders. By this method the reaction of the sulphonic acid with the soda ash causes the fatty acid, if used, to be saponified completely and eliminates the need for the pre-saponification of fats or fatty acids prior to their incorporation into powders of this sort. To obtain this 'auto-saponification' the fatty acid should be of the distilled grade with a maximum of $1\frac{1}{2}$ per cent neutral oil (triglyceride) or unsaponifiable matter present. If neutral oil is present it will not be saponified under these conditions and will tend to make the powder sticky if present in too large quantities.

If solvent detergent powders are to be manufactured, the solvent must be chosen with care for the particular operation involved.

Solvents incorporated in detergent powders made by dry neutralization

---

* A small quantity of the powder is dissolved in water in a test-tube, 5 ml of dilute HCl is added and then a few crystals of potassium iodide. If no brownish-yellow colour appears, there is no free active chlorine present.

should have a boiling range of 125–260°C, such as deodorized kerosine (p 111). The finished product should be marketed in metal or other containers non-permeable to vapour, or in high-density polyethylene bags which may be used as liners for jute or paper sacks.

If the solvent is mixed with the alkyl benzene sulphonic acid prior to neutralization with builders, the process should preferably be carried out so as to dispense with grinding after mixing. During processing, the mixture must be protected against exposure to electric sparks or flame.

The solvent may be incorporated in the detergent powder by two different methods. In the first, the alkyl benzene sulphonic acid is mixed with the solvent and this mixture is added to the builders. (Fatty acids are not recommended for incorporation in solvent/detergent combinations, because the foaming power of such products is too low.) A solvent must be chosen which does not react with alkyl benzene sulphonic acid. Pine oil, for instance, does react and can be used only if diluted with non-reacting solvents to the point where little or no reaction with alkyl benzene sulphonic acid occurs.

As an alternative, the detergent mixture can be neutralized dry as described earlier and the solvent added separately to the finished detergent mixture. A low percentage of alkyl benzene sulphonic acid should be used in this procedure, otherwise the final product may not feel dry enough. A modification of this method calls for pre-mixing of the solvent with colloidal silica. In this case, a non-ionic detergent must be added to the solvent prior to premixing with the silica. The non-ionic renders the silica/solvent combination more readily dispersive in the wash liquor, thus increasing its efficiency.

The first method can dispense with the use of the rather expensive air-floated silica either entirely or in part.

If pine oil is added to alkyl benzene sulphonates, it acts as a synergist towards wetting speed. This effect can be demonstrated and measured by the Draves test method. First observed by one of the authors in combinations of pine oil with liquid alkyl benzene sulphonic acid and sulphated fatty alcohols,[3] this synergism is evident also in combinations with powders based on alkaline builders and alkyl benzene sulphonic acid.

As mentioned above, if solvents are to be incorporated into a powder, grinding must be dispensed with as the current of air through the mill will carry with it some of the solvent, which is both a loss and a fire hazard. This therefore calls for a type of mixer as described on pp 228ff.

One of these types uses the fluid bed principle and again for the same reasons, this cannot be considered for solvent/detergent powders.

It is often required to dye powders to a colour other than white. This can be done by dissolving a water-soluble dye in the water naturally being added to the powder or to disperse with the powders finely divided lakes or pigments, eg, Monastral blue. In general, if a light-coloured sulphonic acid is used, it will not be necessary to bleach the powder prior to dyeing, so that no complication about the use of sodium hypochlorite need arise. However, if the combination of the natural colour of the sulphonic acid and the colour of the dye or pigment produces a 'muddy' appearance, it is necessary to bleach the powder before adding the colouring matter, to give a brighter colour. Inorganic pigments are, in general, not affected by sodium hypochlorite, but lakes can be, so again it is necessary to treat the lake in the same way as the optical brightener when sodium hypochlorite is used.

Some powders, both spray-dried and non-spray-dried appear on the market with coloured spots dispersed throughout the mass. Elaborate arrangements are available for 'dotting' the powder but a simple one is to charge into any dry-powder mixer anhydrous metasilicate or borax and to add to this, while mixing, a water solution of the dye or dispersion of the pigment, the amount to be of the order of 0·5 per cent of the inorganic salt. Mixing takes only a minute or two. This dyed powder is stored until required and is added in the same way as, or together with, perborate to the finished powder, the rate of addition to be between 0·1 and 0·5 per cent of the finished powder.

Perfume is generally added to household powders. If sodium hypochlorite is not used, the desired amount of perfume can be sprayed into the powder towards the end of the mixing process. Sodium hypochlorite can destroy perfume material; so, if the powder is being bleached, it is advisable to add the perfume in the same manner as the optical brightener.

On occasion it is required to manufacture powders containing more than the practical limit of 20 per cent active matter. This is achieved by the use of highly absorptive special silica. This silica will tend to adsorb on to its surface all the stickiness that a high active matter will cause. The powder is manufactured in the normal way as described above without the addition of the silica and when completed prior to its being discharged for ageing 1–2 per cent silica is added. The actual amount is dependent on the actual formulation and can be as low as $\frac{1}{2}$ per cent and should rarely exceed 2 per cent.

To illustrate the above processes we detail below some formulations and procedures using both 100 per cent and 90 per cent alkyl benzene sulphonic acids. It should be pointed out that 100 per cent sulphonic acid never contains 100 per cent of the acid. The term 100 per cent means 100 parts of the sulphonic acid when neutralized to the sodium salt will yield between 99 and 101 parts of active matter. Similarly, for 90 per cent sulphonic acid, 100 parts of the acid will yield 90 parts of active matter after neutralization to the sodium salt. These two figures, however, are very variable. One hundred per cent sulphonic acid made by the sulphonation of alkyl benzene with SO_3 gas does not usually deviate seriously from the practical limits, but for acid or oleum sulphonated material, unless special precautions are taken with the separation of the spent acid, the available active matter very rarely exceeds 90 per cent.

This process can be used to manufacture a cheap, heavy-duty household powder:

Formula 2   Heavy-duty Household Washing Powder

	100% sulphonic acid	90% sulphonic acid
Charge and mix together:		
Light soda ash	58	58
CMC (66% active)	2	2
Sodium tripolyphosphate	15	15
then add with mixing:		
Alkyl benzene sulphonic acid (ABS)	18	20
followed by:		
Water and	2	—
Sodium silicate (40% solution)	5	5

The powder is now discharged on to the floor to age and on the following day is passed through a hammer mill. In practice, several batches can be prepared on the first day and all of these can be combined prior to milling. After milling, the powder is weighed into the same (or another) mixer and sodium perborate and the optical brightening agent are added in the following porportions:

Milled powder	89·9
Sodium perborate	10·0
Optical brightening agent	0·1

This is mixed only for a few minutes and is then ready for packing.

If linear alkyl benzene sulphonic acid is used, the addition of toluene sulphonate and/or commercial very light density silica will obviate stickiness.

For a whiter powder than that produced by this method the water is replaced by sodium hypochlorite solution containing 10 per cent available chlorine. The formula then becomes:

Formula 3

White Household Heavy-duty Washing Powder

	100% sulphonic acid	90% sulphonic acid
Light soda ash	58	57
CMC	2	2
Sodium tripolyphosphate	15	15
ABS	18	20
Sodium hypochlorite solution	2	2
Sodium silicate (40% solution)	5	—
Sodium silicate (54% solution)	—	4

It will be noted that, where the 90 per cent ABS is used, it is necessary to use a more concentrated silicate solution, because extra liquid cannot be added.

After milling, the powder is treated in the same way as the powder from Formula 2.

Recently special disintegrating high-speed mixers, fitted with special rotating knives have been successfully introduced by the authors to produce these dry neutralized powders in one single step, dispensing with ageing and grinding. This type of mixer is produced by the German firm Lödige (Paderborn) and their licensees in the UK and the USA. It is even possible to add perborate immediately after the neutralization action has been completed, even when hypochlorite has been used. If hypochlorite is being used, after the reaction has been completed add any suitable antichlor (p 205) or 1 per cent sodium perborate. This will inactivate any residual hypochlorite. The desired percentage of perborate is now added, and mixing is carried out with the mixing blades only, not using the knives, and at a lower speed than during the neutralization process. A variable speed motor, or a system of gears, is therefore advisable. This procedure diminishes loss of perborate activity.

If a relatively high amount of alkyl benzene sulphonic acid or fatty acids is being used, the mixer should be fitted with a water-cooling jacket to dissipate the heat of reaction.

3. Dry Mixing of Powders

Concentrated spray-dried or drum-dried detergent powders, containing at least 40 per cent active matter, often more than 60 per cent, are available. They are merely blended with the other required ingredients, in a powder mixer, dry blender or similar equipment.

If it is desired to preserve as far as possible the characteristics of a spray-dried powder, spray-dried sodium tripolyphosphate and silicate (p 54) may be used in addition to the spray-dried detergent concentrate. In this way more than half of the ingredients are already of the spray-dried type. When mixing the ingredients care must be taken to avoid (as far as possible) physical breakage of the beads; a tube mixer or a slow rotating drum mixer similar to the mixer used for adding perborate and enzymes to spray-dried powder beads (p 228) is suitable for this purpose.

As this is a dry-mixing process, it is not advisable to incorporate silicate solutions into the powder. Finely divided partially hydrated colloidal silicates are available or sodium metasilicate can be used if the high alkalinity will not adversely affect the final properties of the powder. Alternatively, some concentrated detergent powders are available with sodium silicate already included in the beads.

4. Spray-drying of Powders[4]

The vast majority of household detergent powders are manufactured by the spray-drying process, because spray-dried powders have many advantages over the other types. These advantages can be summarized:

(*a*) The formulation is not limited. Relatively high amounts of active matter can be incorporated; soda ash is not an essential ingredient, and the moisture and bulk density can be varied at will (within definite limits).

(b) The powders present a pleasing appearance and being light have more sales appeal.

(c) Spray-dried powders are dustless and free-flowing and do not tend to lump. For normal formulations no special inner liners are required in packaging.

(d) Because it consists of hollow beads, with a large surface area, the powder dissolves instantly when added to water. This is important when powders are used in machines, where the wash cycle can be as low as 4 min. If the powder takes time to dissolve, valuable washing time is lost. Also, in tub-washing, where no mechanical agitation takes place, portions of powders other than spray-dried powders, can still be undissolved even when the operation has been completed.

(e) Heat-sensitive materials can, within limits, be handled in a spray-drier.

A spray-drier involves large capital outlay, but the results will pay for this investment. Many types are available, both of the jet and of the disc type, and also using both countercurrent and concurrent airflow patterns.

Most, although not all, spray-driers for detergent powders are of the countercurrent airflow, jet-spray type, which produce a large bead with a minimum of dust and a medium bulk density.

Spray-drying techniques depend to a great extent on the type of equipment being used. Spray-driers are equipped with various methods of producing the feed (slurry), which may be continuous (which has many advantages)[5] or batch which allows for easier control of the constituents, since a batch can be prepared, its constituents checked and adjusted if necessary, and then passed for spray-drying. This involves considerable handling and work and also requires at least two batch-preparation tanks, from one of which the slurry is being pumped into the tower while the fresh batch is being prepared in the other. The size of each vessel should be large enough to hold sufficient material to spray during the time it takes for another batch to be prepared and checked in the second one.

If the vessels are comparatively small, this involves frequent preparation of batches during a shift. If the vessel is large, the slurry must be kept warm and stirred during the waiting period (which also includes the time that is taken to pump the slurry into the tower). This waiting period can cause hydration and/or hydrolysis of the phosphate present and may turn the CMC solution into a gel.

With continuous feed, the hold-up time is small and the above complications are largely avoided. More careful control is necessary on the setting of dosing units, but labour is reduced to a minimum. On the whole, the advantages of continuous feeding greatly outweigh the advantages of batch preparation. The source of active matter fed into the spray-drier can be any one of hundreds of alternatives: sulphonates as the acid, or as the already neutralized paste (a concentrated powder can also be used, but this is highly uneconomical, besides causing complications due to the inorganic filler naturally present); sulphates and non-ionics (although the possibilities of adding non-ionics to the powder prior to spray-drying are somewhat limited because of 'pluming' and stickiness, see p 31), and even combinations of the above ingredients and, if desired, admixtures with soap.

It is obvious that if a sulphonic acid is used as the basic material it must be neutralized, either continuously or batch-wise, and it is essential to do this before the acid comes into contact with the rest of the ingredients of the slurry. Otherwise, insoluble silica may be precipitated; the polyphosphate can easily be hydrolysed to orthophosphate; and the optical brightener may also be affected adversely. In continuous operations, the sulphonic acid and caustic soda should be fed into a neutralizing vessel with all the water required for the slurry. This sodium sulphonate paste is then fed into the slurry preparation vessel, where the rest of the ingredients are added. To produce a white powder, sodium hypochlorite can also be added at the neutralization stage, at such a rate that it is completely consumed in this vessel and not allowed to come into contact with the optical brightening agents in the slurry-preparation vessel.

On batch processes it is best to start with the water, caustic soda, sodium hypochlorite (if required), and then the sulphonic acid. After the pH has been adjusted the rest of the ingredients can be added.

For certain formulations, mixtures of two or more active ingredients are used. If a continuous process of slurry preparation is being used and if the dosing arrangements are such that each constituent can be fed separately, there is no problem. Similarly, in batch preparation no difficulties should arise.

However, if, as is sometimes the case, automatic dosing arrangements are such that only one unit can provide the active ingredient and two or more materials are to be used, various pre-mixing arrangements need to be made.

If, as is often the case, a mixture of soap and alkyl benzene sulphonate is used as the active ingredient, the soap can be fed in as a ready-made powder with the other powder ingredients. This again, however, is uneconomical, as it requires previous work to make the soap into a powder. A more usual and efficient arrangement is to pre-mix fatty acids with the alkyl benzene sulphonic acid prior to neutralization, and to feed this mixture into the neutralization vessel in the same way as alkyl benzene sulphonic acid alone. The intimate mixture of fatty acid and the sulphonic acid allows instantaneous neutralization (saponification) in the water solution without any heating. If this pre-mix is used, and if the fatty acid contains any proportion of unsaturated fatty acids (for example, the oleic acid normally present in tallow fatty acids), a reaction between the sulphonic acid and the unsaturated fatty acid can take place. This produces an insoluble material (possibly a sulphone) which can be adsorbed on to the cloth during the washing process, causing an unpleasant odour. The reaction can be inhibited by the addition of 5 per cent water to the sulphonic-fatty acid mixture. However, although alkyl benzene sulphonic acid in the 100 per cent form is normally inert against metals, if water is added, the acid will become reactive. For this reason this sulphonic acid-fatty acid-water mixture should be made and stored in a stainless steel or plastic-coated vessel.

The fatty acids and the alkyl benzene sulphonic acid can of course be introduced into the neutralization system in two parallel streams. The heat of neutralization of the sulphonic acid is sufficient to saponify (neutralize) the distilled fatty acids as well.

If a mixture of long-chain fatty alcohol sulphates and alkyl benzene sulphonates is being used as the active ingredient, the alcohol sulphates are available only as the neutralized salts; and if these are mixed with sulphonic acids (or any acids) the sulphate will tend to decompose at the low pH to produce free insoluble long-chain fatty alcohols. If both the alkyl benzene sulphonate and the long-chain alcohol sulphate sources are the neutralized pastes, they can, of course, be pre-mixed, but if alkyl benzene sulphonic acid is to be used on its own the long-chain fatty alcohol sulphate can only be added to the slurry after the alkyl benzene sulphonic acid has been neutralized.

When alkyl benzene sulphonate and a non-ionic are to be used, there is in general no objection to pre-mixing the non-ionic with the alkyl benzene sulphonic acid or sulphonate. It is advisable to carry out small-scale trials before committing large quantities of material to production, because, if the non-ionic is a fatty acid or amine ethylene oxide condensate (manufacturers of non-ionic detergents do not always disclose the composition), there is again danger of acid hydrolysis of the non-ionic. It should be noted, however, that the bulk of non-ionics are either long-chain alcohol or alkyl phenol ethylene oxide condensates, which are perfectly stable under these conditions.

The CMC should be granular or powdered, so that it is easily dispersed and not necessarily dissolved. In fact, it is not desirable that it should dissolve at all, since this might produce thixotropic gels which cause trouble in pumping.

The incorporation of sodium tripolyphosphate poses a special problem. Sodium tripolyphosphate exists as the anhydrous material in two crystalline (or one crystalline and one meta-crystalline) forms, normally known as Phase (or Type) I and Phase II. Phase I is produced when the calcination temperature (see p 49) is of the order of 650°C and Phase II at a temperature in the neighbourhood of 400°C. It is obvious that very rarely can only one or other of these two types be produced, the resultant is usually a mixture. When dissolved in water both give the identical material $Na_5P_3O_{10}$, or when hydrated form the crystal $Na_5P_3O_{10}.6H_2O$. However, Phase I hydrates very rapidly with the evolution of ten times more heat of hydration than Phase II, which also takes a considerably longer time to hydrate. In figures the heat of hydration for Type I is 13,000 calories and for Type II 1041 calories. (For this reason Phase II is sometimes described as 'slow hydrating'.) Spray-drier feed systems, particularly the continuous system, take this property into account in their design. No hard and fast rules can be given as to which type is best suited for a particular spray-drier, but it should be mentioned that when this property of the phosphate was first observed all spray-drying plants demanded 100 per cent Phase II, because Phase I, when coming into contact with water, hydrated rapidly on the outside surface and formed lumps, and did not allow the inside of the lump to be dispersed. These lumps, at the worst, clogged filters, and at the best, were retained by the strainers and reduced the amount of phosphates fed to the jets. Nowadays it is recognized that for some designs of spray-drier feed systems, a certain portion of Phase I is desirable and some detergent manufacturers are specifying 20 and even 40 per cent Phase I content in their sodium tripolyphosphate.

A more recent tendency is to use pre-hydrated sodium tripolyphosphate which has now become available on the European market.

Another problem in the incorporation of sodium tripolyphosphate into the detergent slurries prior to spray-drying is that this material when dehydrated from solution or from its hexahydrate at a temperature of the order of 100°C can lose only five of the six molecules of crystallization. To 'remove' all the water, at temperatures of 100°C the last molecule of crystallization re-enters the molecule to form a mixture of pyrophosphate and orthophosphate according to the formula:

$$Na_5P_3O_{10} + H_2O = Na_4P_2O_7 + NaH_2PO_4$$

Thus, if the polyphosphate dissolves in the water of the slurry, or if it is hydrated at all, when the slurry is fed into the tower and all the water driven off, no polyphosphate will appear in the finished powder, but instead there will be a mixture of pyrophosphate and orthophosphate.

To overcome this difficulty, the slurry preparation should be organized so that the amount of phosphate which enters into solution is minimal. This can be achieved if the solids concentration in the slurry is not allowed to drop below 60 per cent. The phosphate in solution can be dehydrated, without any danger of decomposition, to its hexahydrate (or even monohydrate, but this is sailing close to the wind). Drying through the tower should thus be arranged so that the powder does not appear as 'bone-dry', but contains sufficient residual moisture to allow the phosphate to exist in the hydrated form. The other inorganic constituents generally present in powders either do not form crystalline hydrates or, if they do, the water of hydration can be driven off at a much lower temperature. Sodium sulphate, for example, has a transition point from the decahydrate to the anhydrous material at 32·4°C. In figures, the finished powder should therefore retain moisture at the ratio of at least 30 per cent of the sodium tripolyphosphate present. Keeping the slurry solids concentration high has an added advantage. Spray-driers are designed to evaporate water (they are normally rated by 'water evaporative capacity'). Concentrated slurries are therefore more economical because more dried powder is discharged per unit time. In general, also, concentrated slurries yield less friable beads with a smaller proportion of dust (ie, a lower particle spread).

MANUFACTURE OF FINISHED DETERGENTS

By virtue of their innate composition, detergent slurries can entrain air. This air entrainment can play havoc with a smooth flow of a slurry through the high pressure lines. De-aeration devices are available as optional additions to spray-driers.

All the above factors and possibly many more due to local conditions, need to be taken into account in designing a slurry unit, whether batch or continuous. Spray-drier manufacturers almost invariably offer slurry preparation units as adjuncts to their towers. One such continuous system is shown in Fig 6.1. The old method of weighing into a rotating drum has been improved by the use of load cells with constant and automating taring through a memory, thus each portion is dosed by difference.

When standardizing a new formulation it is advisable to take a small constant volume vessel of, say, 1 litre to determine the specific gravity of the slurry with a normal desired solids content and to check this specific gravity in the plant from time to time during the actual running operation.

Many household washing powders contain sodium perborate or one of the substitutes for perborate. The introduction of these persalts into the slurry is impossible, as the peroxide will decompose, so this builder is

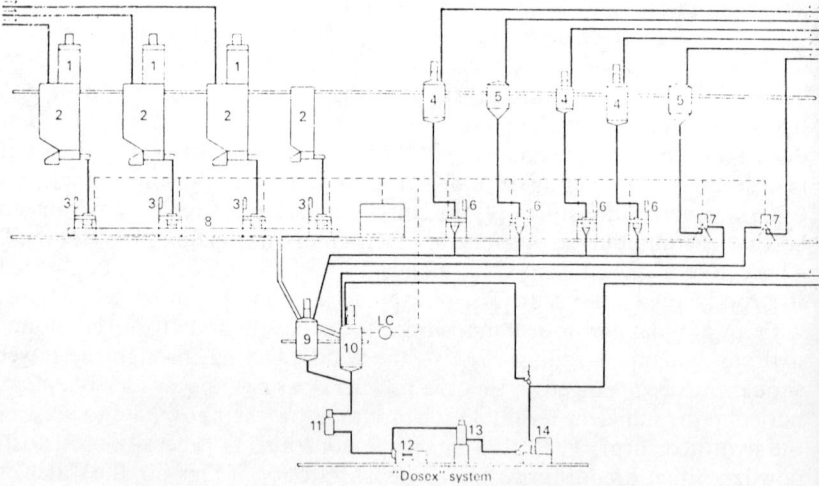

Fig 6.1 Ballestra continuous slurry preparation system

1. Sleeve filler
2. Powder silos
3. Proportional weighing scales or load cells for powders (builders)
4. Tanks for liquids or pastes with agitators
5. Tanks for liquids without agitators
6. Proportioning weighing scales for liquids and pastes
7. Volumetric dosing unit
8. Premix screw conveyor
9. Slurry mixer (crutcher)
10. Ageing vessel
11. Feeding filter
12. Homogenizing pump
13. Self-cleaning filter
14. High-pressure pump

added after the spray tower. Most spray-driers have elaborate perborate dosing arrangements for providing a constant and uniform flow of perborate into the exit stream of the finished powder. This system does not, however, take into account the fact that the bulk density of the perborate is approximately double that of a spray-dried detergent powder and this difference can cause segregation of the perborate in the package; and more particularly, if pneumatic arrangements are used to move the powder, the two constituents will travel at different speeds.

This problem has been overcome by feeding the perborate into a position near the base of the tower where the powder is already dry but still plastic and the perborate will adhere to the individual granules of detergent powder.[6]

The addition of enzymes to powders poses the same problem as that for perborates. The enzyme cannot be added to the slurry, prior to spray-drying, nor can it be added to the base of the tower when the powder is hot. In addition, enzymes are added in quantities of the order of $\frac{1}{2}$–2 per cent of the powder so the problem of mixing and segregation is more involved. Modifications of the physical state of the enzymes are being made (see p 101) to overcome some of these problems and the enzymes are best fed continuously into the powder stream generally via a continuously revolving drum mixer.

The bulk of household powders are perfumed. Perfume can be added in two ways, either as a powder (concentrated powdered perfumes are available for this purpose) in which case the powder is treated like perborate or enzymes, or as a fine spray on the stream of powder passing over a baffle, or by spraying into the rotating drum mixer used for the addition of perborate and/or enzymes to the beads coming from the tower. To spray the perfume it is sometimes advisable that it be diluted with a volatile solvent. This allows better dispersion of the perfume over the powder stream and prevents local build-up of the perfume oil, which may cause lumping of the powder.

The choice of the perfume for detergent powders is important and the following points should be borne in mind:

(a) The type of perfume. Perfumes have both an immediate and residual odour and both need to be considered.
(b) Whether the contents of the powder will adversely affect the perfume on storage.
(c) The volatility of the perfume under the particular conditions of packaging and storage.
(d) Whether the perfume will leave a residual smell on the washing. This is not always a requirement but account must be taken of this possibility.

In the case of perfumes, as part of the sales appeal of a powder depends on this, it is best to collaborate with the manufacturer of the perfume and not to buy the cheapest type available.

The actual manipulations involved in operating a spray-drier are usually laid down by the manufacturer of the plant and it is difficult to give generalized directions for all the possible types of spray-driers in use. However, some guides as to quality control of spray-dried powders are worth while recording:

An acceptable powder is characterized by:

1. Good colour properties.
2. Desirable particle size and spread.
3. Correct bulk density.
4. Correct residual water content.
5. Absence of stickiness.
6. Uniform composition and appearance.

Some of these characteristics are interrelated. For example, if the colour is bad owing to scorching the powder will also be bone dry with a minimal residual water content. If the residual moisture is excessive the powder might become sticky.

1. Colour

Bad colour may be due to off-colour ingredients in the slurry, such as excess iron in one of the inorganic powders or dark-coloured active matter. If discoloration is caused by some inorganic component of the slurry very little can be done about it. However, if an organic colouring matter is at fault, this can frequently be corrected by bleaching of the slurry with sodium hypochlorite or hydrogen peroxide solutions.

In the presence of optical brighteners, bleaching agents must be used judiciously. Not all optical brighteners are stable to oxidizing agents. It is therefore necessary to arrange the bleaching of the dark-coloured constituents before the optical brighteners are added and to add these only after the oxidizing agents have been effectively destroyed. This holds good for both continuous (see Fig 6.1) and batch-wise slurry preparation. A point to remember is that, due to the optical properties of fine opaque particles, a spray-dried powder always looks whiter than the appearance of the slurry would lead one to believe. This is particularly true of slurries containing optical brighteners.

The risk of scorching is greater in a countercurrent plant because the powder reaches the hottest point in the tower when it is practically dry. In a concurrent operation the powder is exposed to maximum heat when it contains most moisture.

Scorching may be counteracted by adding to the volume (ie, the weight) of inlet air, which causes a drop in the temperature of the incoming air without reducing the amount of heat. In operating a spray-drier it is well to remember that the work is not being done by the temperature of the inlet air, but by the quantity of heat, ie, the product of specific heat, mass, and temperature. The volume of the heat is therefore roughly proportional to the product of temperature and mass; roughly because the specific heat varies with temperature.

Any considerable build-up of matter on the walls of the spray-tower is a hazard to the quality of the final product. Such accumulations consist of material in contact with the hot air over extended periods of time and must be scorched. Furthermore, such build-up reduces the effective diameter of the tower and increases the tendency of the particles to impinge on the walls. Eventually the accumulation becomes too heavy for the detergent-metal bond and large chunks of material will break off and may temporarily overload the conveyor system. If the installation includes a screening system, the 'fall out' from the walls will be diverted to one side. But such overdried material is brittle enough to be broken up by the conveyors and the screens and yellow and brownish fragments (not beads) will find their way into the white beaded powder. Unfortunately, this build-up is often due to basic faults in tower design and in many instances little can be done to remedy the situation. However, finer atomization will yield particles with less momentum which, in turn, will ensure that the particles are dried before they reach the walls. Being dry they will have less tendency to stick.

If build-up cannot be avoided, the walls of the tower need to be cleaned at frequent intervals, either by scraping by hand or mechanically or by a jet of water. Spray-driers are often equipped with an auxiliary system for continuous removal of build-up. Ballestra spray-driers are fitted with scraping rings which can be worked either continuously or intermittently during operation, and are shown in Fig 6.2 which gives a general flow sheet of a spray-drier. In this instance the build-up is removed as it is produced, eliminating the production of scorched particles. Another method is to fit to spray-driers an air broom, which prevents build-up. The air broom is a pipe fitted with nozzles which rotates slowly around the chamber. Compressed air blown through the nozzles removes any material adhering to the walls and also cools them slightly.[7]

Sometimes the powder is grey when leaving the tower. This condition can be traced to soot from a poorly adjusted burner, being absorbed into the bead. This problem is more acute in a concurrent drier where the soot comes in contact with the jet of product when it is still wet and able to absorb it. In a countercurrent drier most of the soot is exhausted by the air removal equipment and the fines are then discoloured. Adjustment of the furnace will cure this trouble at once. The most common cause of soot formation is a dearth of primary air in the burner, which causes incomplete combustion. Usually a slight cut in the fuel supply will give complete combustion which in turn will yield a higher temperature using less fuel.

2. *Particle Size and Spread*

Particle size of the powder is dependent on the particle size of the atomized slurry which in turn is inversely proportional to the pressure applied at the jet or to the speed of the atomizing disc and approximately proportional to the viscosity of the slurry. One of the possible sources of abnormal particle spread, all other factors being equal, is a faulty jet or a disc with slurry encrusted on it. (This can also cause build-up on the walls.) A jet can

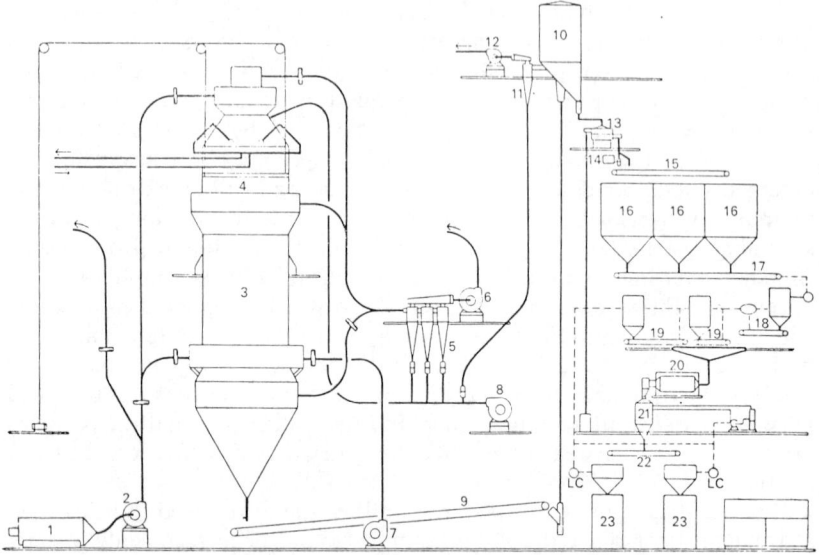

Fig 6.2 Ballestra spray-drying unit and finishing system

1. Furnace
2. First blower
3. Spray tower
4. Cleaning ring
5. Cyclones
6. Suction fan
7. Cold air fan
8. Fan for recovery of fines
9. Belt conveyor
10. Airlift head (separator)
11. Cyclones
12. Airlift fan
13. Screening sieve (filtration)
14. Density indicator
15. Belt conveyor
16. Detergent powder silos
17. Belt conveyor
18. Continuous weighing belt for spray-dried powder
19. Continuous weighing belt for additives
20. Rotating drum mixer
21. Perfume addition unit
22. Belt conveyor
23. Packaging machine

appear to be working satisfactorily but if the orifice has been eroded away from the circular, it will atomize unevenly. For this reason jets should be stripped, cleaned and examined at regular intervals.

A narrow spread in particle size is desirable from the viewpoint of product appearance. Furthermore, if the spread is wide it may range from large particles to dust. The housewife, the ultimate consumer of the product, is particularly sensitive to dust. Finally, a wide spread will add to the percentage of fines produced by the process. In most instances fines can be considered nothing but a necessary evil to be recycled or disposed of—usually uneconomically.

However, where bulk density of a powder is too low, this may be remedied by a slight widening of the particle size range, since the small particles will tend to fill the voids between the large beads.

3. Bulk Density

Bulk density is the key factor in the economics involved in the production of a spray-dried washing powder. Containers are of standard size and shape. It is not practical to sell a half-empty carton if the powder is too heavy and if it is too light the carton will not accommodate the requisite weight. Furthermore, since use directions are always given by volume, insufficient active material will be used by the housewife if the powder is too light and the product will appear inefficient.

The bulk density is dependent on the 'hollowness' of the bead assuming all the beads to be of the same size.

The mass of a hollow sphere is expressed by the formula

$$M = \frac{4}{3} (r_1^3 - r_2^3) D\pi$$

where M = mass
r_1 = external radius of sphere
r_2 = internal radius of sphere
D = absolute density of the material of the bead.

The weight of each sphere is therefore proportional to the difference between the cubes of the external and internal radii of the sphere.

The bulk density of each bead is given by the formula

$$BD = \frac{(r_1^3 - r_2^3) D}{r_1^3}$$

and therefore the closer r_2 approaches to r_1 the lower the bulk density. In other words, the thinner the wall the smaller the bulk density.

These calculations do not yield the bulk density of the final product which is a function of absolute bulk density and particle size spread and an inverse function of the mean diameter of the beads. However, the thicker the walls of the beads the greater will be the bulk density. This would suggest that lower feed concentration will give thinner walls and lower bulk density, and conversely, higher concentration will result in higher bulk density. This relation does hold good for solutions. For slurries, however, any change in slurry concentration affects the amount of material in solution and therefore the density and viscosity of the slurry which in turn play an important part in atomization.

Rather contradictory reports on the effects of changes in slurry concentration have appeared in the literature. In actual practice when spraying solutions a drop in bulk density will result from a decrease in solids content of the slurry. Conversely, in spraying slurries containing powders in suspension, a cut in solids content will result in finer atomization, smaller particles, and increased bulk density (within limits). This phenomenon is due to reduced slurry viscosity and density. Consider a slurry to be a uniform mixture of a solution and finely divided powders. When this slurry is atomized each single droplet will consist of a drop of solution with particles of a powder dispersed throughout the volume of the sphere. The solution will be dried into a hollow sphere and the powder which was present originally will be fused into the wall of the sphere. The powder, having not entered into solution, will in no way change its physical shape, whereas the liquid will give a very thin-walled sphere. The ratio of undissolved powder to solids in true solution will materially affect the bulk density of the finished product. By altering slurry concentration the percentage of undissolved matter is changed and thereby the amount of solid (undried) matter fused into the spheres is also altered. A drastic cut in this undissolved solids content will cause the slurry to approach a true solution. In this case lowering the solids content of the feed will lower the bulk density of the dried material.

Feed temperature has a similar though converse effect on slurry characteristics. Elevated temperatures bring a drop in viscosity and density and an increase in the amount of material in true solution (within limits). Thus an increase in feed temperature gives lower bulk densities. However, increasing the feed temperature causes a drop in Δt, which in turn will cause an increase in bulk density. (Δt is the difference in temperature of the atomized liquid and the ambient temperature. If this difference is large it causes 'puffing' of the bead.)

Actual hollowness of the beads depends on the rate of evaporation of water from the droplets. If the exterior surface of the droplet is dried rapidly to a semi-solid state, it will form a membrane which prevents further escape of moisture. The water retained inside the globule will inflate this sphere into a balloon which will eventually rupture. The size of the balloon depends on the rate of evaporation of the water inside the sphere.

The rate of heat transfer is proportional to the temperature differential between the atomized liquid and the surroundings (Δt). Narrowing of this differential slows and reduces evaporation and thereby reduces puffing of each individual particle. Bulk density is therefore increased. To accelerate the rate of evaporation, the rate of heat transfer must be stepped up by increasing Δt. This in turn will cause a drop in bulk density.

Adjustment of the inlet air temperature will give any desired changes in Δt in a concurrent tower. In a countercurrent installation, adjustment of the inlet air temperature will also change Δt but less so.

A temperature reading even more pertinent to Δt can be taken at the jet. It is general practice to observe the inlet air temperature in the inlet duct immediately prior to its entry into the tower, and the outlet air temperature in the outlet duct at some point between the tower and the exhaust fan. A very important further point of reference is in a plane cutting through the atomization tower about 60 cm (2 ft) below the plane of the jet. Temperature readings at this point will give an accurate idea of any changes in Δt, and are more sensitive than the inlet air temperature because some heat from the inlet air is dissipated in warming up the dried powder and the walls of the chamber.

A change in Δt does not necessarily mean a change in fuel supply to the furnace.

The temperature of the inlet air can be modified by altering the volume of fuel fed to the furnace. In this case the amount of heat from the furnace is changed, which changes the drying conditions. As an alternative, the volume of air entering the tower may be modified while the amount of fuel remains constant. This does not change the quantity of heat but has a material effect on Δt.

In an installation permitting infinite variations in the rate of feed (in a spinning disc atomizer or by the use of a return flow jet) changes in the feed rate can be used to modify Δt. For instance, by stepping up the rate of feed Δt is increased. Here the total evaporation of water is greater but the amount evaporated from each individual particle is smaller. Product moisture content and therefore density are increased.

By virtue of the design Δt must be smaller in a countercurrent spray-drier than in a concurrent installation. Therefore, as a general rule countercurrent spray-drying will yield heavier powders than concurrent operations.

The equation

$$x = \frac{dk\, e^{0.705v}}{2P^{0.375} \sin \tfrac{1}{2}\sigma}$$

where x = mass median drop size
  d = orifice diameter
  k = constant
  v = kinetic viscosity
  P = nozzle pressure
  σ = spray cone angle

shows the drop diameter to be an inverse function of pressure. The greater the pressure on the nozzle (or the faster the revolutions of a disc) the smaller the drop size. The same amount of material dispersed into more and smaller drops means a larger surface area giving more evaporation and lower residual moisture. Increased evaporation means a lower Δt. This in turn results in less and slower evaporation per droplet and therefore less puffing and greater density. Increasing the pressure on a jet or the revolutions of a disc will increase bulk density.

As will be noted from the above equation, the viscosity of the slurry plays a very important part in the atomization. The viscosity of the slurry is dependent on the actual formulation (ie, the types of ingredients), the pH, the degree of hydration of the phosphate, the concentration of solid matter, the amount of material in true solution and in suspension and the temperature. It is suggested that continuous viscosity metering be done with a dynamic viscometer to control any change in slurry structure. It may even be arranged with an automatic alarm if the viscosity changes radically from pre-set conditions.

The amount of residual water retained in the powder is another factor affecting density which is discussed below.

The actual formulation of a product plays an important part in bulk density. Generally, the higher the percentage of active ingredients the lower the bulk density, and the higher the alkaline builder content the higher the bulk density. Sodium sulphate rather surprisingly tends to reduce density. Bulk density is increased by sodium chloride.

4. Residual Moisture

For economic reasons it may be necessary to retain a certain amount of residual water in the powder. As the powder must of necessity be free-flowing, this water can only be present as water of crystallization and not as surface moisture. Account should also be taken of the fact that certain inorganic salts which have more than one physical form tend to assume the stable form. For example, washing soda crystals $Na_2CO_3.10H_2O$ will effloresce down to $Na_2CO_3.H_2O$, whereas anhydrous sodium carbonate will take up moisture from the atmosphere to its stable form $Na_2CO_3.H_2O$. Therefore if a formula containing soda ash is spray-dried to be bone dry, the soda ash will tend to absorb moisture up to 17 per cent of its weight. This phenomenon occurs with most inorganic salts used in detergents but it is difficult to give hard and fast rules because, in virtually every case, the problem may be complicated by double and co-crystals. Generally a spray-dried powder containing a fair proportion of inorganic salts can be expected to pick up moisture on storage. Table 6.1 lists some crystalline inorganic salts used as fillers or builders in detergent powders and the percentage of water absorbed by the anhydrous salts to form the stable molecule.

TABLE 6.1

Percentage of Water Absorbed by Anhydrous Salts to Form Crystals

	Per cent water absorbed by anhydrous salts	Temperature of instability of crystal
$Na_2B_3O_7.5H_2O$	44	—
$Na_2B_4O_7.10H_2O$	89	60
$Na_2CO_3.H_2O$	17	100
$Na_2CO_3.7H_2O$	105	32
$Na_2CO_3.10H_2O$	170	32
$Na_2CO_3NaHCO_3.2H_2O$	17	85
$Na_2HPO_4.7H_2O$	79	—
$Na_3PO_4.10H_2O$	110	—
$Na_3PO_4.12H_2O$	130	100
$Na_4P_2O_7.10H_2O$	67	93
$Na_5P_3O_{10}.6H_2O$	29	—
$Na_2SiO_3.5H_2O$	74	—
$Na_2SiO_3.9H_2O$	133	100
$Na_2SO_4.7H_2O$	89	—
$Na_2SO_4.10H_2O$	126	32·4

Storage conditions determine the rate at which water of crystallization is taken up. If the powder is packed immediately in 'airtight' containers, the rate of absorption will be comparatively slow. (No container commonly used for packaging powders is truly air-tight.) Storage in an atmosphere of fairly high relative humidity will result in fast moisture take-up and a tendency to agglomeration. However, the absorption will be slower the closer the moisture content of the powder is adjusted to that of the stable form. If a powder must be stored in ambient relative humidity exceeding 60 per cent, the risk of lump formation can be minimized by discharging the powder from the spray-tower with at least three-quarters of the stable moisture content. It is economically unsound to dry a powder to 1 per cent moisture content when the powder must revert to, say, 7 per cent. The operation would be a lot more efficient if the powder were dried to 7 per cent in the first place.

The stable moisture content can be determined only by prolonged storage tests. These call for small samples of the powder to be kept in desiccators designed to expose the samples to varying degrees of relative humidity.

Temperature of the outlet air and product residence time in the tower are the two factors used to control moisture content of the finished powder.

Contrary to general belief the outlet air temperature need not be above 100°C to obtain a desirable product. Spray-driers can work efficiently at temperatures well below 100°C. However, for a bone dry powder the temperature must be in the vicinity of 100°C or near the boiling point of water at the altitude of the particular plant.

To keep water of crystallization in the finished powder the outlet air temperature must be low and residence in the tower must be extended to permit crystallization (which has both a time and heat factor) to take place. Some spray-driers are designed to discharge the powder while slightly wet and then to transport it through a pneumatic conveyor or air lift to the packing department. This conveyor system serves a useful purpose by cooling the product and effecting crystallization (for example $Na_2SO_4.10H_2O$ exists only below 32·4°C; to form the crystal the wet powder must be cooled below this temperature). However, the extra cooling step is not essential. A dry free-flowing powder of 10–15 per cent moisture content can be obtained from the tower without this aid. The amount of water of crystallization retainable in a powder is determined by the type and quantity of the inorganic salts incorporated in the particular formulation.

Spray-drier manufacturers and operators sometimes do not pay sufficient attention to residence time in the tower. Disadvantages attendant upon excessive time in the chamber have been discussed above. But there are disadvantages also resulting from too short a time in the tower. It is not enough to expose the spray to a quantity of heat sufficient to dry it. The spray must be in contact with the hot air long enough to permit all necessary heat to be transferred. In a countercurrent installation operating at a low outlet temperature the temperature at the point of atomization will be only a few degrees higher than that of the slurry. Therefore Δt will be small and the rate of heat transfer slow. The residence time in the tower must then be increased. This is a very important point.

Air pressures in spray-drier towers are usually balanced to range between 0 and 5 Pa vacuum water gauge. This is accomplished by means of dampers on either the inlet air duct or exhaust duct, or both. The rate of exhaustion of air plus vapour is adjusted to be either equal to or slightly greater than the rate of ingress of hot air into the chamber. Positive pressure inside the tower is not advisable since it may cause blowing out of dust. The vacuum in the chamber can be raised, however, quite easily to 125 Pa water gauge provided that the tower is sealed off at its base with a rotary valve or other sealing-discharging device. The increased vacuum will retard the rate of fall of the dried particles and thus extend the residence in the tower. At the same time the higher vacuum will tend to add to the amount of fines produced, the actual increase being dependent on the particle spread of the powder.

Thus if the powder exits slightly wet, this may be remedied by increasing the vacuum in the tower (by opening the outlet air duct slightly). By the same token, if more residual water is required, a drop in outlet temperature (by increasing the rate of feed or decreasing inlet heat) and a higher vacuum in the tower will have the desired effect. Operating variables are tabulated in Table 6.2.

The following range of parameters shows typical tower operation conditions:

Slurry concentration	60–70 per cent
Hot air inlet temperature	300–450°C
Average air velocity	0·3–0·6 m/second
Exhaust air outlet temperature	90–95°C
Powder temperature at outlet	60–70°C
Powder bulk density at outlet	200–400 g/litre
Retained moisture in powder at outlet	2–15 per cent

5. Stickiness

Stickness is a phenomenon difficult to define but absolutely unacceptable in a spray-dried powder. It may be caused by poor drying or faulty atomization leaving large particles incompletely dried or by some ingredient or lack thereof in the formulation. Addition of sodium silicate will somewhat improve a slightly sticky powder. However, certain materials tend, *per se*, to yield a sticky product. In particular the linear alkyl benzene sulphonates yield stickier products than the branched chain materials. This stickiness is usually easily overcome by the addition of one of the hydrotropes mentioned on p 97, particularly sodium toluene sulphonate. Certain alkylates, for all that and depending on their isomerism, do, when sulphonated, yield powders which remain sticky even when toluene sulphonate is added. This can be overcome by the addition of small percentages of fumed silica or calcium silicate (p 78). Magnesium silicate, added either as such or, more commonly, formed *in situ* by the reaction of magnesium sulphate (added as a stabilizing agent for perborate) and sodium silicate, also adds to the free flowing properties of problematic powders.

TABLE 6.2

Effect of Variation of Operating Conditions on Powder Characteristics

	Particle spread		Particle size		Bulk density		Water content	
	Incr	Decr	Incr	Decr	Incr	Decr	Incr	Decr
Orifice diameter	+	—	+	—				
Pressure	—	+	—	+	+	+		
Slurry concentration (solids)	+	—	+	—	—	+*		
Active matter			+	—	—	+		
Alkalis				+	+	—		
Viscosity			+	—				
Rate of feed			+	—	+	—	+	—
Feed temperature					+	—†		
Δt					—	+†		
Outlet air temperature					—	+	—	+‡

+ sign indicates increase of the function in the left-hand column, and — sign the converse.
* Within limits.
† These two factors tend to neutralize each other.
‡ In conjunction with increased residence time in the chamber.

6. Product Uniformity

After careful adjustment of all processing variables one may attain all desired characteristics in the product, but this is not enough. For uniform product quality, optimum conditions must be established, maintained and repeated from run to run. Trial runs need to be made to establish the best processing conditions for each individual product and these must be recorded and repeated for each production run. Accurate measuring instruments located at all vital points in the spray-drying installation are absolutely essential. Readings should be made and recorded at frequent intervals or better still the instruments should be of the continuously recording type. A recording instrument will predict a trend which can only be derived from ordinary indicators by very frequent readings and recording of results.

7. Separation of Powder

Separation of the powder from the exhaust air is an integral part of the function performed by a spray-drying installation. Factors pertaining to the formation of fines in the process have been discussed earlier. A second and sometimes third separation step is performed by external cyclone separators and/or bag filters or by wet scrubbing. The manufacturer is faced with a problem of collecting the resulting fines and using them.

8. Wet Scrubbing

Even the best of cyclones or bag filters do not completely trap all the solid particles entrained in the exhaust air, particularly those of micron size. In addition due to intermittent faulty operation of the tower, the separation system can become overloaded.

The detergent manufacturer has a moral responsibility to ensure that he does not contaminate or pollute the air he discharges, this apart from any legislation which has been or will be enacted.

As dry separation systems are known to pass, at best, a few per cent of the entrained particles, recourse may be made to wet scrubbing, when the solid material is scrubbed out by water. Wet scrubbers can bring the concentration of solid particles in the outgoing air down to 15–25 ppm.

One such scrubber is shown in Fig 6.3. The water is circulated through the unit until the concentration reaches a predetermined amount, when the solution is pumped to the slurry preparation.

To avoid troublesome foam formation, particularly when the active content is high, a salt or preferably a silicate solution rather than water can be used. If the fuel used for hot air generation contains an appreciable amount of sulphur, caustic soda can also be injected to neutralize acidity.

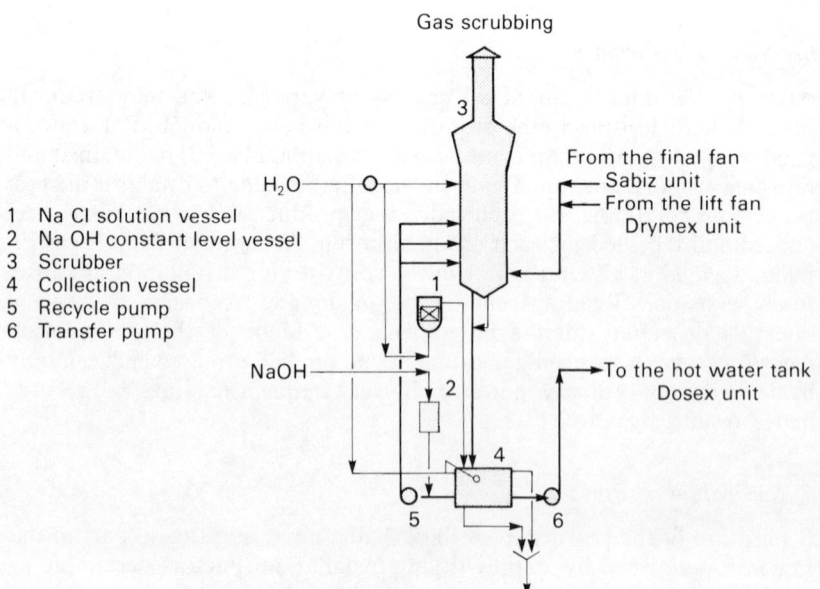

1 Na Cl solution vessel
2 Na OH constant level vessel
3 Scrubber
4 Collection vessel
5 Recycle pump
6 Transfer pump

Fig 6.3   Ballestra wet scrubber system

9. Use of Fines

In some plants detergent fines can be used in the manufacture of scouring powders or other related products. In that case the fines can be collected in bags at regular intervals from the cyclones or filter bags. If no other use can be made of the fines they must be reintroduced into the slurry. Where the slurry is made up continuously this reintroduction should be continuous also. In plants where the slurry is made up in batches, reintroduction of fines simply calls for transporting them from the collecting points to the slurry vessels.

Where a wet scrubber is used the reintroduction of fines is easiest. Water is pumped through the scrubber to form a solution of fines and this solution is recirculated continuously until it reaches a predetermined concentration when it is pumped to the slurry vessel where the primary water must be adjusted accordingly.

Another method (Ballestra system) used is to blow the fines continuously to the top of the tower and to reintroduce them concurrently with the atomized slurry (see Fig 6.2). As they come into contact with the wet atomized slurry they tend to agglomerate with the atomized particles and in this way all the problems of handling fines are disposed of.

There is one prerequisite for the automatic reintroduction of fines into either the slurry or the tower: there must be no soot whatsoever and careful attention must be paid to the accurate adjustment of the furnace. The slightest trace of carbon in the fines will discolour a large amount of powder. Freedom from soot is not easily attained since some of the factors involved are beyond the operator's control. A temporary stoppage of electric current, for instance, can cause soot. An outside use for fines is therefore usually preferable to their reintroduction, if freedom from soot cannot be guaranteed. In a scouring powder for example, a trace of off-colour powder will not affect the appearance or performance of the final product but it is axiomatic that white linen cannot be washed with a powder containing soot.

5. Combination of Spray-dried and Dry-mixed Powders

The introduction of linear alkyl benzene led to problems of stickiness which were rapidly overcome. Nowadays more and more powders have certain amounts of non-ionic detergents incorporated into them and this is again leading to problems of stickiness. The use of non-ionic detergents in spray-dried powders causes difficulties as soon as the percentage (depending on the type) of the non-ionic reaches a certain figure. The difficulties encountered are stickiness of the powder, clogging of cyclones, build-up on walls of the tower, 'pluming', etc (see however p 31). Thus the development of chemically advanced powders can be hindered by technical problems.

Some of the above factors can be overcome by mechanical means such as wet scrubbers and automatic cleaning rings; but for all that, some of the newer additives are heat sensitive and in using certain non-ionic detergents the solution from a wet scrubber will become black due to scorching of the material.

A method used to overcome the problem was to dose powdered non-ionics on to the spray-dried powder. These are available in various grades, often with 25 or more ethylene oxide units which lower the biodegradability, and alter the hydrophile-hydrophobe balance, thus reducing the efficiency of the non-ionic as a detergent. In addition, technical problems were encountered when the non-ionic powder was mixed with a spray-dried powder while the latter was still hot.

All of the above problems have been overcome by the use of a combination of dry-mixing and spray-drying techniques.[8] The principle is basically that the alkyl benzene sulphonic acid and the soap (fatty acid) portions of the active matter are spray-dried in the normal way with phosphates, CMC, etc. The non-ionic component, ethanolamide, and other difficult-to-spray constituents are adsorbed on to a portion of the inorganic filler, with or without the use of colloidal or fumed silica (p 78) in a mixer such as described on pp 229, 231.

Ballestra 'Combex' System

This system has been developed and described by Davidsohn[9] and consists of a special mixing/dispersing unit in which a powder containing ingredients that are difficult to spray-dry is produced. It has two advantages in that it can be a fully integrated system linked to a spray-drier or can be run separately to make 'heavy powders' and it may be adjusted to produce powders of a granular or fine structure.

This process consists of a dispersion system which distributes the active matter components on to the detergent builders. This very homogeneous dispersion of the surfactants on to the builders is accomplished in a mixer which has specially designed plough type horizontal mixing blades rotating at high speed, and passing very close to the mixer walls. This design guarantees very efficient blending of all the components. In addition, ultra high speed rotating desintegrators are inserted to prevent any lump formation.

In contrast to 'spray-mixing', 'fluid bed', and 'agglomeration' processes, fixation of the active matter components is accomplished by mechanical disintegration of the solid builder material on to which the active matter components are added by means of a rather simple dispersion system. In most spray-mixing systems special high absorption types of builders are used, whereas in a system based on mechanical dispersion and powerful disintegration a much wider variety of builders in normal powder form may be used. The mixing-disintegration unit is of a relatively small size in relation to its output. Loading, discharging and residence times are short. While most spray-mixing systems are designed to obtain powders approaching the structure of spray-dried beads with relatively low density, the system described here aims at a powder with relatively high density, but with a structure to make it adhere strongly to the surface of beads coming from a spray tower.

The mixing unit may be used either for the incorporation of high percentages of non-ionics or for the incorporation of sulphonic acid and/or distilled fatty acids, with or without non-ionics. In this case the LAS and fatty acids are neutralized immediately on contact with the builders. Fatty acids are transformed in a single step operation into soap without leaving any unreacted fatty acids. Certain processing details are to be observed to guarantee the complete neutralization of LAS and the complete saponification of distilled fatty acids.

This 'dry neutralization and saponification' is accomplished by using a small percentage of an alkaline aqueous phase (representing only a small fraction of the stoichiometric amount of NaOH) as a kind of accelerator or 'trigger' for the reaction. The aqueous phase, generally only 2–4 per cent on the total mix, is added at the same time, but not pre-mixed with the surfactant components. To give a general indication of the efficiency of the mixing system: only about 3–4 minutes are required for even distribution of non-reacting non-ionic components on to the builder mix, and about 4–6 minutes for 'reacting' components, ie, AB sulphonic acid and/or distilled fatty acids. Automatic charge and discharge takes approximately an additional minute or two, depending on the size of the mixing system. The process can be run completely automatically with weighing and dosing similar to continuous slurry preparation.

Depending on the bulk density of the builders used, the amount of the surfactants added, the bulk density of this powder from the first phase of the process is in the range of 0·6–0·9. The product from this process may be used as such for detergent powders, eg, washing powders for commercial laundries, metal cleaning detergents, etc. The powder is less dusty than otherwise produced 'heavy powders'. The process may be carried out in such a manner that powders, either with a granular or a finer structure, are produced. The process is flexible enough to produce powders in a wide range of particle size.

The dry-mixed powder can also be fed continuously on to the powder being discharged from the spray drier while it is still hot and the two streams of powder are passed through a slow-acting baffle mixer, a slowly rotating cylinder mixer or by weighing belts. From this mixer the powder is taken to storage or packing.

The entire system is completely co-ordinated and features automatic controls. Being a closed system, there are no problems with dust. Proportions of builders and liquid components, as well as those of the final mixture, are set on the central control panel and may be changed quickly; the density of the final mixtures of 'heavy' powder and beads from the spray-drier is determined more by the low density of the beads than by the high density of 'heavy' powder. Table 6.3 gives some examples of the density of various mixtures of 'heavy' powder with spray-dried beads having a density of 0·3.

TABLE 6.3

Density of Mixed Powders Made by the Combex System

Density of 'heavy' powder	Density of mixture (Proportion of heavy to light powder with a density of 0·3)		
	50:50	40:60	30:70
0·9	0.52	0·43	0·38
0·8	0·45	0·41	0·37
0·7	0·42	0·40	0·36
0·6	0·40	0·39	0·35

The approximate expected density of the resultant mixtures may be calculated as follows:

$$\text{density of mixture} = \frac{100}{\frac{a}{d'} + \frac{b}{d''}}$$

where a = percentage of heavy powder
b = percentage by weight of beads
d' = density of heavy powder
d'' = density of beads

This does, of course, give only an approximation of the expected density, but may serve as an indication.

Using this method the detergent chemist has a great amount of latitude in formulation and the output of the spray-tower is enhanced.

In certain countries of the world there is a swing away from light density spray-dried powders to medium density powders. Two of the factors influencing this swing are high packaging costs and disposal problems with packages. This combination system can be arranged to produce powders of intermediate bulk density, regardless of whether non-ionic detergents are used, by producing a powder as described on pp 202, 228 and mixing it with a spray-dried powder in a proportion to give the desired formulation and density. This again will vastly increase the production of powder without the necessity of increasing the capacity of the spray-drying plant.

A further development is offered by Ballestra Lugano (Switzerland). This company has introduced the 'Turbo Drymex' plant. The alkaline powders are fluidized in a cooled air stream and the liquid components are sprayed on to the fluidized powders by special nozzles. Intimate mixing allows of the production of *expanded granules* by the explosive release of CO_2 from the reaction.

These expanded granules are agglomerated, cooled and aged in a second stage by the use of another turbo-reactor, with or without the addition of further liquid ingredients, from additional nozzles.

If no neutralization occurs (when only non-ionics are used), agglomeration and granulation can still be achieved by spraying through four different nozzles, using agglomerating agents.

Perfume is added to the cooled powder by a further nozzle.

Patterson–Kelley Systems

The Patterson–Kelley Company of the USA offers its V-shaped blender (Fig 6.4) for both powder mixing and liquid/solids blending and agglomeration. As can be seen from the drawing, this method achieves a precision blend through a divergent flow and by the intermeshing action when two inclined cylinders combine their flow. Material is rotated close to the axis, reducing power requirement. As the blender rotates, liquid is sprayed into the material through a high-speed liquid dispersion bar located concentric to the trunnion axis. The blades on the bar aerate the material to increase the speed and thoroughness of the blend. Liquid is dispersed through disc apertures in a controlled pattern and extended to all solids throughout the blend.

The blender is used for manufacture of detergents by combined neutralization/absorption processes, for the hydration of sodium tripolyphosphate, and if colloidal sodium silicate solution is added, agglomeration can be achieved.

An improvement of this system is their Cross-Flow Blender, where the two legs of the V are of unequal length, giving unequal displacement in the powder mass in each revolution of the shell, producing an axial exchange from each leg, thus faster blending.

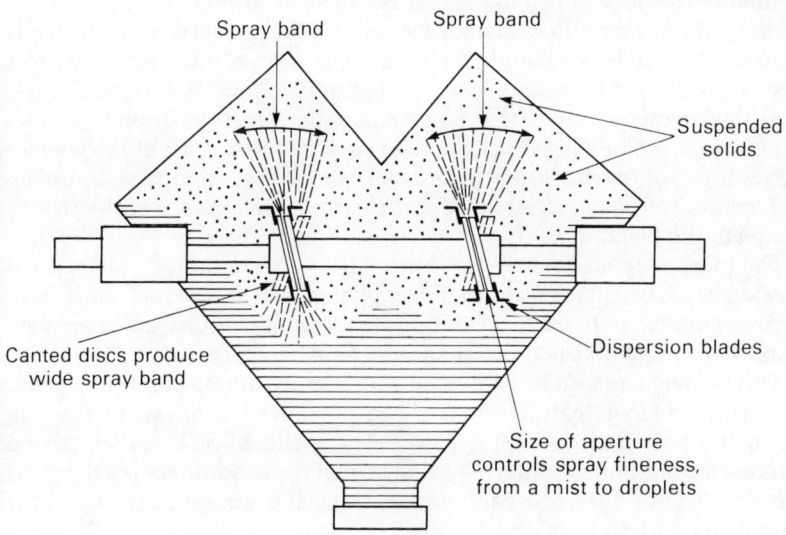

Fig 6.4   Patterson–Kelley V-shaped blender

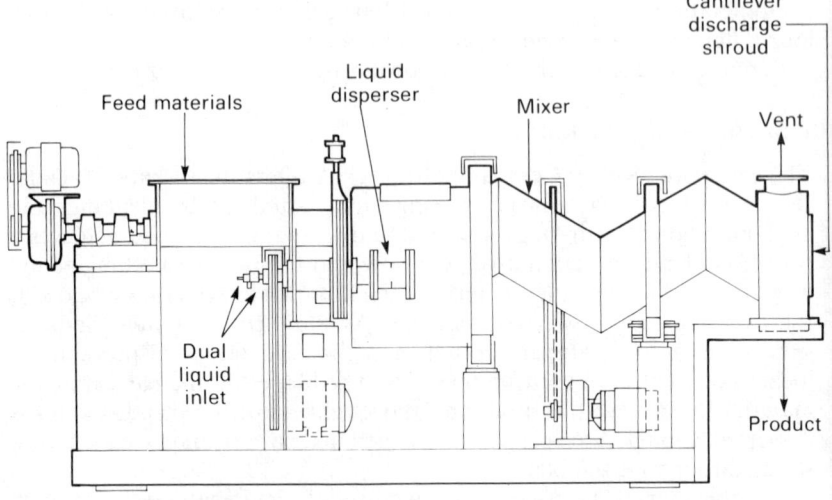

Fig 6.5 Patterson–Kelley zig-zag continuous blender

The above two blenders are of the batch type, the solid material needs to be added while the blender is stationary and the machine needs to be stopped for discharge. There is no cooling so if perborate, enzymes and perfume are to be added, this can only be done after ageing and cooling, when the powder needs to be reintroduced into the same or another blender.

A development of this principle of uneven flow is the P–K zig-zag continuous blender shown in Fig 6.5. The blender combines the action of a rotating, eccentric drum with multiple recycling to produce uniform blends of both solid/solids and liquid/solids materials. The zig-zag completes most blends within two minutes. Two liquids can be added simultaneously.

As the blender revolves, the blended material leaves the drum and flows into the legs, where recycling begins. At each half turn, part of the powder moves forward, part of it backwards within the legs. Splitting, tumbling and merging of the material bring particles into contact with each other to complete the blending.

The blender typically recycles thirty parts of material for every part it discharges. Recycling serves as an averaging device to level off short-term feed variations. With each revolution, the machine discharges a uniform quantity of material of constant volume and weight.

This blender can also be combined with a spray-drying plant to increase capacity.[10] STP is hydrated in a zig-zag blender (as shown in Fig 6.6) mounted in series with a spray tower. The hydrated STP is then passed through a conditioner/cooler. The first part of the conditioner is a fluidized bed which both ages and cools the hydrate, the second part brings the power down to 30°C.

On passing out of the conditioner the material is split into two parts, the

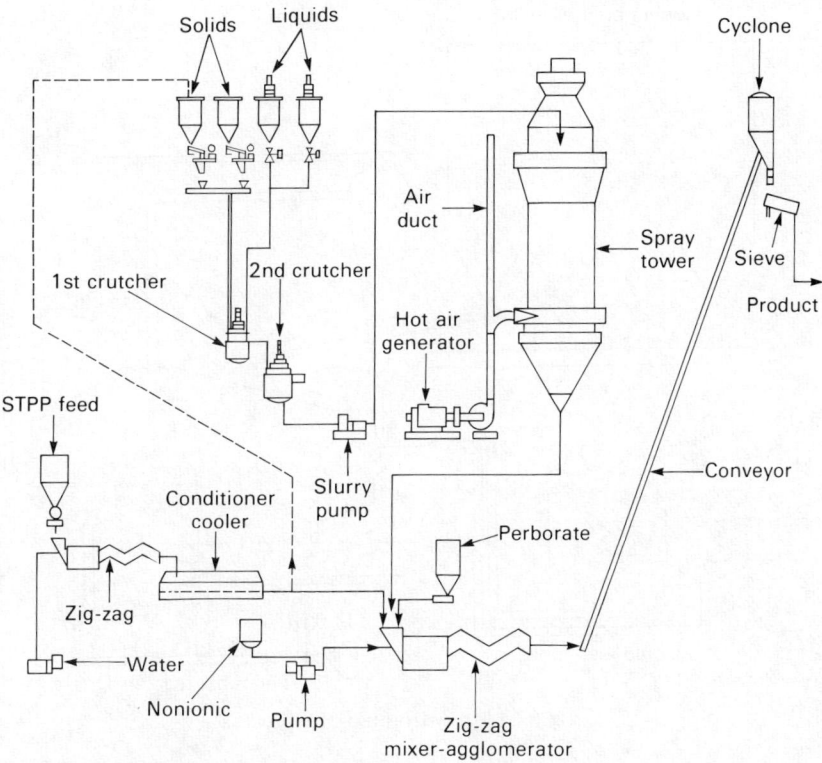

Fig 6.6 Patterson–Kelley system for increasing spray-drier capacity

smaller part is fed to the crutchers to become part of the spray-dried matrix, while the larger part moves on to the second stage.

The second stage is a second zig-zag blender which is used to combine the spray-dried matrix, the STP hexahydrate and the non-ionic. Perborate, perfume and enzymes can be added at this stage. The non-ionic is dispersed through a liquid dispersion bar, designed for gentle but thorough mixing so as not to destroy the bead structure of the matrix. The non-ionic acts as a binder to form a soft agglomerate and to prevent segregation. Bulk densities of 410–430 g/litre are obtained.

Anhydro System

The Anhydro fluid-mix system, offered by APV Anhydro, Denmark is illustrated in Fig 6.7.

Mixing, neutralization and agglomeration take place in a fluid bed reactor. Powders are fed continuously from the day silos one of which is fed by a pre-mixer to dilute small quantity powders with one of the other powdered ingredients so that they can be measured efficiently in the volemetric measuring device.

MANUFACTURE OF FINISHED DETERGENTS

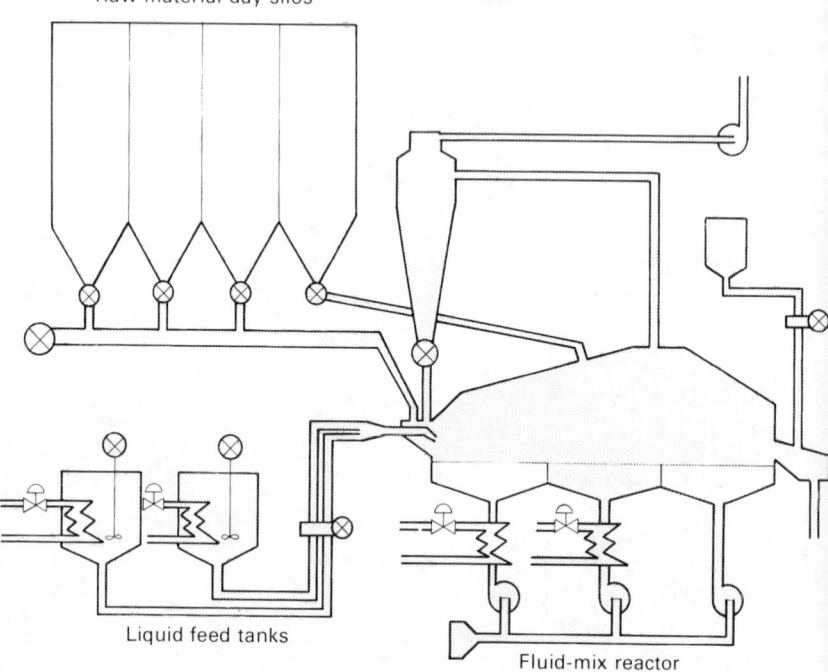

Fig 6.7 Anhydro fluid-mix process

Materials discharged from the silos, after being measured by the volumetric measuring devices at the base of the silos, are transported to the fluid bed by belt conveyors.

Liquids are fed into the reactor by metering pump(s) and atomized by compressed air directly on to the bed.

The fluid bed reactor consists of a mixing chamber with a fluidized area, fines separating area and a perforated plate for air distribution. The air supply is divided into a number of compartments each with a guiding vane to allow differential fluidization to take place.

The reactor bed outlet is connected to a series of cyclones to separate fines from the exhaust air. The fines separated thus are continuously recycled and agglomerated in the reactor bed.

Heat sensitive materials, perborate, enzymes, etc, are added in an area adjacent to the fluidized bed, where the powder has cooled sufficiently.

As no mechanical mixing is used, the fluid-mix reactor can be easily combined with a spray-drying unit, the discharge from the spray-drier being connected to the entrance to the fluidized bed, thus the beads from the tower are combined in the agglomerization process.

Bulk densities, depending on the raw materials used, the liquid/solids ratio and whether spray-dried powder is added can vary from 0·35 to 1·00 g/ml.[11]

All the above plants therefore can be operated on their own or can be used to complement the spray-drier to double its capacity. With the growing use of non-ionics in powders, one of these plants is the ideal solution, by itself or in combination with a spray-drier, to produce powders suitable for the present-day requirements.

6. Drum-drying of Powders

Occasionally detergents are manufactured by drum-drying. Historically the first dry alkyl benzene sulphonate, in a concentrated form, appeared on the market as drum-dried flakes containing 40 per cent active matter and filled with sodium sulphate.

This process of manufacture does not commend itself, as the flakes produced are not very attractive and have a relatively small surface area. Compared to spray-dried powders, and even to well-made powders from the simultaneous absorption and neutralization process, drum-dried powders are slow to dissolve in water.

The operation of a drum-drier is expensive (compared with a spray-drier) and there is always the risk of scorching the finished product. Furthermore, unless the drums are made of stainless steel, the detergent slurry will corrode them quickly.

Drum-driers have one advantage over spray-driers in that they can be started and stopped in a relatively short time, allowing for the possibility of short runs.

For the above reasons drum-driers are not normally used for the manufacture of built powders, but when a highly concentrated powder of alkyl benzene sulphonate or fatty alcohol sulphate needs to be made in relatively short runs, drum-drying has a definite economic advantage over spray-drying.

Liquid Detergents[12,13]

The very first detergents sold in large quantities for household use were merely simple solutions of anionic detergent in water of active concentrations varying from 5 to 20 per cent. They rapidly became standard household requisites for dishwashing and for the washing of fine articles of clothing such as wool and silk. Of recent years, these light-duty detergents have become more sophisticated and liquids (or lotions) have now appeared with builders for heavy-duty washing, both by hand, and more particularly, in household washing machines.

At first glance, it might be considered that liquids are cheaper in raw material costs than powders, as the filler (or diluent) is water rather than soda ash or sodium sulphate. This might hold true for light-duty liquids, but where the heavy-duty product is to be considered, it is by no means axiomatic. For heavy-duty liquids hydrotropes, potassium rather than sodium salts and ethanolamines need to be used. These in no way add to detergency but increase the price somewhat. However, to offset this plant for liquid production is in no way comparable in cost with powder plants.

The advantages to the manufacturer of liquids are that they can be made in relatively simple plant and also that the diluent is water which costs virtually nothing, compared with sodium sulphate or soda ash, which, although cheap, are obviously far more expensive to use as diluents.

To the user, the advantages are that liquids are instantaneously dispersed in water, the material can be perfumed and be given a very attractive appearance and the container—a glass or polythene bottle, for example—can be designed in such a way as to catch the eye.

Liquid detergents can be made from a variety of starting materials, but in every case the plant is the same. A vessel equipped with a slow-speed stirrer is all that is required and the stirrer should be positioned so that it is well under the surface of the liquid, so as not to cause foaming. The authors know of a manufacturer who processes 4·5 m^3 at a time of liquid detergent, stirring by hand with a wooden paddle, in a matter of hours. It is, however, necessary that the vessel be of a non-corrodible material. Stainless steel is satisfactory, but expensive; concrete or asbestos-cement vessels are eminently suitable, and so are those of glass-fibre reinforced plastics.

The choice of active ingredient depends on the requirements of the finished liquid, but here particularly, the versatility of alkyl benzene sulphonic acid made by sulphonating with SO_3 is of particular value. Other anionic raw materials are available as the already neutralized salt, with or without the presence, whether incidental or not, of various non-detergent materials. Alkyl benzene sulphonic acid based on oleum sulphonation, when neutralized, will also contain a fairly large proportion of inorganic sulphate ions. Alkyl benzene sulphonic acid sulphonated with SO_3 and sold as the 100 per cent material, contains a minimum of free sulphuric acid and thus gives the manufacturer the greatest scope in formulation.

This 100 per cent material can be neutralized by any base to give various types of liquids, from low viscosity to gels. The solubility of the alkyl benzene sulphonic acid salt of the alkali metals is opposite to that of the regular order, in that the most soluble of the salts is that of lithium, followed by sodium, with potassium the least soluble. When manufacturing heavy-duty liquid detergents, this must be taken into account if potassium silicate and pyrophosphate are to be used, since the potassium ion of the inorganic filler can precipitate the sulphonic acid ion unless the formulation is carefully controlled.

Table 6.4 gives the physical properties of 10 per cent solutions of dodecyl benzene sulphonic acid, neutralized by various bases. In every case 100 g of the sulphonic acid was neutralized to pH 7 by the particular base and the final solution made up to 1 litre.

The figures in Table 6.4 will vary with the type of ABS used, as small amounts of free sulphuric acid in the original material will greatly influence the behaviour of the neutralized salt. Furthermore, these figures

TABLE 6.4

Physical Properties of ABS Solutions Neutralized by Various Bases

Base	Appearance of sol at 20°C	Viscosity at 25°C (Pa s)	Surface tension of 0·1% active material N/m	Foam height of 0·1% active material solution (Ross Miles)	
				Initial cm	After 5 min cm
Ammonia	Separates into 2 phases	—	0·0310	17·5	17
NaOH	Clear liq	0.0045	0.0300	18	17
KOH	Gel	0.0285	0.0310	16	15
LiOH	Clear liq	0·0024	0·0310	17	16
Monoethanolamine	Clear liq	0·0285	0·0310	16·5	15
Diethanolamine	Clear liq	0·0870	0·0300	17·5	16·5
Triethanolamine	Clear liq	0·0635	0·0310	18	17
Morpholine	Gel	0·4350	0·0300	12	11·5
Monoisopropanolamine	Gel	0·3100	0·0310	15	15
Di-isopropanolamine	Clear liq	0·3450	0·0300	15	14
Tri-isopropanolamine	Clear liq	0·0910	0·0310	15	14·5
Mixed isopropanolamine	Clear liq	0·3750	0·0305	16	15
Diglycolamine	Clear liq	0·1180	0·0310	16	15
Diethylethanolamine	Clear liq	0·3050	0·0320	15	14
Triethylamine	Liquid*	0·1350	0·0320	14	13
Trimethylamine	Liquid*	0·3100	0·0360	13	12
Cyclohexylamine	Liquid*	0·0240	0·0310	6·5	6
Iso-butylamine	Gel*	0·7000	0·0340†	5†	5†
n-Butylamine	Cloudy liq*	0·0078	0·0310	5	4
sec-Butylamine	Gel*	0·4450	0·0305	5	4
ter-Butylamine	Gel*	0·2250	0·0310	5	4
Diethylamine	Gel*	0·5250	0·0310	12	11
Dimethylamine	Gel* separates	0·1250	0·0310	12·5	12

* With characteristic smell.   † Not completely soluble.

were given for propylene tetramer DDBS. Linear DDBS in general has a lower viscosity, both for the alkali and ethanolamine and ethylamine salts, but these figures give a relative idea of the differences according to the base used.

If partial neutralization of the sulphonic acid is to be done with caustic soda (or potash) as is possible when using the DDBS produced from SO_3, a slight amount of ferrous iron present in the caustic soda can cause unpleasant complications in the final product. (The same holds good for sulphonic acid produced by oleum or sulphuric acid sulphonation and where iron pick-up is possible.) Traces of the ferrous salt of the sulphonic acid will in time precipitate as a black sludge. This can be eliminated by oxidizing the ferrous to ferric ions, and is easily accomplished by the use of sodium hypochlorite thus:

All the water necessary for the formulation, together with the caustic soda solution, is run into the vessel. Then 1 per cent of a 10 per cent solution of sodium hypochlorite is added and the solution mixed. The sulphonic acid is now added and mixing continued until all the sulphonic acid has been dissolved, and the solution allowed to stand for at least 20 min longer to permit the oxidation process to go to completion (see p 204).

The ethanolamine or other organic basic which is to be used for the completion of the neutralization is now added, together with all the other additives.

The sodium hypochlorite will have been partially destroyed in the initial oxidation process; then, when the sulphonic acid is added, the excess hypochlorite will completely bleach all the colouring matter if any is present in the sulphonic acid. After this process a small residue of active chlorine will still be present, but this need not be eliminated as it will react with the organic amines to produce a chloramine, which is unstable. The amines will thus reduce the balance of the active chlorine left and render it innocuous. (It is for this reason that sodium hypochlorite cannot be used to bleach amine-containing solutions.)

If for any reason it is not possible to use sodium hypochlorite for this oxidation process, the same effect can be achieved by the use of hydrogen peroxide, either as such, or in the form of sodium perborate. It is general practice to add to detergents based on alkyl benzene sulphonic acid one of the alkylolamides as a foam stabilizer, thickening agent and skin-protecting agent. No special procedures are involved. Opacifying agents, especially for solutions based on linear alkyl benzene sulphonates are often used to give the liquid a 'different' appearance. Aqueous dispersions of an alkali insoluble polymer of styrene, substituted styrene, a co-polymer of styrene with acrylamine, PVC or polyvinylidene chloride are used. (Such opacifying agents are offered, eg, by Hüls, Monsanto and GAF Corp.) All these above materials produce an opacity in the liquid which enhances the appearance. Stability and freeze-thaw studies should, however, be made on all trial formulations.

Liquid formulations can be light sensitive. This can however be overcome either by the addition of ultra-violet absorbers or by packing in opaque bottles.

Liquid formulations allow the incorporation of solvents (solvents are however being incorporated successfully in powders). These solvents need not necessarily be of the water-immiscible type, glycol ethers being examples of excellent fat solvents which are completely water-miscible.

LIQUID DETERGENTS

Light-duty liquid detergents can be based on alkyl benzene sulphonic acid, alcohol sulphates, non-ionics as such, or sulphated non-ionics, or amphoteric detergents, or combinations of two or more of these materials.

Except when sulphonic acids are used, all the other components are available as concentrated (from 40–100 per cent) neutralized material which needs only to be diluted down to the required activity. With sulphonic acid, however, the manufacturer has the choice of varying his neutralizing agent to give his liquid any property he wants, from a low viscosity to a thixotropic gel (refer to Table 6.4).

If non-ionics of the ethylene oxide condensation type are being used in liquid detergents, either alone or in combination, it should be borne in mind that this type of material has a peculiar cloud point, in that it forms a cloudy solution when the temperature is raised and not (as is usually the case) when the temperature is lowered. It is thus important that products should be tested extensively, taking into account the highest possible temperature that the liquid may encounter on storage.

Heavy-duty liquids are coming more and more into use and require special formulation techniques. The problem here is to incorporate into the liquid active matter, sequestering agents, silicates, anti-redeposition and optical brightening agents. To achieve all this and to retain the finished material as a clear liquid with a low cloud point is not easy. If at a moderately high temperature the liquid 'clouds', it is an indication that separation into at least two phases is taking place and each phase will contain different proportions of each of the ingredients.

This problem is best overcome by producing the heavy-duty liquid as a lotion or gel, but this must be formulated so that it will not separate into two phases at the highest ambient temperature to which it may be subjected.

TOILET PREPARATIONS

As hair shampoos, shaving creams and bath preparations are all basically detergents, more and more detergent manufacturers are manufacturing either ready-made cosmetic preparations of this type or a concentrate for supply to the cosmetic manufacturer. The alcohol and ether sulphates have proved themselves over the years to be dermatologically safe for this type of preparation.

For hair shampoos the choice depends on the desired physical form of the product. For a cream the sodium salt of one of the fatty alcohol sulphates is recommended. For a liquid an ether sulphate is used or else an ethanolamine salt of a fatty alcohol sulphate. The latter has the disadvantage in that it cannot normally be obtained with as light a colour as the ether sulphate and it also foams less, but this can be enhanced by the use of a foam booster. The pearlescent type of shampoo can use as a base either an ether sulphate or sodium salt of the alcohol sulphate.

In addition to the well-tried sulphated alcohols and ethers, more and more use is being made of the amphoteric detergents (Miranol, p 43) as either a portion or all of the active ingredient in shampoos, etc. If they are used as the sole ingredient an additional advantage is that cationic detergents can be incorporated as bactericides.

Perfumes used in shampoos should all be dermatologically tested and in addition preservatives need to be added, as fungi and yeasts grow in detergent solutions. All formulations should be made up to include perfume dyes and preservatives and subjected to extensive storage tests as no hard and fast rule can be laid down for the interaction of the micro constituents. Furthermore, it is best to add a small portion of a chelating agent (0·25 per cent or so) as traces of iron may have an adverse effect on dyes in storage.

Dyes themselves or the constituents of certain dyes have come to be considered health hazards and are now subject to certification for use in toiletries or domestic preparations.

Some basic formulae for shampoos are given below:

Formula 4

Cream Shampoo

Sodium fatty alcohol (C_{12-14}) sulphate (50% paste)	25
Coconut monoethanolamide	2
Thickening agent	1
Sodium chloride	2
Chelating agent	0·25
Perfume, dye, preservative	qs
Water	to 100

The thickening agent (see p 84) is predissolved or dispersed in a portion of the water in accordance with the manufacturer's instructions. The sodium chloride is added to increase the viscosity if desired.

Formula 5

Liquid Shampoo

Sodium ether sulphate (C_{12-14}, 2 mol ethylene oxide) (100% basis)	12·5
Thickening agent	qs
Sodium chloride	2
Perfume, dye, preservative	qs
Chelating agent	0·25
Water	to 100

The thickening agent is added as required because hard and fast rules cannot be given as very often the concentrated ether sulphate contains ethyl alcohol or propylene glycol both of which drastically lower the viscosity.

Formula 6

Liquid Shampoo

Triethanolamine fatty alcohol sulphate (40% concentrate)	35
Coconut diethanolamide	3
Chelating agent	0·25
Perfume, dye, preservative	qs
Water	to 100

In place of the triethanolamine salt, the monoethanolamine neutralized fatty alcohol sulphate can be used. This will give a higher viscosity.

A pearlescent effect can be obtained by adding to the paste ethylene glycol stearate, diethylene glycol stearate, polyethylene glycol stearate, magnesium stearate or myristic diethanolamide.

Formula 7

Pearlescent Shampoo

Sodium fatty alcohol sulphate (50% concentrate)	25
Palmitic monoethanolamide	2
Ethylene glycol stearate	1
Chelating agent	0·25
Perfume, dye, preservatives	qs
Water	to 100

The alcohol sulphate, the glycol stearate, the monoethanolamide and the water are charged into the vessel and warmed with gentle stirring until a clear solution free from dispersed particles is obtained. This is then cooled to 50°C and the other ingredients added and passed to packing. When cold the pearlescence will have developed.

Formula 8

Bubble Bath Preparation

Sodium ether sulphate (100% basis)	20
Triethanolamine fatty alcohol sulphate (100% basis)	8
Chelating agent	0·25
Perfume, dye, preservative	qs
Water	to 100

Paste Detergents

In certain parts of the world detergents are used in a finished or semi-finished form as pastes of varying concentrations. Certain detergent raw materials are sold as pastes: for example, the sodium salt of lauryl alcohol sulphate in a concentrated form without any addition of inorganic fillers.

The detergent manufacturer who wishes to manufacture a paste can do so very easily from alkyl benzene sulphonic acid, again either from the 100 per cent or 90 per cent varieties, depending on which is available to him. To produce a snow-white paste from 90 per cent sulphonic acid can only be achieved if the acid is neutralized immediately after sulphonation, and not stored. The 100 per cent sulphonic acid can produce a snow-white paste even though stored for many months.

The manufacture of the paste can be done in a soap crutcher, or in a specially built neutralizer, equipped with a cooling jacket as shown in Fig 6.8. The neutralizing vessel whould preferably be built of stainless steel, but if precautions are taken so that the pH of the paste does not fall below 7 during the process of manufacture, a mild steel vessel can be used.

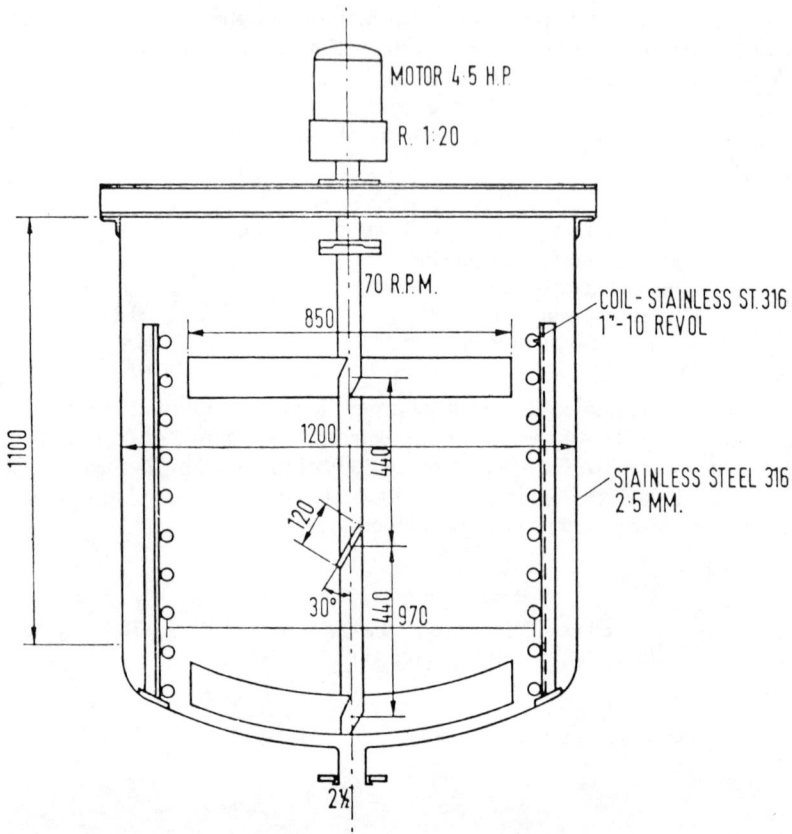

Fig 6.8 Mixer for liquid and paste-form detergents
Reproduced by permission of Zohar Detergent factory, Israel

Because sodium alkyl benzene sulphonates are not completely homogeneous in concentrated solution, especially in the presence of electrolytes, various materials must be used to prevent separation into two phases. This can be done by the use of hydrotropes, small proportions of CMC or alkylolamides. However, hydrotropes will tend to produce thinner pastes and this will mean adding inorganic salts to make them stiffer again, depending, of course, on the requirements of the market. In general, tridecyl benzene sulphonate will produce the stiffest paste; dodecyl benzene sulphonate a paste of medium consistency; and the biodegradable linear alkyl benzene sulphonates, the thinnest pastes. Consistency can again be varied by the addition of inorganic salts. In some countries high concentrations of borax are added to produce a stiff paste.

These inorganic salts can be sodium chloride, sodium sulphate, tetra sodium pyrophosphate or sodium silicate solution, or a combination of these. To achieve a stiff paste, it is very rarely necessary to add more than 3 per cent in total. Sodium chloride is the cheapest and gives the greatest stiffening effect, but is corrosive to iron or steel. Sodium sulphate and tetra sodium pyrophosphate suffer from the disadvantage that in cold weather they might crystallize out from the paste into needle-like crystals, while both the tetra sodium pyrophosphate and the sodium silicate raise the pH of the paste, but add to the detergency. The high pH can be regulated by using some of the sulphonic acid to neutralize a portion of the alkalinity of these materials. In particular, to improve the stability of a paste based on SO_3 sulphonated LAS the 1:3·3 ratio sodium silicate is used for partial neutralization (see also p 204). Colloidal silica is formed in the neutralization and it serves as a kind of absorbent skeleton to prevent separation. The LAS is thus neutralized by sodium only, the exact amount of caustic soda being dependent on the acid value of the sulphonic acid and also whether partial neutralization by an inorganic salt has been done. The acid value will vary not only with the average molecular weight of the alkyl benzene (tridecyl benzene has a lower acid value than dodecyl benzene), but also with the method of manufacture (ie, oleum, sulphuric acid or SO_3) and finally according to the individual manufacturer.

For this reason, although the acid value is usually expressed as the amount of milligrams of caustic potash required to neutralize 1 g of acid, it is better to consider the acid value as the number of grams of caustic soda required to neutralize 100 g of acid. This can readily be calculated from the internationally accepted acid value by the formula:

$$\text{Acid value in g NaOH}/100 \text{ g acid} = \frac{\text{Acid value in mg KOH/g acid}}{14}$$

The method of manufacture depends on whether a stainless-steel or mild-steel vessel is being used.

If CMC is to be included in the formulation, it is necessary to dissolve it separately. If a separate mixing vessel with a slow-speed stirrer is available, water to the amount of ten times the weight of the CMC is charged into the container and the CMC poured in slowly with stirring. The stirring is continued until all the lumps of CMC are dispersed, concurrently with the process of the manufacture of the paste. If a mixing vessel is not available, the solution can be made in the same way in a 200-litre drum and the CMC added while stirring with a hand crutcher. This dispersion is best prepared the day before it is required and left to swell and dissolve by itself overnight.

The CMC dispersion, if in contact with iron or steel, will tend to corrode the vessel and also will become discoloured by the rust formed. It is therefore essential to produce this dispersion in a non-corrodible vessel—stainless steel, asbestos cement or steel clad with plastic or rubber. The stirrer itself should also be non-corrodible.

However, some manufacturers are able to achieve rapid wetting, swelling and solution of the CMC in a matter of half an hour by the use of a high speed turbine mixer. The addition of alkali can also help in this dispersion.

If a mild steel neutralizing vessel is being used for the manufacture of the paste, all the remaining water called for in the formula (all the water if no CMC is to be added) and the requisite amount of caustic soda solution, with inorganic salts if they are to be used, are run into the vessel. Mixing and cooling is started and the sulphonic acid added at such a rate that the temperature does not rise above 65°C. (If 100 per cent acid is used the heat of reaction is low and the temperature will not rise very high, provided that an efficient cooling coil or jacket is available. If 90 per cent acid is used, the heat of reaction is considerably higher and the rate of addition of the sulphonic acid must be adjusted.)

When about 80 per cent of the sulphonic acid has been taken up and the paste is still distinctly alkaline, the requisite amount of sodium hypochlorite solution is added and mixed in. The balance of the sulphonic acid is now added, taking care that the pH of the paste does not fall below 7. Mixing is continued until the paste is uniform and the CMC solution (if called for) is added, or, if an alkylolamide is to be used, this can now be put in. If a hydrotrope is employed, this can be added with the sulphonic acid.

When a stainless-steel neutralizing vessel is being used, the procedure is similar, except that the sodium hypochlorite can be put in at the beginning with the water and caustic soda and no special precautions need be taken about not passing to the acid pH. If an acid pH is reached, the desired pH of 7·5–8 can be achieved by the addition of a litre or two of caustic soda solution.

A slight variation of this paste procedure occurs when a highly concentrated liquid is required. This is done by neutralizing the sulphonic acid with an ethanolamine.

In this instance, the requisite water (which depends on the final concentration desired) and the ethanolamine are charged into the vessel. The sulphonic acid is added slowly with intermittent stirring so as not to generate foam. The reaction is instantaneous. The only precautions that need be observed are to avoid the production of foam and to keep the temperature from rising above 60°C.

Calcium Sulphonates

Although efforts of formulators are concentrated in the direction of neutralizing the effects of calcium ions by chelation, the calcium salts of alkyl benzene sulphonic acid play an important role in the production of agricultural emulsifiers because of their solubility in the semi-polar solvents used as diluents for insecticides. The calcium ion, being divalent, yields a sulphonate which has the pictorial formula

$$\text{Ca} \diagup_{\displaystyle SO_3}^{\displaystyle SO_3-Ar-R} \diagdown_{\displaystyle Ar-R}$$

that is, the hydrophilic portion is in the centre of the molecule with two hydrophobic tails. For reasons not readily apparent this allows of spontaneous emulsification, an absolute prerequisite in the field. The sulphonic acid is not necessarily of the same molecular weight as that used in cleaning materials, quite often it can be greater, increasing its oil-solubility. The calcium sulphonate is not the only constituent of these emulsifiers whose formulation is beyond the scope of this volume.

To manufacture calcium sulphonates, good quality SO_3 sulphonated alkyl benzene is essential with an absolute minimum of free sulphuric acid. To attain this completeness of sulphonation is often sacrificed so that (almost) no surplus SO_3 is needed for the sulphonation. As can be understood, the presence of free sulphuric acid will produce insoluble calcium sulphate.

The neutralization is performed with good quality lime in the presence of a polar solvent as a diluent, and after completion the diluent is distilled off.

Solid Detergents

Solid detergents are usually manufactured in three forms for the following different applications:

(*a*) toilet cakes;
(*b*) household scrubbing;
(*c*) 'one-shot' additions to washing machines.

1. Detergent Toilet Bars

The manufacture of a detergent toilet bar comparable with soap is fraught with many difficulties, and is the subject of many patent applications throughout the world. Not many of these patents have been successfully exploited commercially for technological or economic reasons, or both. A few toilet detergent bars have been put on the market, but not all of them have captured the sales that the manufacturers had hoped for.

One hundred per cent of detergent active matter is considerably more expensive than 100 per cent soap. This does not matter when detergents in liquid or powder form are being marketed, as 20–25 per cent active matter in a well-made powder can compete very favourably with a 40–60 per cent fatty acid soap powder. To be sold in the same price range as soap a detergent bar cannot have much more than 40 per cent active matter. If only 40 per cent active matter can be used the mass needs to be filled to 100 per cent with relatively inexpensive materials and this causes the problem. The use of soluble inorganic salts is precluded because the cake will be heavy compared to soap and will tend to effloresce. Talc, although not efflorescent, will again make the cake extremely heavy and hard to the feel.

A detergent used as a soap will tend to defat the skin overmuch and 'superfatting' or retarding agents need to be incorporated into the bar.

One of the big disadvantages to overcome is the behaviour of a detergent tablet when wet compared to a soap tablet. When soap is wetted, water is absorbed over the whole surface of the soap and the viscosity of the water-soap mixture drops dramatically and then increases with further dilution to form the viscous 'middle soap'. Thus when a wet tablet of soap is placed in a soap dish this viscous middle phase prevents water from penetrating into the tablet and the soap remains dry inside, the outside surface gradually drying out by evaporation. This middle phase does not occur with detergents and therefore the water in a wet detergent tablet under the same conditions will penetrate right to the centre of the mass, making it soft and slimy.

For all that, synthetic toilet tablets have one very big advantage in that because of the chemistry of synthetic detergents they do not make a scum or leave a ring in the bath tub. A toilet detergent tablet should therefore not be readily soluble or should not disintegrate, and should have the feel of soap when wet; it should lather freely and when stored should not crack or effloresce. Finally, it should be possible to work the mass in normal soap-milling and stamping machinery.

The bulk of these problems have been overcome by the use of a soap/detergent mixture, but although this improves the physical behaviour of the cake, it does not do away with the disadvantages of soap. Tolerance to hard water can be improved by the addition of lime soap dispersants[14] or even by using them wholly as the synthetic part of the 'combo' soap as it is sometimes called.

There are a few detergent cakes on the market and these are based on sulphated monoglycerides, fatty alcohol sulphates, alkyl isethionates, sulphosuccinates and sulphated fatty acids. Very often the potassium salt is preferred to the sodium because of its better rheological properties. To these active raw materials are added plasticizers such as fatty alcohols, the phosphate and phthalate esters, alkylolamides, particularly coconut or palmitic mixtures, and polymerized ethylene glycols and binders such as starch and talc.

Manufacture is the subject of many patents and trade secrets, but the process consists of using raw material in as nearly an anhydrous form as is possible to maintain, melting it with plasticizers, adding the binders, and then milling, plodding and stamping in the normal soap-making machinery.

Martin Hollstein (*Proceedings of the XVth Congress of the Spanish Detergent Committee, 1984*) gives the general composition for a syndet toilet soap:

Surfactant	30–70%
Plasticizer/binder	10–30%
Filler	10–30%
Additives	0–20%
Water	3–10%

A specific formula we suggest can be

Coco fatty alcohol sodium sulphate (70–80% AM)	50%
Monoalkanolamide of palmitic/stearic acid derived from either hydrogenated tallow or palm oil methyl ester or fatty acid	20%
White dextrin	30%

The dextrin serves not only as a binder but also as a skin protectant. This formulation contains no insoluble matter.

2. Household Scrub Bars

Historically, in this field again the first attempts were based on a detergent which had been available as a water solution. Binding the water was done with materials such as bentonite and sodium stearate, ie, soap produced *in situ*. Because the starting material contained at least as much water as active matter, the final bar or cake contained the same ratio of water to active detergent matter. This limited the amount of active matter that could be incorporated, because the more active matter added, the more water appeared in the final product. Nowadays a bar of this type can be manufactured at concentrations between 20 and 60 per cent, as desired, by the use of alkyl benzene sulphonic acids.

The sulphonic acid is mixed with the dry ingredients in a dough-mixer. A plough-mixer or amalgamator is also suitable provided that the arms and motor are strong enough to withstand the resistance of the material and the neutralizing agent is either soda ash or a concentrated caustic soda solution. If caustic soda is used, a percentage of water is automatically added to the material, but the final product can be controlled to have a neutral pH. When soda ash is used (as in the manufacture of powders) a surplus of soda ash needs to be present and it is thus impossible to have a neutral pH. Although a neutral pH is not essential, as it would be for bars meant for toilet use, nevertheless the bar will still come into contact with the hands, and must not be too alkaline. The pH can be lowered by having a large amount of bicarbonate in the formula or by the addition of sodium acid pyrophosphate after the neutralization has been completed.

After neutralization, and the addition of any special ingredients, the mass is allowed to age. This ageing process varies with the formulation and further manipulation. If a bar of high active-matter content (60 per cent or so) is being made, the mass should be aged for at least 12 hours. If a low active-matter material (40 per cent) is being manufactured, the mass can be transferred to the next stage immediately. In fact, it is not advisable to allow it to age, as it will set hard and become unworkable. (If by any chance a low active-matter bar has set hard before it can be worked, it is possible to regenerate the plasticity by warming it to 50–60°C in a soap-drying room.)

Low-active concentration bars or tablets are not recommended because they are both inefficient and uneconomical in use. However, in certain of the developing countries it is considered that the unit cost of a bar might be too high if it contains the normal amount of active matter. This is a false consideration because the filling materials as well as the extra manipulations involved all cost money. For all that, bars of 25 per cent active matter or less are being produced.

To make bars of this type filling materials need to be used. Plain water cannot be used alone as this will make the bar too soft. If inorganic salts alone are used the bars become heavy, hard and unattractive in appearance and are also subject to efflorescence on storage. These problems can be partially overcome in various ways.

Paraffin wax at the rate of 2 per cent can be added. The wax needs to be melted separately and poured on the mass while mixing, before the temperature has fallen to 60°C. The addition of the paraffin wax gives a better feel to the tablet and to a certain extent hinders the defatting action of the detergent on the human skin. This skin-protecting action can be enhanced by using a blend of paraffin wax and polyethylene wax. In this instance the polyethylene wax (which has a melting point of the order of 100°C) needs to be dissolved in the molten paraffin wax. Surprisingly enough, this addition of waxes in no way detracts from the detergency of the bar or tablet.

Another method is to use 5 per cent starch and to include in the basic formula a total of 15 per cent water. After the reaction is completed, the starch is added and the mass heated to a minimum of 80°C to solubilize the starch.

Finally, a combination of both the above methods can be used and also on occasion talc is added to improve the feel of the tablets.

After the requisite ageing period, the mass is passed through a soap mill. In certain factories this milling is dispensed with and the aged mass is fed directly into a plodder. However, the appearance and texture of the finished product is definitely inferior if it is not passed through a mill.

For this reason, we stated above that the period of ageing depends on the further processing. If the mass is to be milled, the ageing period can be shorter, as it is desirable to feed the mass into the mill while it is still very plastic. If no milling is to be done, the mass should be less plastic. In some factories the mass is allowed to set hard, then chipped in a soap-chipper, and then passed into a mill or direct into the plodder.

Fabric Softeners

In the USA initially, and afterwards in many other countries, fabric softeners have become a necessary adjunct to washing of clothes particularly when machines are being used.

It was becoming apparent that clothes were coming out of the machine 'hard' to the feel. The wool industry had for years used the term 'handle' to test the feel or softness of the fibres and this in turn was a function of the efficient washing or scouring of the wool.

In the same manner as CMC was introduced to produce whiter washing, fabric softeners filled the gap formed by the 'hardness' of the clothes. The reason for the washing coming out hard is a build-up of salts on the fibres and these salts are deposited in such a manner that they are not easily rinsed away. (It should be remembered that no matter how much chelating or sequestering agents are added to the washing powders or liquids, the final rinse in household washing is with tap, ie, relatively hard, water.)

Certain quaternary ammonium salts (cationic detergents) and also alkylolamides were found to be effective as additives for the softening of clothes. These softeners are adsorbed on to the fibres from the final rinse and the washing comes out softer, probably due to an interfibre lubricating effect.

In addition to softening, the fabric softeners of the quaternary ammonium type also act as bacteriostats (see pp 28, 102, 195) if not bactericides. This is particularly important in the USA, where the average washing machine is set to work at about 55°C for about ten minutes, and this temperature-time factor is not sufficient to destroy germs.

Finally, with the synthetic materials being used nowadays, a very common occurrence is the build-up of electrostatic charges on the fibres. The softener tends to diminish production of this static electricity.

The common cationic detergents in use are di-tallow di-methyl ammonium chloride (or iodide) (I), alkyl di-methyl aryl ammonium chloride (II) or 2-heptadecyl-1-methyl-1-(2'-stearoylamido-ethyl)-imidazolinium methyl sulphate (III) which are pictured below:

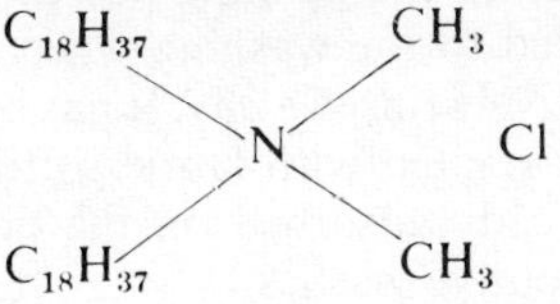

$C_{18}H_{37}$ — N(CH_3)(CH_3) Cl — $C_{18}H_{37}$

I

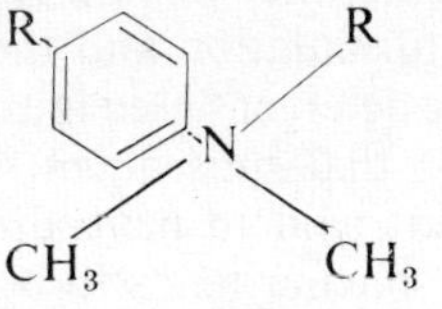

R — C$_6$H$_4$ — N(CH_3)(CH_3) Cl — R

II

$C_{17}H_{35}$—C=N
    | | CH_3SO_4
CH_3—N—CH_2—CH_2
    |
$C_{17}H_{35}CONCH_2CH_2$

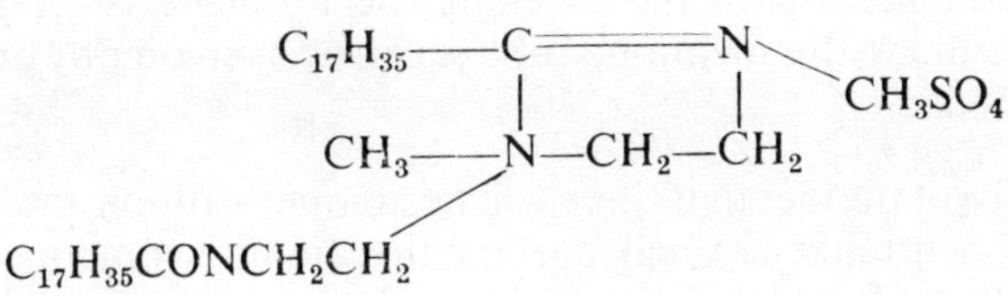

III

It will be observed that III has a quaternary, an ethanolamide and a sulphate group making it an amphoteric. This type has only 70 per cent of the softening efficiency of types I and II.

Viscosity of the softener should of course be constant and here again we come across an anomaly, in that these materials behave differently from other surfactants. If the electrolyte content of the final solution is high the formulation can be almost water-like in its viscosity. If the electrolyte content is very low the solution (dispersion) can be like a jelly. To allow for variations in the hardness of the water supply it is best to use demineralized water for diluting and then control viscosity by addition of electrolyte (sodium chloride or sodium acetate).

These materials are merely dispersed in water to give a convenient concentration so that 0·1 per cent (based on the weight of the clothes) of the softener can be dosed. Thus for a machine taking 5 kg of wash, 5 g of the softener need to be applied. It is therefore advisable to dilute the softener sufficiently so that 5 g (or less for machines taking smaller quantities) of the 100 per cent material can be measured easily.

This concentration is normally in the 5–8 per cent range, but 'concentrated' softeners of double, triple or even greater strengths are available. At the higher concentrations the solubility of the distearyl or di-tallow derivatives is poor and the 'disperson' could separate on storage. To prevent this, recourse to emulsifiers is necessary. Of the emulsifiers suitable for this purpose we mention glycerol monostearate, ethoxylated fatty amines, and fatty alkyl diamines.[15]

Again as the solubility is poor, if too viscous a solution is made, it might not disperse sufficiently in the water during the rinse cycle. For the very concentrated solutions viscosity can be lowered by the addition of small quantities of calcium chloride.[16]

The imidazoline softeners give a thin solution, and the viscosity of this solution can only be increased by using a thickener (p 84) or by combining them in the ratio of 60:40 or 70:30 with a quaternary softener. This has the added advantage of giving enhanced softening.

As the quaternary ammonium salts are incompatible with anionic detergents, they obviously cannot be included in the washing powder or liquid formulation and they need to be added to the washing when the anionic detergent used initially has been effectively rinsed away. It has been found[17] that subsequent washing of clothes treated with fabric softener does not tend to neutralize the softener already adsorbed on the fibre; rather softener tends to build up with consecutive washes.

However, there are on the market washing powders and liquids which incorporate a softener, that is the softening effect is achieved in the wash cycle. The objection to this mentioned above can be overcome by one of the following methods:

1. Acceptance of the fact that there will be anionic–cationic interaction. This will neutralize a small part of the anionic component (and incidentally reduce foaming) to form the complex (or salt). This

complex is still substantive to the fibres, its softening effect will be diminished but will still produce an anti-static function.
2. Use of only non-ionic material as the active matter.
3. Encapsulation of the cationic with a relatively high melting, water-insoluble material (paraffin wax). The softener then becomes dispersed in the water when the temperature in the wash cycle reaches its maximum after the bulk of the washing has been done. There will of course be a competing reaction between the anionic on the one hand and adherence to the fibre on the other.
4. Use of a tertiary amine with two fatty acid groups (di-tallow methyl) for example). These materials are negatively charged at normal wash pH and behave as ordinary bases, although not soluble to any extent in water. This insolubility allows them to adhere to the fibres of the cloth. In subsequent rinse cycles the pH will drop to the neutral range and they become converted to cationic material (they have been called crypto-cationic) and exert the softening effect.

A further tendency has arisen to add optical brightening agents to the softening solution. Care must be taken in the selection of the brightening agents as many of them are incompatible with cationic detergents or unstable in solution. It is best to work in close co-operation with the manufacturers when selecting these dyestuffs.

Fabric softeners for the rinse cycle are always perfumed. The type and quantity of perfume should be chosen with care, preferably with the help of an expert perfumer, as a well-formulated softener can leave the clothes with a pleasant, fresh smell even if they are passed through a drier.

The criteria, apart from the germicidal or bacteriostatic properties, that have been suggested[17] for fabric softeners are:

1. Easy dispersion in a final rinse solution.
2. Substantivity to the fabric.
3. Ability to impart softness and fluffiness (handle).
4. Ability to impart good interfibre lubrication without giving a greasy feel.
5. Ability to impart an anti-static effect.
6. The softener must not adversely affect the colour of the fabric (yellowing or greying) and, optionally, should impart a brightening effect.
7. The fabric must be easily re-wettable after treatment.

Abrasive Cleaners

To this class belong the abrasive hand-cleaners and scouring powders for pots and pans as well as cleaners for tiles and bath-tubs. The vast majority are sold as powders or pastes, but bars and liquids are also available.

Powders, pastes and solids are made in much the same way as the non-abrasive product, with the addition of the abrasive material. In liquids, abrasive matter will always tend to settle out, but, to prevent it from compacting, a gelling agent is incorporated. This increases the viscosity, so that the abrasive can easily be re-dispersed when the liquid is shaken to prevent the abrasive from compacting. The choice of abrasive in the above compounds is one of extreme importance in avoiding undue costs and yet giving the abrasiveness desired without scratching. For example, an abrasive used in a wall-cleaner must be softer than one used on windows, one for aluminium ware softer than one for steelware. Usually the manufacturer employs a compromise type of material. The following materials are listed in order of hardness. Generally, they are used in a size of 200 mesh or finer.

Talc: Hydrous magnesium silicate; very soft; suitable for special scouring agents for tile and porcelain enamel, but not for a general household product.

Diatomaceous Earth: Siliceous skeletons of small aquatic plants called diatoms; also called infusorial earth, kieselguhr, Tripolite, etc; soft; expensive; not recommended in general household products.

Dolomite and Calcite: Natural magnesium carbonate and calcium carbonate.

Marble Dust: Calcium carbonate, mild abrasive, commonly used in household cleaners.

Volcanic Ash: Sedimentary rock composed of volcanic dust, ash, and cinders; suitable for use in household cleaners. Pumice, also a derivative of volcanic ash, is slightly harder.

Felspar: Mixture of various metal aluminium silicates, chiefly $K_2O.Al_2O.6SiO_2$. Its hardness is just about the equivalent of volcanic ash, and it is usually of a whiter colour, so that it is a very desirable product when available.

Quartz: Crystallized SiO_2. It is harder than pumice and is apt to be abrasive on softer metals such as aluminium and copper. Not recommended where milder abrasives are available.

Sand: Fine grains of disintegrated siliceous rock, chiefly quartz which it resembles in hardness.

In many cases the soap plays a minor role. This fact leads to an interesting development, namely, the use of synthetic detergents to replace soap in the formula. A product containing 5 per cent soap will rapidly lose its sudsing value through the precipitation of the soap by the hard water. This destroys the penetrating action necessary to such cleaners. Synthetic detergents such as alkyl benzene sulphates which are not affected by hard water retain this property under all conditions of use and in America have largely captured this type of market.

Abrasive powders or pastes containing about 5–10 per cent active detergents (the DDBS detergents are specially suitable) have been found by the authors to possess excellent detergent and foaming qualities.

The Revised US Federal Specification P-D-221 for this kind of product also specifies cleaners which contain vegetable abrasives. For this purpose sawdust, etc, may be used, and it seems that the fact is taken into consideration that these kinds of abrasive are usually less irritating to the skin than the common mineral abrasives, with the exception perhaps of bentonite or colloidal clay.

Again, it must be emphasized that fineness of particle size of the abrasive material, especially if inorganic, is of the greatest importance; otherwise irritation of the skin may be caused.

The strong defatting action of synthetic detergents and their power of penetrating into the skin may lead to dermatitis. However, the irritant effect of synthetic detergents can be counteracted by adding superfatting agents such as lecithin and lanolin, and also alkanolamides, especially those based on vegetable oils with a high content of double unsaturated fatty acids.

References

1. Davidsohn, A., *Soap*, **36**, 5, 215 (1960), and **36**, 6, 151 (1960), *Soap*, **41**, 9, 73–86 (1965).
2. Technical Brochure: Neutralization of Surface Active Salt Producing Acids by Anhydrous Simet, Rhone-Poulenc Chimie Minerale.
3. Davidsohn, A., *SPC*, **25**, 724 (1952).
4. Milwidsky, B. M., *Soap*, **38**, 10, 59 (1962).
5. Ballestra and Silvis, *J Am Oil Chem Soc*, **41**, 12, 1–5 (1964).
6. Davidsohn, A. and Holtzman, S., *Seife, Oele, Fette*, **91**, 723 (1965).
7. Lyne, C. W., *Brit Chem Eng*, **16**, 370 (1971).
8. Davidsohn, A. and Moretti, G. F., *Sefe, Oele, Fette*, **111**, 444 (1985).
9. Davidsohn, A., *Soap*, 27 (1972).
10. Kuti, S. A., Paper presented at the World Surfactants Congress Munich, West Germany (May 1984).
11. Briggs, J. L. A., *Sefen-Ole-Fette-Wachsen*, **105**, 15 (1979).
12. Davidsohn, A., *SPC*, **33**, 176 (1960).
13. Milwidsky, B. M., *Soap*, **39**, 54 (1963), and **39**, 4, 53 (1963).
14. Linfield, W. M., *JAOCS*, **55**, 87 (1978).
15. Puchta, R., *JAOCS*, **61**, 367 (1984).
16. Fuller, J. G., Paper presented at CSMA Seminar (June 1985).
17. Ginn, M. E., Schemach, T. A. and Jungerman, E., *J Am Oil Chem Soc*, **42** (12), 1084 (1965).

7. Application and Formulation of Detergents

The physico-chemical reactions which take place in the cleaning processes are beyond the scope of this volume. There are excellent treatises on the theory of detergency, some of which are mentioned in the Selected Bibliography (p 309). In this chapter we shall be discussing only formulations for particular applications.

Foam

It is expedient, however, that a few words be said on foam. At one time the amount of foam formed by a detergent was considered to be a measure of its effectiveness. It is true that foam is formed when surface tension is low, but lowering of surface tension is not always a criterion of detergency because this is in actual fact related to the lowering of interfacial tension.

In certain detergent operations high foam is a definite requirement, in certain cases it is immaterial whether the detergent foams or not, and in other cases foam can be considered a nuisance if not a prohibition to the use of the detergent for a particular operation.

It is obvious that a hair shampoo, shaving cream (other than the brushless type) and bubble bath preparations need to produce copious foam. Dishwashing compounds which are primarily meant for washing by hand in a sink full of water also need to foam copiously but the reason is not so obvious. As the plates are dipped in the solution and washed the oil is freed from the surface of the plate and floats to the top of the solution and in time this fatty layer can become appreciably thick. When the plate is withdrawn from the liquid it will pass through this layer last and part of the oil might become redeposited on the surface. If, however, a thick foam is developed in the liquid by the physical actions of washing, the oil will be trapped in the tremendous surface area of the foam and the amount of oil available for redeposition on the plate in its passage out of the solution is greatly reduced. When the foam becomes saturated with oil it collapses and this is an indication that the solution is no longer suitable for washing. For detergents intended for hand-washing of clothes foam is desirable as a sales-appeal factor.

Commercial and automatic household laundry machines are almost without exception 'foam sensitive'. If the detergent foams unduly the foam overflows on to the floor and also can interfere with the free flow of clothes through the water. In automatic machines the foam can interfere with water level pumps and the proper working of the controls in the machine. A small amount of foam is necessary as this tends to trap dirt particles.

Similarly, dishwashing machines cannot tolerate foam, firstly because again the foam might overflow and also interfere with the pumping of the liquid and secondly, once foam forms, bubbles remain on the dishes and leave spots on drying.

Hard and fast rules cannot be laid down as certain household washing machines (usually those with a propeller) can tolerate foam, and in dishwashing machines using a propeller a small amount of foam is not harmful, but in the case of those working with jets the smallest trace of foam leaves spots on glassware.

From our experience we can lay down the following rough guidelines for foam:

(a) Anionic detergents in general produce voluminous foam; somewhat less is produced in hard water but foam is always increased with an increase in temperature.
(b) Non-ionic detergents foam considerably less than anionics, but this depends on the type of non-ionic. There are 'low-foam' non-ionics which can be used alone for most operations, while other non-ionics still need 'foam control'. Non-ionics usually foam somewhat less in hot water than in cold.
(c) When soap is added to an anionic detergent, foam is depressed.
(d) Non-ionic detergents do not depress foam of anionics; they can even enhance it.
(e) Alkaline builders in general enhance the foaming power of all types of detergents.

We shall later in this chapter include formulations for both powders and liquids with foam control but cognizance should be taken of a Proctor and Gamble patent[1] whereby a foam control agent is added to a ready-made powder.

The patent describes a combination of a white mineral oil, paraffin wax or microcrystalline wax and/or glycerol monostearate with microfine precipitated hydrophobic silica in the proportions:

Mineral oil	80 parts
Paraffin wax or microcrystalline wax and/or glycerol monostearate	2–12 parts
Microfine hydrophobic silica	0·3–1 part

The solid waxes are dissolved in the oil with warming and the silica dispersed in it.

To give a non-foaming detergent, 100 parts of finished powder are sprayed with 1 part of this solution/dispersion.

Household Cleaning

The days of one bar of soap, which served as a hair shampoo, toilet bar, laundry soap, for general washing and (mixed with sand from the garden) for pot-scouring, have long since passed. Nowadays, materials are manufactured for special purposes, such as heavy-duty laundering (the washing of cotton goods which by their very nature are usually heavily soiled), fine wash (the laundering of delicate fabrics such as silks, nylons, woollens), general purpose, dishwashing, floor-washing, window-cleaning, tile-cleaning, etc.

Heavy-duty Laundering

Detergents should be formulated with the type of clientele in mind. Werdelmann[2] has made a study of the difference between European and American washing habits.

The obvious differences are that Europeans use a front-loading, rotating drum washer whereas the Americans use a top-loading agitator machine. The Europeans also wash at a higher temperature than the Americans. These two factors require European powders to be low-foaming, while American machines can tolerate a moderate-foaming washing powder or liquid.

Without going into details we also quote his figures for the range of hardness of water.

	Hardness parts/10^6		
	0–90	90–270	270 and higher
USA	60%	35%	5%
Europe	9%	49%	42%

Although these figures can be meaningless for the continents as a whole, they do provide some indication.

Again washing habits both personal and for clothes differ greatly in the two continents. Table 7.1, compiled by Dr Werdelmann, is quoted.

TABLE 7.1

Comparison of Formulation of Heavy-duty Detergents

	USA	Europe
Surfactant	10–20	20–15
STP	35–60	35–45
Sodium silicate	4–10	3–5
Sodium perborate	—	20–35
Optical brighteners	0·1–1·0	0·7–0·8
Dosage	1–2 g/litre	7–8 g/litre
Clothes : liquid ratio	1:15–1:25 kg/litre	1:5 kg/litre

Since this report was published, however, perborate together with an activator are gradually appearing on the US market (see also p 88). Enzymes which had their ups and downs in the USA are also increasing in popularity.

Non-ionics perform better than other detergents on fabrics made from synthetic fibres and also under cold water conditions[3] (see p 279). They also find use as 'rubbing agents' for badly soiled collars.

Cotton goods, which are still the bulk of the household wash, require a moderately high alkalinity. In contact with a solution of a pH of 10 or more, the cellulose fibre swells slightly, allowing the water to penetrate into the fibre and thus loosen adhering dirt. The other ingredients of these cleaners have already been described in Chapters 2, 3 and 4.

Household heavy-duty washing powders are generally of two types, for hand washing and for fully automatic washing machines. The hand-washing type of powder requires a copious lather for psychological reasons. Fully automatic washing machines, which have a drum which revolves around a horizontal axis, cannot use a powder with a copious foam. Because of the action of the drum, the foam can spill out of the machine, and also the foam interferes with the action of the automatic floats which adjust the level of the water. However, a small amount of foam is necessary, as some of the dirt is trapped by this foam. For these reasons the powder has to have 'controlled foam'. Semi-automatic machines, ie, those that have propellers or impellers to do the agitation, or with drums rotating about a vertical axis, can in general, utilize either of the two types of powders.

A formulation for a hand-washing or semi-automatic machine, manufactured by simultaneous absorption and neutralization, has already been discussed in Chapter 6 (p 208), Formulae 2 and 3. The addition of sodium perborate is, of course, optional and depends on washing practices in the particular area. We recommend the addition of at least 10 per cent sodium perborate to all laundry powders. In countries where this is not normal practice the addition of sodium perborate might make the powder expensive, but the extra whiteness achieved will compensate for the increased cost.

If the powder is being manufactured by the spray-drying process, the formulation can, if desired, be the same as that given in Formula 2 or 3, but an infinitely better powder can be made by using Formula 9:

Formula 9
    Spray-dried Heavy-duty Household Hand-washing Powder

Sodium alkyl benzene sulphonate (or equivalent)	25
Sodium tripolyphosphate	25
CMC (66% basis)	2·5
Sodium silicate (1:2·4 ratio)	15
Optical brightening agent	0·2
Sodium sulphate	32·3

and to this powder after spray-drying add sodium perborate at the rate of

Spray-dried powder	90
Sodium perborate	10

Foam Control

The above formula cannot be used in foam-sensitive washing machines. To overcome the foaming problem several methods of foam control are used. One is to use a low-foaming non-ionic detergent as the active matter. An alkyl phenol condensed with not more than eight molecules of ethylene oxide will not foam unduly, but the physical characteristics (free-flowing, stickiness) both of spray-dried powders and those made by other methods will suffer when using low ethylene oxide content non-ionics and therefore the amount of active matter it is possible to incorporate in the powder is limited.

A typical powder to be made by either spray-drying or spray-mixing is given in Formula 10.

Formula 10

Heavy-duty Fully Automatic Washing Machine Powder

Non-ionic detergent (low-foaming)	10
Sodium tripolyphosphate	30
CMC (66% basis)	1·5
Sodium silicate (1:2·4 ratio)	15
Optical brightening agent	0·2
Sodium sulphate (can be partially replaced by soda ash)	43·3

Another trend is to use a moderately high-foaming non-ionic such as iso-tridecyl alcohol with 15 molecules of ethylene oxide, tallow alcohol or alkyl phenol both with 10–12 molecules of ethylene oxide. All of these detergents foam quite highly, and to depress the foam a nonyl phenol with $1\frac{1}{2}$ molecules of ethylene oxide is used. The addition of this foam-depressor is critical and if proportions different from the optimum are used the foam is likely to be enhanced. The amount suggested is normally 15 per cent of the total active matter, but this figure should be checked against the particular type of active matter with the exact proportions of builders being used.

Sodium perborate can be added in the same manner as in Formula 9, if the powder is spray-dried. If the powder is manufactured by a spray-mixing process which has arrangements for continuous discharge of the powder, the sodium perborate can be fed continuously on to the conveyor belt which receives the discharged powder. If not, the perborate can be incorporated batch-wise in a powder mixer. On pp 205, 215 we describe methods for the incorporation of perborate both continuously and batch-wise.

To incorporate appreciable quantities of non-ionics into spray-dried powders, recourse must be made to the 'Pluronics, some of which are available as flakes, prills or flakeable solids. Alternatively all the methods described on p 227 and p 233 should be considered.

A simpler and cheaper 'controlled foam' powder can be made by the use of a mixture of soap and alkyl benzene sulphonic acid. This will give both lower foam and higher detergency than when either is used alone in comparable proportions. It has been found by the authors that the foam is at a minimum when the soap proportion of the total active matter is between 30 and 60 per cent. Outside these limits, foam begins to form. If hard water is prevalent in the area, the detergent/soap ratio should be richer in the synthetic portion. If there is predominantly soft water in the area, more soap can be used if required. A word of caution needs to be given in the use of soap or soap mixtures in automatic washing machines. Some automatic machines have their heating elements protruding directly into the water, while others have the elements protected by a sleeve, which disperses the initial heat over a greater surface area. If the element is itself immersed in the water, it has been found that in areas of moderate to very hard water, a layer of insoluble limesoap can form on the element, decreasing the efficiency of the heat transfer. This can to a certain extent be eliminated by the inclusion of NTA or EDTA in the formula, but this leads to a further complication in that both NTA and EDTA in the presence of oxidizing agents (perborate) are corrosive to copper and zinc. Inhibitors have been developed to minimize this corrosion to an acceptable level.

A powder of this type can be made by the absorption and neutralization process according to Formula 11.

Formula 11

    Low-foaming Machine Powder for Soft-water Areas

Mix together with warming to 45°C

(a) LAS (100 per cent)	11·6
Distilled tallow fatty acid	6·4

In the powder mixer, mix together:

Sodium tripolyphosphate	15
(b) Soda ash, light	55
CMC (66% basis)	2

When uniform, add (a) with mixing. When the LAS/tallow fatty acid mixture has been dispersed, add immediately in order

Sodium hypochlorite	2
Sodium silicate 40%	8

This is mixed for a few minutes until the reaction is seen to have been completed, and then discharged for ageing. Next day the powder is ground and recharged to the mixer in the following proportions:

Milled powder from Formula 6	90
Sodium perborate	10
Optical brightening agent	0·15

and mixed until uniform and then sent to packing.

APPLICATION AND FORMULATION OF DETERGENTS

If one of the mixers described on p 208 is used the ageing and grinding can of course be dispensed with.

To manufacture the same formula using ready-made soap powder the formulation is:

Formula 12

    Low-foaming Machine Powder for Soft-water Areas using Ready-made Soap Powder

Mix together:
Sodium tripolyphosphate	15
Soda ash, light	53·7
CMC (66% basis)	2

add in order:
LAS (100%)	11·4
Sodium hypochlorite	2
Sodium silicate	8

when the reaction is completed add:
High titre soap powder 82 per cent fatty acids	7·9

Thereafter proceed as in Formula 11.

The same powder can of course be made by spray-drying, but a vastly superior one can be made in a spray-drier according to Formula 13. Modern tendencies are to manufacture 'ternary' powders, that is, powders having the active matter made up of three constituents, soap, an anionic and a low-foaming non-ionic detergent, a formula for which is given in Formula 14.

Formulae 13–14

    Spray-dried Household Low-foaming Laundry Powders

	13	14
Sodium dodecylbenzene sulphonate (100% basis)	15	7
Tallow soap, soda based (100% basis)	10	5
Low-foaming non-ionic detergent	—	7
Sodium tripolyphosphate	15	15
Tetrasodium pyrophosphate	10	10
(optional, can be completely replaced by sodium tripolyphosphate)		
Sodium silicate (anhydrous basis, ratio 1:2·5)	7	7
CMC (66% basis)	2·5	2·5
Optical brightening agent	0·2	0·2
Soda ash	0–20	0–20
Sodium sulphate	to make 100	

Sodium perborate is added at the rate of 10 per cent (or more if desired) after drying.

In place of tallow soap, the soap of behenic acid (a C_{22} saturated acid) is being used. Sodium behenate in the formulation acts mainly as a foam control agent (a very efficient one) and at a maximum concentration of 2 per cent. The synthetic active agents can then be increased to give a more effective wash, even at low temperatures.

Behenic acid has a high melting point (80°C), therefore special precautions need to be applied for its introduction and saponification.

For dry neutralized powders it is best dissolved in the LAS and the liquid non-ionic (if being used) or a solvent with the aid of heat (see p 206 for solvent detergent powders) and the mixture is sprayed hot on to the powders.

For spray-drying it can be dissolved in the LAS, and the mixture kept at a higher than normal temperature in the supply tank. The neutralized paste or slurry, which should also be kept at a higher temperature than normal, will probably have a higher viscosity than normal. Contrary to generally accepted principles, lowering the water content of the slurry slightly will, in this case, also decrease the viscosity.

As mentioned on p 227, non-ionics can cause trouble on being spray-dried, pluming and also auto-oxidation.

Pluming can be minimized by the use of the narrow range ethoxylates (p 31). Auto-oxidation, which at best can discolour the fines and at worst will discolour the powder, can be inhibited according to a Lever patent[4] by the inclusion of about 1 per cent of charge transfer agent, two examples of which are stannic chloride or tetrachlorobenzoquinone. The most efficient methods of incorporation are of course the methods already described. These methods also lend themselves to the incorporation of enzymes, perborate, etc.

Even in areas where there are no restrictions (total or partial) on the use of phosphates, consideration should be given to the use of zeolites, either the zeolite itself or the zeolite/silicate blend described on p 61. This will have the added advantage, particularly in dry mixed or agglomerated powders, of surface adsorption. The zeolite can adsorb on its surface larger amounts of liquid surfactants than the normal builders are able to, thus drier, more free-flowing powders with higher active matter can be produced.

Where there are bans or restrictions on the use of phosphate, the STP suggested in the formulations can be replaced by mixtures of zeolites on the one hand and NTA or one of the polycarboxylic acids (p 92).

Enzymatic laundry powders for automatic washing machines are now being produced in large quantities. It is difficult to give directions as to the amount of enzyme concentrate to add to powder as the concentration of the 'net' enzyme varies from manufacturer to manufacturer, but they give explicit instructions as to usage. Enzymes for washing machine powders are commonly used together with perborate. The enzyme operates while the temperature of the water is low and it is rapidly inactivated when the temperature reaches 60°C, perborate taking over. Although the manufacturers state that the enzymes are stable to perborate, this stability is only relative and if both enzymes and perborate are included in a powder, it is advisable to increase the enzyme concentration over that normally recommended. Formula 14 with little or no soda ash can serve as a useful base for enzymatic powders.

In all of the above and subsequent formulae which are to be spray-dried, consideration should be given to the addition of 3–5 per cent of sodium toluene sulphonate. This does not serve the purpose that it does in a liquid detergent. In this instance the function of the toluene sulphonate is twofold: it lowers the viscosity of the slurry, allowing a higher solids content, and also promotes the free-flowing characteristics of the finished powder, particularly when linear alkyl benzene sulphonate is used. Also when powders are made by processes other than spray-drying, the inclusion of toluene sulphonate is essential if linear alkyl benzene sulphonate is being used.

Heavy-duty liquid detergents are now making headway for household laundering. The problem here is that the solution for efficient laundering needs a certain amount of alkali, sequestering agent and soil-suspending agent. To incorporate sufficient of these with a detergent into a solution which will stay bright and clear is not easy. As a result, new materials are coming on the market specially for this purpose and new techniques are being employed. Many patents for heavy-duty liquids are daily appearing in the literature and as a result the trend is to move away from conventional materials.

The active ingredient can be chosen from one or more of the alkyl benzene sulphonates, olefin sulphonates, paraffin sulphonates or ethylene oxide condensates. For reasons given below the anionic should preferably not be neutralized with a sodium ion, rather potassium or an ethanolamine. This condition cannot always be observed, particularly if paraffin or olefin sulphonates are bought as the already hydrolysed/ neutralized solution.

The linear alkyl benzene sulphonates show better solubility in water than do the branched. For the production of liquids the choice should be in the lower register of molecular weight (C_{10-12} rather than C_{11-13}). EniChem of Italy is fractionating its detergent alkylate into high and low 2-phenyl fractions. Again the high 2-phenyl fraction is more suitable for the production of liquids for reasons of solubility. True the detergency is somewhat lower (p 126) than for the low 2-phenyl isomers, but the production of liquid detergents does require some compromises and the detergency can be enhanced in different ways.

On this score it should be pointed out that, contrary to expectation, the potassium salt of LAS is not more soluble than the sodium salt; it is less soluble. However, the potassium salts are used on occasion because the inorganic constituents are often potassium salts and to introduce a sodium salt would result in precipitation of the sodium salt of the inorganic compound by double decomposition. This also works the other way in that the potassium ion of the salt could precipitate the LAS. More often than not, however, the LAS is neutralized with an ethanolamine to obviate this problem.

The problem in heavy-duty liquid detergents (HDLD) formulations is in incorporating the builders into the solution in such a way that the liquid has enough of all the necessary ingredients for efficient washing without detracting from the appearance of the product.

The builders in question are sequestering agents, which are usually phosphates; alkalis, which are invariably silicates; anti-soil redeposition agents (CMC) and optical brighteners.

Spray-dried heavy-duty powders invariably contain sodium tripolyphosphate as the sequestering agent, sometimes with the addition of other sequestering agents. Polyphosphates hydrolyse rapidly in neutral or acid solutions, first to a mixture of ortho- and pyrophosphates, and the pyrophosphates in turn hydrolyse further, albeit at a slower rate, to the simple phosphates. If the pH of the solution is 9 or higher, the hydrolysis of polyphosphates is slow[5] so much so that a solution containing polyphosphate at this pH can be stored for a year at 25°C with no appreciable hydrolysis having taken place.

The solubility of sodium tripolyphosphate in pure water is 15 per cent at room temperature, but this figure will be considerably lower in the presence of other dissolved materials, particularly anionic, due to the common ion effect. The solubility of tetrapotassium pyrophosphate in water is over 60 per cent whereas the corresponding sodium salt is only soluble to the extent of 5 per cent. It is for this reason that tetrapotassium pyrophosphate is often used as a constituent of liquid detergents, however sodium salts need to be excluded almost completely because, due to the common ion effect, tetrasodium pyrophosphate can be precipitated from a solution containing both sodium and potassium ions. Again the use of pyrophosphate is a compromise as it does not have the sequestering and peptizing properties found in the polyphosphates.

Potassium tripolyphosphate, with a solubility in water of 55 per cent, has appeared on the market and this does not suffer from the same defect in that its sodium salt might precipitate due to the increased solubility of its sodium analogue.

Alternatively, phosphates are completely dispensed with and the organic sequestering agents, such as sodium (or potassium) ethylene diamine tetraacetate, or nitrilo triacetate[6] are employed.

The alkali required is obtained from the colloidal silicates. Potassium silicate has been found to be superior to sodium silicate in this respect and silicates act as corrosion inhibitors against the action of phosphates on stainless steels.

However, to get sufficient of all three of the above ingredients into solution is virtually impossible, as the presence of one affects the solubility of the others. For this reason it is necessary to use a hydrotope, ie, a solvent aid, a product which is itself soluble in the medium and aids in the solution of other products. There are a variety of these, but the most commonly used are potassium (or sodium) xylene, toluene, cumene or ethyl benzene sulphonates. These have to be used in relatively large quantities, of the order of 5–10 per cent of the finished product. They can either be sulphonated (or bought) as such, or else co-sulphated in the correct proportion with the original sulphonate, if the basic detergent is one of the sulphonate types.

Urea, apart from its other uses in the detergent industry, is an efficient hydrotrope, as efficient as the lower alkyl sulphonates. However, it suffers from one disadvantage. If one considers its method of manufacture, the combination of carbon dioxide with ammonia:

$$CO_2 + 2NH_3 \rightarrow NH_4CONH_2$$
Ammonium carbamate

$$NH_4CO_2NH_2 \rightarrow NH_2CONH_2 + H_2O$$
Urea

it is conceivable that industrial urea can contain small proportions of unreacted ammonium carbamate. In water solution this can hydrolyse:

$$NH_4CO_2NH_2 + H_2O \rightarrow (NH_4)_2CO_3$$

to ammonium carbonate. If the solution is alkaline, which it almost invariably is, a smell of ammonia will become apparent, which might be objectionable to some people. Thus if urea is to be considered as the hydrotrope, the purchase specification should stipulate no ammonium carbamate to be present.

Berol of Sweden has developed potassium salts of lower molecular weight phosphate esters specifically as hydrotropes for the production of liquid detergents based on alkyl phenol ethoxylates in admixture with potassium tripolyphosphate. These phosphate esters have a further advantage in that they have detergent or wetting properties in their own right, thus aiding in the washing process, a property which none of the other hydrotropes can claim.

CMC now has to be brought into this solution. It is best incorporated by making a 10 per cent 'solution' separately, when it swells and forms a gel. This swelling and gelling, however, is influenced by ions already in solution and, if present, the CMC in a very short time precipitates out and sinks to the bottom. This problem can be overcome in three ways. One method is to adjust the density of the solution so that it is the same as the density of the precipitated CMC ($\pm$ 1·37) and the precipitate will therefore not settle.[7] This is achieved by the addition of sodium sulphate or chloride. In another method carboxymethyl cellulose and methyl cellulose are used.[8] Both cellulose compounds tend to precipitate from the solution. By itself, the methyl cellulose would normally rise to the top and the CMC by itself would normally sink to the bottom. By using this pair in equimolar proportions, however, they precipitate, but neither rise nor fall. Finally, Hercules Powder Co has brought on the market a new type of low molecular weight CMC which has excellent stability to HDLD solutions with high percentage of electrolytes.[9]

With the incorporation of CMC as outlined above, one tends to get an opaque type of suspension. It is felt that once the solution tends to be thick and opaque, it should be made to look like a lotion, and the incorporation of sodium nitrate[10] is said to give a better appearance and easier density control. Alternatively one of the opacifying agents mentioned on p 238 can be used.

If it is decided to dispense with phosphates, one of the organic chelating acids (p 88) neutralized preferably in this case by monoethanolamine, can be used. Alkalinity can also be obtained by excess monoethanolamine.

In the manufacture of these liquids, very often a sludge of insoluble material settles out. This has been identified as traces of iron contamination in the ingredients or even the water. The addition of 1 per cent triethanolamine, over and above any base needed for neutralization will eliminate this sediment (p 92).

Finally, it is desirable to incorporate an optical brightener in the solution. This is the least of the producer's problems, as optical brighteners are now available which are stable in water solution. The amount is so small that this has no effect on the solubility of the other materials, nor have they any bearing on the solubility of the dye.

Thus, to summarize, the following formulae could be used as heavy-duty liquid detergents.

Formulae 15, 16, 17, 18

Heavy-duty Liquid Detergents

	15	16	17	18
Alkyl aryl sulphonic acid (ABS)	10	20	9	12
Diethanolamine	3·6	7·2	3·3	4
Non-ionic (100%)	2	—	3	—
PVP (100%)	0·7	—	—	0·7
$K_4P_2O_7$ (100%)	12	12	10	—
Potassium silicate (100%)	4	3	4	—
Monoethanolamine	—	—	—	3
EDTA	—	—	—	5
CMC (100%)	—	1	1	—
Potassium xylene sulphonate or other hydrotope	5	5	4	4
Optical brightening agent	0·1	0·1	0·1	0·1
Water to	100	100	100	100

All the above formulae are of the medium- to high-foaming type and thus unsuitable for fully automatic washing machines.

A formulation for a heavy-duty liquid detergent with a 'controlled foam', using the normal type of CMC is given in Formula 19:

Formula 19

Heavy-duty Liquid Detergent with 'Controlled Foam'

Disperse with gentle mixing:
CMC (66% basis)   2
in water   18

Into a stainless-steel vessel equipped with a slow-speed stirrer charge:

Distilled coconut oil fatty acid	8
Water	25
Caustic potash (40% solution)	5

Stir and warm gently until the solution attains a clear and homogeneous appearance, then add with stirring in order:

Non-ionic detergent	4
Monoethanolamine	1·7
LAS	4
NTA or EDTA (acid form)	1
Hydrotrope	8
Tetrapotassium pyrophosphate or Potassium tripolyphosphate	10
Potassium silicate (40% solution, 1:2·5 mol ratio)	10
Optical brightening agent	0·2

The CMC gel prepared above is now added, mixing continued, and water added to make this solution up to 100.

This will produce a lotion type of liquid. It is not necessary to 'saponify' by boiling, because if all the above directions are adhered to, the coconut fatty acid will be completely neutralized and no free unsaponified oil will be present.

PVP is being used with success in place of the CMC. The PVP is added as a 5 per cent pre-prepared solution, and the water adjusted accordingly. This will produce a clear solution, possibly with a slight haze due to insoluble matter sometimes found in poly- or pyrophosphates.

As mentioned on p 265 trace quantities of iron contamination in the raw materials can cause a sludge of ferric hydroxide to separate. This separation is not always immediately visible and an insurance against this is the addition of 1 per cent triethanolamine to chelate the ferric ions. The iron will still be present in the solution, but inactivated.

On p 94 the Gantrez resins were mentioned as chelating agents. GAF, the manufacturer, has published a detailed description of an additional use, that as a stabilizer for liquid detergent solutions containing non-ionics.

These resins are high-molecular-weight polymers with anhydride rings, which on hydrolysis with water form dibasic acids. Thus for a typical molecular weight of 40,000, the hydrolysed resin can have 250 dibasic acid groups per molecule. These need to be neutralized and in fact the maximum chelating action is attained at a pH of 10 minimum. These acid groups can also be esterified[11] and in our particular field the esterification agent they suggest is a non-ionic detergent, which normally has a terminal —OH group available for reaction with a carboxylic acid. If the major portion of the carboxylic groups is esterified, an insoluble mass is obtained, but if a partial ester is formed (1 per cent non-ionic based on the weight of the resin), this serves as a stabilizing agent for non-ionic liquids.

The non-ionic of choice is an alkyl phenol with 15 ethylene oxide units and the procedure is important. A stock solution of the resin is made, the type suggested is Gantrez AN-149 (medium viscosity). Dissolve in water

Non-ionic	10
Water	890

and raise the temperature to 90°C. Add slowly with stirring

Gantrez AN-149	100

keeping the temperature at 90°C, and continue stirring till a clear solution is obtained. (Caution, a large amount of foam might be formed in this esterification.)

A formulation we have found to give excellent results as a heavy duty liquid is

Water	4·0
Gantrez stock solution	10·0
NaOH (45% solution)	2·0
CMC	0·5
Nonyl phenol 9-ethoxylate	3·5
NaOH (45% solution)	29·0
Potassium silicate (1:3 mol ratio, 36%)	51·0
Optical brightener, dye, perfume	qs

The solution is made at 60°C, every ingredient to be dissolved completely before the next one is added.

The first addition of caustic soda is to neutralize the Gantrez resin. The second is to convert the silicate to metasilicate. We have found that a mixture of potassium/sodium metasilicate gives better freeze–thaw characteristics than pure sodium metasilicate as recommended by GAF. This can also be reversed, to use potassium hydroxide and sodium silicate to achieve the same purpose.

Finally, mention must be made of the fact that in the United States, one of the giant soapers, after ignoring the liquid household heavy-duty market almost completely, introduced its new liquid heavy-detergent towards the end of 1984, and from all accounts, at the time of going to press, this liquid is making remarkable inroads into the existing liquid market and also taking a large portion of the powder business. From the press release, this liquid is more concentrated than normal liquids on the market, being termed $\frac{1}{2}$ cup (rather than 1 cup for existing brands) and contains 12 active ingredients; four surfactants, anionic, non-ionic and cationic (thus it seems to have a softener built in), two builders stated to be citrate and laurate (the laurate is obviously a soap, they use it as a chelating agent), three 'stain fighters' (two enzymes and one chemical never before used in heavy duty liquids), two brightening agents and a 'revolutionary molecule that makes all the 12 ingredients work together', apparently a hydrotrope. The problem of inactivation of enzymes (the calcium ions are chelated) has obviously been overcome by a method known only to this company.

An interesting and new development in the liquid heavy-duty household laundering field is a liquid containing a mixture of active chlorine and anionic detergent. It had been generally considered that chlorine (derived from sodium hypochlorite) could not be stable in the presence of considerable amounts of organic matter. However, sodium toluene or xylene sulphonates seem to stabilize sodium hypochlorite, and mixtures of sodium hypochlorite with sodium ether sulphates can now be produced with the chlorine reasonably stable for over three months.[12]

Formula 20
    Heavy-duty Liquid Detergent and Bleach
Sodium lauryl ether sulphate (60% concentration)	20
Sodium toluene sulphonate	5
Sodium hypochlorite solution (10% available chlorine)	75

Even better results have been obtained in using Dowfax 2A1, one of the range of alkylated diphenyloxide disulphonates of the generic formula

$$\underset{SO_3^-X^+}{\underset{|}{C_6H_4}}-O-\underset{SO_3^-X^+}{\underset{|}{C_6H_3}}-R$$

produced by Dow. In the 2A1 version 'X' is the sodium ion. This novel surface active material is stable in moderately strong alkali and acid solutions, in the presence of large amounts of inorganic materials and in solutions of oxidizing agents.

The high stability of hypochlorite in admixture with di-sulpho-diphenyl type of detergent is due to the inertia against chlorination of the multi-substituted benzene ring, one of the rings being tri-substituted, the other di-substituted.

To achieve maximum stability, it is suggested that the diluted Dowfax 2A1 be heated to 70°C together with 2 per cent of a 12 per cent active chlorine hypochlorite solution, then cooled to 30°C and 20 per cent of a 12 per cent active chlorine hypochlorite solution added. This will give a 2·5 per cent active chlorine solution stable for a reasonable period, when packed in a plastic bottle. For a 6 per cent active detergent solution the formulation could be:

Soft water	65·4
Dowfax 2A1	13·0
Sodium hypochlorite, 12% active	1·6
heat to 70°C, then cool to 30°C and add	
Sodium hypochlorite, 12% active	20·0

Free alkalinity should then be adjusted by the addition of caustic soda solution to 0·5–1·0 per cent. This free alkalinity, in addition to stabilizing the hypochlorite, reacts with the water hardness lowering the pH for maximum disinfecting and bleaching action of the active chlorine.

Light-duty Household Products

In the field of light-duty detergents, or (as it is sometimes called) fine wash, are the materials for washing delicate fabrics such as wool, nylon, silk, etc. Also, in general, this type of detergent is suitable for household dishwashing by hand as well.

Synthetic detergents first took a hold in the household for this purpose, and liquids consisting only of detergents diluted with water very quickly achieved a large sale and large quantities are still being sold. Nowadays, liquid detergents are more sophisticated than the plain solutions that were originally sold. They can be based on anionic detergents only, usually dodecyl or tridecyl benzene sulphonate, or can utilize a mixture of the above anionics with a non-ionic, or a sulphated ether. In addition to foam boosters, these liquids can have viscosity increasers and cloud-point depressants. Furthermore, they need not necessarily be transparent liquids, but can be opaque lotions.

The concentration of active matter present in a liquid detergent of this type varies from country to country, but an average figure is 12 per cent for the simpler liquids, but this can go up to 40 per cent for the more sophisticated market.

The cloud point of a detergent is an important factor in its sales appeal, as it is axiomatic that no housewife will buy a liquid which tends to deposit a precipitate or to cloud over on storage. Requirements for the actual cloud point will vary from place to place and no hard-and-fast rules can be given. The cloud point naturally obtained from a particular concentration of active matter depends on the type of material used and the method of neutralization.

In general, the sodium salt of the alkyl benzene sulphonic acids, at a concentration of 12 per cent active matter, gives a cloud point too high for commercial use. The diethanolamine and triethanolamine salts of the alkyl benzene sulphonic acids give a very low cloud point, lower than is needed generally. Thus for economic reasons, neutralization is usually done partially with caustic soda and partially with an ethanolamine.

Among the alkyl benzene sulphonates, all other things being equal, the cloud point of the neutralized solution rises in the following order:

linear dodecyl benzene sulphonate;
linear tridecyl benzene sulphonate;
branched dodecyl benzene sulphonate;
branched tridecyl benzene sulphonate.

The viscosity of the final solution, all other things being again equal, rises in the same order.

In addition to the above factors, the method of sulphonation is important, since on this method depends the amount of free sulphuric acid present. When neutralized, the acid produces inorganic sulphates, both of which raise the cloud point and the viscosity. SO_3 sulphonated material will give the lowest free sulphuric acid present; and normal oleum sulphonation will produce the highest. (The figure is normally as high as 7-8 per cent.)

The factors involved in producing an acceptable liquid detergent are therefore dependent on:

(1) the type of alkyl benzene sulphonate;
(2) the method of manufacture of this sulphonate;
(3) whether any other active material is used.

To take an average case, if conventional dodecyl benzene, sulphonated with SO_3, is to be used as the base material for the manufacture of a 12 per cent liquid, a cloud point of below 5°C can be obtained by neutralizing half of the active matter with caustic soda and the other half with diethanolamine or triethanolamine. From Table 6.4, Chapter 6, on p 237, it will be seen that neutralization with diethanolamine will give a slightly higher viscosity than if neutralization had been done with triethanolamine. Diethanolamine is a solid, except in very hot climates, and therefore is more difficult to handle. On the other hand, when alkyl benzene sulphonates are wholly or partially neutralized with triethanolamine, which is a liquid, a buffer is formed at a pH of approximately 6. To pass this buffer, it is necessary to use a large excess of base. To preserve a low cloud point, it is not advisable to use too much caustic soda to overcome this buffer, so a fair excess of triethanolamine must therefore be used. The manufacturer must consequently consider the disadvantages involved in handling diethanolamine against the extra cost of triethanolamine.

If in the examples of alkyl benzene sulphonates given above, one of the sulphonates which normally gives a low cloud point were to be used (p 269) to maintain the same cloud point level, the proportion of caustic soda can be raised and that of the ethanolamine correspondingly reduced.

Furthermore, if an acid sulphonated material is used as the base material, the proportion of ethanolamine needs to be raised considerably (and the caustic soda to be lowered) to maintain the low cloud point.

If it is required to raise the viscosity of the solution, a simple but limited method is the addition of an inorganic salt, like sodium sulphate or sodium chloride. This is limited in that it raises the cloud point. A better and more efficient method of increasing the viscosity is the use of an alkylolamide, preferably a diethanolamide.

Because of the various factors involved in the formulation of a liquid, it is not possible to describe formulations for all possible permutations and combinations.

Formula 21 gives a basic formula to be used as a starting-point. The manufacturer is advised to modify this according to the conditions he requires and the materials available to him.

Formula 21

                Light-duty Household Liquid Detergent

LAS (SO_3 sulphonate)   10

Triethanolamine	2
Caustic soda (45% solution)	1·7
Sodium hypochlorite (10% solution)	0·6
Lauric acid diethanolamide	1
Sodium sulphate	1
Water	83·7

Full and detailed instructions on the method of production of this solution are given on p 235.

An opaque lotion type of liquid can be made (taking into account the same factors as mentioned above) from Formula 22.

Formula 22

Lotion-type Light-duty Liquid Detergent

LAS	19·5
Monoethanolamine	4
Lauric acid monoethanolamide	1·5
Sodium hypochlorite (10% solution)	0·6
Sodium sulphate	0·9
Water	73·5

In this case the dodecyl benzene sulphonic acid, the sodium sulphate and the water are mixed in the neutralization vessel, the sodium hypochlorite added and mixing continued for at least 20 min. The lauric acid monoethanolamide, being solid, is best dissolved in the monoethanolamine with heating. The mixture of the two is then added to the vessel, after the bleaching has if necessary been completed. If an opaque solution is required an opacifier can be added. The product can now be dyed and perfumed.

To manufacture a liquid from an alkyl benzene sulphonate neutralized only with caustic soda (in practice, a concentrated paste of the sodium sulphonate is used as the starting material), a low cloud point can be achieved by the addition of urea. With the sodium sulphonate, however, urea lowers the viscosity considerably. Addition of a fatty acid dialkylolamide will restore the viscosity.

Modern tendencies are to use both stronger solutions and mixtures of active matter. One of the main uses for light-duty liquid detergents is hand dishwashing, and for this particular application foam is important other than from the point of view of sales appeal. Grease released from the dishes floats to the top of the washing solution and forms an oily film which can be redeposited on the plate when it is withdrawn from the sink after having been washed. Due to the enormous surface area in a foam, this grease is held in a thin, easily rinsable film on the bubbles, preventing redeposition. When the foam collapses due to saturation with oil, the solution is considered to be exhausted and methods of test for the efficiency of dishwashing liquids are based on this foam collapse.

Ether sulphates, both of the alcohol and alkyl phenol types are finding more and more use as constituents of dishwashing liquids. If the ammonium salt of the alkyl phenol type is used care must be taken not to raise the pH of the final liquid over 8 or otherwise a smell of ammonia will appear. If a high pH is desired without the ammonia smell, the ammonium ether sulphates can be heated with a stoichiometric amount of caustic soda until the ammonia is distilled off. Alternatively an 'ammoniated' cleaner, which has some sales appeal, can be made at the high pH.

The choice of the type of ether sulphate is rather wide, as the base material can be any one of the alcohols available on the market or any of the alkyl phenols and then again the amount of ethylene oxide can be varied from between 1·7–4 molecules per molecule. Gohlke and Bergerhausen[13] have investigated viscosity and foam height with different alcohols and different degrees of ethoxylation. They have come to the conclusion that the best foamers are a C_{12-14} alcohol with 2 molecules of ethylene oxide. Another important fact in the make-up of the ether sulphate is the amount of polyethylene glycol present (see pp 31, 187). The authors have found that relatively large (of the order of 3 per cent) amounts of polyethylene glycol can act as a solvent or 'thinner' on the final sulphate to reduce the viscosity of the solution.

As the wetting properties of ether sulphates are low they are not recommended as dishwashing agents alone, rather in conjunction with nonionic or other anionic detergents.

Alcohol sulphates, neutralized with ethanolamines can, of course, be used alone, or with an alkylolamide as a foam booster. It is interesting to note that alkylolamides have no effect on the foaming properties of ether sulphates. Some typical basic formulations are given in formulae 23–27.

Formulae 23–27

Light-duty Liquid Detergents

	23	24	25	26	27
LAS (SO$_3$ sulphonated)	10	15	20	12	—
Caustic soda (40% solution)	3·2	3·2	3·2	3·2	—
Triethanolamine	—	2·2	4·4	9	—
Sodium alcohol ether sulphate (100% basis)	—	3	3	—	—
Ethanolamine alcohol sulphate (100% basis)	—	—	—	—	24
Coconut diethanolamide*	2	1	1	—	2
Sodium sulphate or chloride	qs	qs	qs	qs	qs
Dye, perfume	qs	qs	qs	qs	qs
Water	to make 100				

It will be observed that the first and fourth formulations are relatively cheap, the others more sophisticated.

SO$_3$ sulphonated LAS can be considered the work-horse of liquid detergent formulations. It is good practice, when mixtures of LAS with alcohol or ether sulphates are used, to ensure that the LAS is neutralized completely before the sulphate is introduced into the reactor. This will

* The ethanolamide might contain some free ethanolamine, in which case this can be used as a partial neutralizing agent for the LAS, reducing the amount of NaOH or triethanolamine.

eliminate hydrolysis of the acid-unstable sulphates. To facilitate formulation the following data show the alkali requirement of a typical LAS.

100 kg LAS requires for complete neutralization:

12·8 kg NaOH (100%)
18·8 kg KOH (100%)
45·5 kg Triethanolamine
33·6 kg Diethanolamine
19·7 kg Monoethanolamine

The inorganic salts are included only if it is desired to increase the viscosity. The diethanolamide acts as a viscosity booster but particularly where the ether sulphates are included coconut monoethanolamide will be more effective. To incorporate the monoethanolamide, the solution is merely warmed to 60°C with mixing, when all the monethanolamide will melt and dissolve and will not be thrown out of solution on cooling.

Household fine-wash detergents are, of course, not limited to liquids. They can be made as spray-dried powders, sometimes only with 20–30 per cent active matter and the balance sodium sulphate as the inert filler, but more often than not they include sodium tripolyphosphate, some silicate, occasionally CMC, and an optical brightening agent, substantive to the fibres for which the powder is being used. The phosphate, of course, helps in the detergency, and the silicate acts in this case as a corrosion inhibitor. Here, the action of the CMC is not as pronounced as it is on cottons, but on woollen fabrics it does aid in preventing redeposition, and as these powders are meant for hand-washing, it does tend to give a protective colloidal action on the skin.

A powder of this type can be manufactured as detailed in Formula 28.

Formula 28

Household Fine-wash Spray-dried Powder

	Minimum	Maximum
Sodium alkyl benzene sulphonate	10	25
Sodium tripolyphosphate	15	25
Sodium silicate (preferably 1:3 ratio) anhydrous	3	5
CMC (100%)	0	1
Lauric acid monoethanolamide	0	2·5
Optical brightening agent	0·2	0·2
Sodium sulphate	Balance	

By virtue of the nature of manufacture, fine-wash powders of this type cannot be made by the absorption methods, as the surplus of soda ash will prove detrimental to the operation of the powder. However, a dry-mixed type of powder can easily be made to Formula 28, using as a base a concentrated powder of about 60 per cent active matter.

In certain parts of the world, neutral pastes of sodium alkyl benzene sulphonate are sold also for fine-wash purposes. These pastes vary in concentration between 20 per cent and 50 per cent active matter. The consistency of the paste varies with the ingredients and the method of manufacture of the alkyl benzene sulphonic acid. Stiffer pastes are obtained with tridecyl benzene sulphonic acid, sulphonated with oleum; and the softest paste is linear dodecyl benzene sulphonic acid, sulphonated with SO_3 gas.

The alkyl benzene sulphonates have limited solubility in water. They, however, can absorb water to form a natural paste. If the amount of water present is above that which can naturally be absorbed, the mass will separate into two phases on standing: a concentrated phase of alkyl benzene sulphonate on top and a weak solution of alkyl benzene sulphonate and inorganic salts at the bottom. The concentration of the sodium alkyl benzene sulphonate in the upper phase varies with the material, but is of the order of 55 per cent active matter. This means that pastes of 55 per cent can be manufactured with no special additions. If lower concentrations are required, it is necessary to add certain ingredients to prevent this separation. Materials that can be used are: hydrotropes, which will make the final paste thinner in consistency, so that stiffening materials have again to be added; alkylolamides, which have the added advantages of foam boosting and skin protection (these pastes are to be used by hand); and CMC. Both the CMC and the alkylolamides tend to increase the viscosity of the mass in such a way that separation cannot take place.

If only CMC is used as the thickening agent, a general rule is to manufacture the paste to a 55 per cent concentration and then to add sufficient of a 10 per cent CMC solution in water to bring the concentration down to the required amount. Sodium sulphate is then added to increase the consistency. The method of manufacture is described in detail on p 241.

A 40 per cent active paste, using both conventional dodecyl benzene and linear dodecyl benzene sulphonic acids, both sulphonated by SO_3 gas, is given in Formula 29.

Formula 29

40 per cent Detergent Paste

ABS (100%)	40
Caustic soda (45% solution)	11·4
Sodium sulphate	2
Sodium hypochlorite solution	0·6
Water	29·3
⎰ CMC	1·7 ⎱
⎱ Water	15 ⎰

This paste may be reduced to 30 per cent by increasing CMC to 2·5 or 3·5 per cent and adding water.

General-purpose Detergents

Powders of this type are the most popular for household use. They are not harshly alkaline, and contain relatively large quantities of both active matter and sodium tripolyphosphate. These achieve the action on cottons without the addition of further alkalis, and the alkalinity naturally present from the phosphate does not harm delicate fabrics.

By virtue of their formulation these powders are truly general-purpose and can be used for virtually every household job. They suffer from the disadvantage, however, of being unsuitable for fully automatic household washing machines, as, in general, they foam too much.

A generally accepted formula for this type of powder to be manufactured by a spray-drier is given in Formula 30.

Formula 30

Spray-dried General-purpose Powder

Active matter as alkyl benzene sulphonic acid neutralized with caustic soda	25
Sodium toluene sulphonate*	2·5
Lauric monoethanolamide	3
Sodium tripolyphosphate	30
Sodium silicate (1:2 ratio) anhydrous	10
CMC (100% basis)	2
Optical brightening agent	0·2
Sodium sulphate	27·3

* Optional. Its purpose is to lower the viscosity of the slurry and increase the hardness of the beads (see pp 98, 224).

The addition of 10 per cent sodium perborate is, in our opinion, advisable, but again is dependent on washing practices in the particular country. As a guide, in the United Kingdom, powders of this sort have at least 8 per cent sodium perborate added, and in most of Europe even more.

If sodium perborate is to be added, we suggest that 2 per cent of the sodium sulphate be replaced by magnesium sulphate on any anhydrous basis. This magnesium sulphate is added to the slurry prior to spray-drying.

A general-purpose powder can be made by the absorption and mixing process as well, but in this case, due to the limitations of the process, the active matter is limited. The formulation suggested is given in Formula 31.

Formula 31

General-purpose Powder

Dodecyl benzene sulphonic acid	18
Sodium tripolyphosphate	25
Sodium silicate (1:2 ratio) 40% solution	5
CMC (100% basis)	5
Sodium bicarbonate	28
Soda ash	20
Optical brightening agent	0·15
Water (or sodium hypochlorite solution)	2

As in Formula 31, the same remarks apply regarding the addition of sodium perborate. If this is to be introduced, it is suggested that the sodium bicarbonate be lowered to 25 per cent and 3 per cent magnesium sulphate crystals added. If sodium hypochlorite is used to bleach the powder and if sodium perborate is to be incorporated, the precautions as detailed on pp 204–205 must be observed.

A general-purpose powder can be manufactured by dry-mixing according to Formula 32.

Formula 32

General-purpose Powder

60% alkyl benzene sulphonate powder containing silicate	45
Sodium tripolyphosphate	30
CMC (100% basis)	2
Optical brightening agent	0·2
Sodium sulphate	22·8

Here, if sodium perborate is to be used, no special precautions need to be taken and it can, of course, be added initially with all the other ingredients.

In all the formulations given hitherto, we have included relatively large amounts of sodium tripolyphosphate, this despite the fact that in certain countries there is a complete or partial ban on the use of phosphates. To date no 'plug-in' replacement for this excellent builder has been developed. Where this is a partial ban, the above formulations to include STP up to the limit allowed and to add a zeolite to make up the difference is suggested. Where there is a total ban, the available alternatives must be considered; increasing the silicate, use of NTA, use of zeolite (the zeolite/silicate co-crystal mentioned on p 61 might be eminently suitable), or one of the polymers specially mooted for this purpose. The final formulation will probably be a combination of any two or more of the above.

For fully automatic, front-loading household washing machines, foam control needs to be applied. This can be by the use of soap as for heavy duty powders, but infinitely better performance will be attained if the ternary mixture of the anionic/non-ionic/soap is used as described on p 260. With the new processes and mixing equipment for the inclusion of non-ionics into powders (pp 228, 231), there need be no practical limits to the formulation.

CHOICE OF NON-IONIC

As can be appreciated from the descriptions of the possible non-ionic surfactants, the choice is wide and often confusing. Much work has been done on the optimization of the molecule. Cox and Matson have compared various molecules for their cleaning efficiency in hard-surface cleaning.[14] Kravetz has done the same for soil removal on cloth.[15] It is true that the results cannot be compared directly because of the different substrates, but the results are revealing.

Their results can be summarized:

1. From Cox and Matson's work on hard-surface cleaning it appears that linear alcohols with a low molecular weight and containing 50 per cent ethylene oxide (average chain length 8·6 carbons with 3·1 moles EO) in a relatively high concentration (5 per cent), are better in cleaning efficiency for grease, wax and particulate soil than the molecules containing C_{10} and higher chain lengths with the appropriate amount of ethylene oxide to maintain the balance.

    This was explained by the fact that the low-molecular-weight hydrophobe acts as a solvent and this was confirmed by adding a glycol ether when no improvement of soil removal was noted.
2. For low dilutions of the surfactant the optimum chain length was shifted to the C_{8-10} range. Under these conditions of dilution the hydrophobe can no longer act as a solvent and performance was dependent solely on the surface active effect.
3. Comparison of the low carbon number ethoxylated alcohols with a built formulation containing nonyl phenol with $9\frac{1}{2}$ EO units again showed a better performance for the low-molecular-weight alcohol ethoxylates.
4. Kravetz compared the performance of a linear primary alcohol ethoxylate, a linear secondary alcohol ethoxylate, both of approximately C_{13} average chain lengths, and a branched octyl phenol ethoxylate, each with varying amounts of ethylane oxide and for both cotton and cotton/polyester blends.
5. The results show that 7–9 EO units on each give optimum performance for sebum, oil and clay removal on the blend but for sebum removal from pure cotton 12–15 EO was superior.

Further work by Rosen[16] confirms the above in that he found that short-chain non-ionics are better in removing water repellent soils.

The above facts and figures do not take into account the added complications of the presence of anionic detergents and builders but they do indicate a trend.

From the above it appears that any of the commercially available non-ionic types with approximately 8 EO units will give good all round performance. For powders, for technological considerations, the choice of non-ionic has been in a somewhat higher register hitherto.

With the development of systems for incoporation of non-ionics into powders without spray-drying (p 228) the problems associated with making powders using these (relatively) low-molecular-weight materials no longer apply.

The use of this type of non-ionic in liquids will tend to give a low cloud point but this can be overcome by the judicious use of co-solvents and hydrotropes.

The term cloud point can be somewhat confusing. Normally it is the temperature at which a cloud forms on cooling. This demonstrates the lowest temperature at which a liquid can be stored and still remain clear. When applied to non-ionic solutions it indicates the temperature at which the solution becomes cloudy on heating (p 30), at this temperature the solution separates into two phases, one of which will be richer in non-ionic than the other.

CONCENTRATED POWDERS

A new approach to the manufacture of powders is what is called the $\frac{1}{2}$ or $\frac{1}{4}$ cup concentrates, where a minimum of inert filler, if any at all, is used. A patent by Colgate[17] describes the technology for the production of these powders.

In brief the process is to mix STP with water and sodium silicate in a crutcher to allow the STP to hydrate to its hexahydrate, then to add a further quantity of STP and water under conditions that this second addition does not hydrate and to spray-dry the mix. To the spray-dried beads non-ionic detergent is added in a special mixer.

The details are:

Mix together	
STP	14·5
Sodium silicate (1:2·4 ratio, 50% solution)	15·2
Deionized water	21·0
Maintain this slurry at 60°C for hydration and then raise the temperature to approximately 90°C, add	
STP	28·3
Deionized water	21·0

At this higher temperature no hydration takes place. This slurry is then spray-dried to a bead containing 10% moisture, with a bulk density of ± 0.55 g/ml.

The finished beads are then sprayed, either continuously or batch-wise with a non-ionic, of the alcohol ethoxylate type mixed with minor ingredients; optical brighteners, dye, perfume, etc. The post-mixer mentioned in the patent is one of the Patterson–Kelley types but it is conceivable that any of the mixers described on pp 228, 233 can be used. The finished powder has a bulk density of 0·68 g/ml and contains

Base bead	78·0
Non-ionic	19·7
Minor ingredients	2·3

The granules are attractive and dustless and a further claim by the patentors is that they are sufficiently pourable to be packed in a transparent specially designed bottle.

Further embodiments of the patent include adding fillers other than STP to the hydrated STP. No mention of the inclusion of CMC is made but it is envisaged that it could quite easily be added to the slurry prior to spray-drying.

Cold Water Washing

A generation or so ago all washing was done at the boil or near it. With the advent of synthetic fibres wash temperatures came down drastically and now in an attempt at energy conservation householders are beginning to use cold water only, for washing of clothes.

The industry is facing and meeting the challenge, and a challenge it is because the highly complicated washing process requires, among other things, energy to break the bond of the dirt to the substrate. This energy can come from lowering of the interfacial tension, the mechanical energy imparted by the motor of the machine and from heat.

For cold washing the thermal energy needs to be replaced, and one method is by increasing the amount of active matter in solution and using more non-ionics which are better for soil removal from synthetics. Zweig and her co-workers have studied the parameters involved in cold washing with non-ionics and have come up with some surprising and novel findings.[18] The conclusions can be summarized:

1. Non-polar soils are difficult to remove from polyester/cotton fabrics at low temperatures, thus the inherent detergency needs to be enhanced.
2. Optimum detergency for this system is found with non-ionic mixtures which have cloud points in the range of 15–25°C *below* the wash temperature.

   This is explained by the fact that a surfactant-rich pseudo-phase separates at temperatures above the cloud point. This can be considered to be globules of concentrated surfactant which are attracted to the soil (note also the remarks of Rosen).[16]
3. Low cloud points were obtained by blending lightly ethoxylated alcohol or unethoxylated alcohol with a normal detergent non-ionic.

   In figures, a blend of 3 mol ethoxylated alcohol with an 8 mol non-ionic to give a cloud point close to 0°C enhanced the mineral oil detergency of the 8 mol alcohol ethoxylate by 50 per cent and a blend of a 9 mol non-ionic with decanol, again to give a cloud point close to zero enhanced the detergency of the alcohol-9 ethoxylate by 20 per cent.

These are the bare figures, but as the authors state there is a window through which efficient detergent systems for cold water washing can be seen. The technological problems of formulating these low cloud point non-ionics into liquid detergents need to be investigated.

As can be seen from the foregoing, possibilities of formulating are varied and later in this chapter we also discuss solvent powder detergents. The actual formulations to be used depend on many factors, mainly the

APPLICATION AND FORMULATION OF DETERGENTS

TABLE 7.2

Typical Formulations for Powders Produced by Dry Neutralization

	All purpose non-machine	All purpose machine	Light-duty hand	Heavy-duty non-machine	Heavy-duty machine	Solvent powder
LAS	14–15	5	15	12–15	5	5
Dist. fatty acids	—	4	—	—	4	3
Non-ionic	2–3	6	2	—	4	3
STP*	30	30	20	15	20	20
Metasilicate ($5H_2O$) or spray-dried disilicate	3	3	—	5–6	5–6	5
Soda ash	15–20	15–20	20 max	to 100	to 100	to 100
Sodium bicarbonate	—	—	30	—	—	—
Sodium sulphate	to 100	to 100	to 100	—	—	—
CMC	2	2	—	1·5	1·5	1·5
Optical brightener	0·2	0·2	0·2	0·2	0·2	0·2
Perborate	10	10	—	10	10	—
Enzymes as 300,000 DU	—	0·5	—	0·5	—	—
Perfume	qs	qs	qs	qs	qs	qs
NaOH (30% solution)	2–3	2–3	2–3	2–3	2–3	2–3
Solvent deodorized kerosine	—	—	—	—	—	4

* See p 61.

materials easily available and the trends in the area where the powder is to be sold. The above statement also holds good for liquids.

To sum up we give in Tables 7.2 and 7.3 formulations which can be used as starting points for the manufacture of both dry-mixed and spray-dried powders. These formulations do not necessarily parallel those already described in the text, but do give an indication of what is being made.

Hard-Surface Cleaners

A new development in both the household and the institutional cleaning fields is the all-purpose liquid meant specifically for hard-surface cleaning

TABLE 7.3

Typical Formulations for Powders Produced by Spray-drying

	Heavy-duty hand	All-purpose cold water hand	Heavy-duty machine	Light-duty hand
LAS (Na salt)	12	16	6	18
Non-ionic	3	4	4	2
Soap or distilled fatty acids (see p 258)	—	—	6	—
CMC	2	3	2	2
Sodium silicate, 1:2·45 ratio (100% basis)	3	3	4	—
Soda ash	10	10	—	—
STP (see p 61)	30	35	30	15
$MgSO_4$	1·5	—	1·5	—
Optical brightener	0·15	0·15	0·15	0·1
Na_2SO_4	22	28	30	60·1
Perborate	15	—	15	—
Sodium toluene sulphonate	1	1	1	1
Perfume	qs	qs	qs	qs
Enzymes	—	—	0·5–0·8	—
Residual moisture (approx.)	10	5	10	5

Note
1. Part or all of the LAS can be replaced by sulphonated methyl esters, see p 168.
2. Perfumes for this type of powder should be of low volatility and added after spray drying (see p 215).
3. All the above formulations lend themselves to production by the combined systems described on pp 228–234.

—by hard surfaces are meant those surfaces that cannot be immersed in a bath or basin for cleaning, and the operation needs to be done *in situ*.

These cleaners are usually liquid, alkaline and often contain solvents. Their purpose is to clean grease, mud, and atmospheric grime from walls, doors, glass, tiles, etc. Alkalinity can be derived from alkaline salts, or ammonia (occasionally caustic soda or potash) can be added. The solvent is added because it has been found that oil stains, although theoretically saponifiable by the alkali, cannot always be removed without it. This solvent needs to be soluble in water (unless an emulsion is to be made), reasonably odourless, non-toxic, with a fairly high flash-point and last but not least, a good fat solvent. The glycol ethers, in particular ethylene glycol monobutyl ether or dipropylene glycol methyl ether answer to all the above requirements admirably. Silicates are added both to buffer the alkalinity and to minimize corrosion on metallic surfaces. A typical formulation is:

Formula 33

Hard-surface Cleaner

Sodium alkyl benzene sulphonate (100% basis)	15
EDTA sodium salt	4
Ammonia (100% basis) (optional)	5
Butyl cellosolve	7
Sodium silicate (1:2 ratio, 100% basis)	3
Water	to 100

It goes without saying that if ammonia is to be added the solution cannot be used on copper surfaces.

A generalized formulation which is becoming popular in household use is:

Syndet	5–8 wt %
Hydrotrope	6
Alkali	5
Dipropylene glycol methyl ether	4
Pine oil	2
Water	Balance

The syndet can be non-foaming (non-ionic) or medium foaming (LAS) or high foaming (mixture of LAS and ether sulphate). The alkali can be ammonia, trisodium phosphate, tetrasodium pyrophosphate or sodium silicate, or one of the organic amines such as monoethanolamine. If organic alkine salts are not used the hydrotrope can be dispensed with. It is necessary to use softened water for this formulation as otherwise the hardness salts will precipitate or form a haze due to the high pH.

A particular instance of hard-surface cleaning is the oven cleaner. The above formula will theoretically clean ovens, but as grease is normally heavily encrusted on the surfaces of ovens more alkalinity than that provided in this formula is needed. A typical formula could be:

Formula 34

Hard-surface Cleaner

Alkyl benzene sulphonate, sodium salt	4
Phosphoric acid (85%)	4·5
Caustic potash (100%) or monoethanolamine	9
Tetrapotassium pyrophosphate	4·5
Ethylene glycol monobutyl ether	6
Isopropyl alcohol	2
Sodium silicate 1:2 ratio (100%)	2
Water	68

It will be noted that the amount of caustic potash suggested is in excess of that required to neutralize the phosphoric acid, the surplus is needed to provide alkalinity. In the above formula the glycol ether is not completely soluble but the isopropyl alcohol acts as a coupling agent. Contrary to general belief, isopropyl alcohol is in itself a good fat solvent.

For certain institutional purposes it might be necessary or desirable to produce an oven cleaner as a paste or viscous liquid so that it will not drain down vertical surfaces. Formulae 34 and 35 can be modified by the addition of either a thickening agent (p 84) or colloidal silica. If a viscous liquid is to be made it can be conveniently packed as an aerosol. Vanderbilt Corporation suggests the following formula for an aerosol oven cleaner:

Formula 35

Aerosol Oven Cleaner

Veegum-T*	1·5
Ammonia solution	6
1, 1, 1-Trichlorethane	18
Water	24
Ethanol	7·5
Tergitol NPX†	18
Propellant:	
Dichlorodifluoromethane	17·5
Dichlorotetrafluoroethane	7·5

* Vanderbilt Corporation.   † Union Carbide.

A low temperature oven cleaner has been described in a patent[19] where mannitol or sorbitol is used for alcoholysis of the fat deposited, with sodium or potassium bicarbonate as the alcoholysis catalyst and salts of low molecular weight organic acids to esterify the alcoholysis product. The formulation could be:

Sorbitol	2–5
Potassium bicarbonate	0·1–4·0
Eutectic mixture of	
Sodium acetate	
Lithium acetate	1–5
Potassium acetate	

Thickener	qs
Precipitated chalk	20–30
Wetting agent	qs

The eutectic mixture is to lower the melting point of the salts to allow them to react easily. Other salts recommended could be sodium tartrate, Rochelle salt or sodium glycolate.

Machine Dishwashing

Until a few years ago household dishwashing was done by using one of the fine-wash or general-purpose formulations mentioned above. Of recent years the household dishwashing machine has come into popular use and this has requirements of its own. The mechanical cleaning action is done by means of jets of water. These jets are produced by a high-pressure pump or by the whipping action of a fast revolving propeller. In either case it is essential to the operation of the machine that the detergent added be completely non-foaming (and not even with a 'controlled foam' as in household laundry).

Non-ionics, in general, foam considerably less than anionics, but even they do foam slightly, so the amount of detergent is therefore kept to a minimum and the cleaning effect is achieved by the use of alkalis and phosphates. Almost completely non-foaming non-ionics have now been developed by blocking the terminal —OH group. One of the common methods is to add a methyl group to the terminal —OH. If, for example, a non-ionic is to be made by the esterfication of a polyethylene glycol and a fatty acid (see p 32), instead of the polyethylene glycol a methoxy-polyethylene glycol is used. Another method is to condense to the finished non-ionic detergent a further molecule of butylene oxide. The formulation also depends to a very great extent on the type of water being used. Household water-softeners are only now beginning to appear on the market; so the formula must take into account whether the water being used is soft, moderately hard or hard.

For soft-water areas, soda ash, besides being cheap, can provide a portion of the alkalinity, but in moderately hard and hard water the soda ash will leave a scum of calcium carbonate on crockery and cutlery.

For moderately hard water tetrasodium pyrophosphate can be used to give both detergency and alkalinity, but for hard water sodium tripolyphosphate in combination with strong alkalis must be employed.

Most dishwashing machines have a drying cycle as well as the rinsing cycles, or at least dishes are left hot and the last traces of moisture on them dry very quickly on contact with the air. It is, therefore necessary to provide for the last traces of water to drain from the dishes in a uniform film to avoid water spots, particularly on glassware.

This effect can be achieved by incorporating solvents or an organic chlorine-releasing compound into the powder. Chlorine-releasing materials react with non-ionic detergents and it has been found by the authors that these powders work very successfully without any detergent at all. If, however, it is desired to incorporate a detergent, a very small amount of non-ionic detergent can be used. FMC has suggested a method whereby non-ionic detergents can be incorporated into these powders in the presence of chlorine releasing agents.[20] It recommends the low-foam modified non-ionic detergents mentioned above. This detergent, at the rate of 1–2 per cent based on the final weight, is pre-mixed with the most alkaline ingredient present (anhydrous metasilicate), the tripolyphosphate added next and then the other ingredients and finally the chlorine-releasing compound.

Highly alkaline materials such as metasilicate can affect the over-glaze on delicate china and cause 'crazing' of the glaze. This can be eliminated by the incorporation of boric acid (not more than 5 per cent), sodium aluminate (2 per cent)[21] or zinc salts in the formula.

Since these powders usually contain large amounts of sodium metasilicate, they are not normally made by spray-drying, as there is a limit to the amount of sodium metasilicate (which is hygroscopic) which can be incorporated into a spray-dried powder. They are produced instead, by a mixing and absorption technique.

To make these powders dustless, granulation techniques are used. The powders and the non-ionic detergent are mixed as described above, without the chlorine-releasing material. On to this mixture a small amount of waterglass is sprayed in a mixer with a revolving or tumbling action, when the waterglass glues the particles together and the revolving action gives them a spherical shape. The chlorine-containing material is then added.

A formula suitable for use in soft-water areas is given in Formula 36.

Formula 36

Machine Dish-washing Powder for Soft-water Areas

Tetrasodium pyrophosphate	49
Sodium metasilicate anhydrous	25
Soda ash	22·5
Sodium dichloro-iso-cyanurate (60% available chlorine)	1·5–4·5
Non-ionic detergent, eg, nonyl-phenol 9 mol ethox	2

This formulation is also suitable for dishwashing machines used in catering establishments.

Because of the small amount of detergent involved, it is not necessary to use any special procedure for the incorporation of the active matter. The easiest method is to blend in with the other powders a concentrated detergent powder. Alternatively, any other already prepared detergent powder can be used as the source of the active matter and due allowance must then be made for the other ingredients in this powder.

Where the water is moderately hard the product should be made according to Formula 37

Formula 37

Machine Dish-washing Powder for Moderately Hard-water Areas

Tetrasodium pyrophosphates	60
Sodium metasilicate anhydrous	38
Trichloro-iso-cyanuric acid	1—3·5
Non-ionic detergent	1

For very hard-water areas the following formulation will be suitable:

Formula 38

Machine Dish-washing Powder for Hard-water Areas

Sodium tripolyphosphate	50
Sodium metasilicate pentahydrate	25
Trisodium phosphate anhydrous	15
Non-ionic detergent	3
Hexylene glycol	2
Isopropyl alcohol	1·5
Water	3·5

The liquids are pre-mixed, and then the powders are charged into the mixer and the liquid sprayed on to the powders while mixing.

For institutional machine dishwashing the tendency is to move away from powders to liquids as these can be dosed either continuously or in accordance with pre-set requirements based on electronic controls.

Liquid for dishwashing machines can be formulated with either a non-ionic component or active chlorine, but not both.

Alkalinity is obtained from caustic alkali (soda or potash) with the addition of the corresponding silicate, together with a condensed phosphate. The same problems of common ion effect occur as described for HDLD (p 239), thus a basic formula could be:

Tetrapotassium pyrophosphate	15
Potassium silicate (1:2 mol ratio, 40% solution)	8
Potassium hydroxide (45%)	10
Soft water	67

Potassium tripolyphosphate, if available, would obviously give vastly improved performance.

To this solution is added a per cent of a non-foaming non-ionic or 2 per cent active chlorine. The chlorine in this instance can be added by direct injection of chlorine gas to form potassium hypochlorite in the solution.

ABRASIVE-TYPE CLEANERS

The most popular household abrasive cleaners are scouring powders. These are usually dry mixes of all of the ingredients. A typical formulation is given in Formula 39.

HARD-SURFACE CLEANERS

Formula 39

Household Scouring Powder

Abrasive	87
Soda ash	5
60% concentrated detergent powder	8

If desired, this can also be manufactured from sulphonic acid by mixing together:

Abrasive (powdered calcite or marble)	85
Soda ash	7

then adding in the mixer:

100% alkyl benzene sulphonic acid (ABS)	5
and water	3

If this scouring powder is to contain active chlorine, as is becoming the fashion now, it is advisable not to use soda ash, as concentrated organic-chlorine-releasing materials are not in general stable in the presence of soda ash, except for the potassium salt of trichloro-iso-cyanuric acid.

Alkalinity can be obtained by anhydrous sodium tripolyphosphate or tetrasodium pyrophosphate. To avoid introducing moisture into the powder, the active matter can best be put in by the use of an already made detergent powder, not necessarily a concentrated one. A chlorine containing powder is best packed in plastic containers.

A suggested formulation is Formula 40 which is a chlorine-containing household scouring powder.

Formula 40

Chlorine-containing Detergent Scouring Powder

Detergent powder containing 20% active matter (without soda ash)	15
Tetrasodium pyrophosphate	5
Abrasive	79
Trichloro-iso-cyanuric acid	1

Where a spray-drier is being operated, Formula 40 is a useful outlet for the cyclone fines produced by the spray-drier.

A scouring liquid can be made according to Formula 41.

Formula 41

Household Scouring Liquid

Disperse	
Bentonite	5
in water	25
Add 12% active ready-made liquid detergent	35
and dissolve in the liquid:	
Sodium metasilicate pentahydrate or water-glass	3
and then disperse in this solution:	
Abrasive	32

Special suspending agents suitable for this type of product are now available. These prevent settling of the abrasives.

APPLICATION AND FORMULATION OF DETERGENTS

MISCELLANEOUS HOUSEHOLD CLEANERS

Window cleaners of the gentle abrasive type, using whiting, were once popular. These have now been superseded by liquids like those of Formula 42.

Formula 42

Household Window-cleaning Liquid

Active detergent matter	0·25–0·5
Isopropyl alcohol	15–35
Water	to 100
Dye	as required

Similarly, for stone or tile floors, although materials such as that given in Formula 1 on p 202 are used, liquid floor cleaners according to Formula 43 are now also being manufactured.

Formula 43

Floor Cleaner

Active detergent matter	2–5
Isopropyl alcohol	8–15
Pine oil	1–2
Water	to 100

COMMERCIAL LAUNDERING

In commercial laundering, powders are used which do not foam and which are in general more alkaline than household powders. Most commercial laundries use soft water, and we suggest to the detergent manufacturer that if he supplies washing powders to a laundry which does not use soft (or softened) water, he should make every effort to convince his customer that it is essential for him to use it.

Laundry powders can be made both by spray-drying and by absorption and neutralization. The two formulations below have been used with success in commercial laundries.

Formula 44

Spray-dried Industrial Laundry Powder

Alkyl benzene sulphonic acid	20 ⎱ Neutralized with
Distilled tallow fatty acid	15 ⎰ NaOH to sodium salts
Tetrasodium pyrophosphate	15
Sodium metasilicate	15
CMC (100% basis)	2
Optical brightening agent	0·15
Soda ash	33

Formula 45

Industrial Laundry Powder not Spray-dried

Alkyl benzene sulphonic acid	11
Distilled tallow fatty acid	7

Tetrasodium pyrophosphate	10
Sodium metasilicate pentahydrate	7
CMC (100% basis)	1·8
Optical brightening agent	0·1
Soda ash	61·1
Water	2

Directions for the manufacture of Formula 45 are given on p 204.

It is obvious that for a given load of washing more of Formula 45 will need to be used than of Formula 44.

Many laundries carry out a pre-wash at a lower temperature.* One of the above powders can be added to the pre-wash, but a more effective method is to use for the pre-wash a solvent detergent.[22]

Solvent Detergents

We have already described on p 205 the manufacture of powders containing solvents.

It is not too difficult to combine non-ionics with solvents, but anionics are not easy to combine, especially those of the alkyl aryl sulphonate type, as most of these are insoluble, or only slightly soluble, in most non-polar solvents such as kerosene or deodorized kerosene which, of course, are generally the least expensive ones.

Detergents often serve only as emulsifying agents for the solvents, and not so much as detergent or cleaning agents. Emulsifiable insecticide concentrates, for example, are based on a solution of the emulsifying agent within the insecticide solvent mixture, forming a clear stable solution which, on dilution, gives a milky white emulsion with water. Here the emulsifying agent is very often a non-ionic detergent acting as an emulsifier, rather than as a detergent proper. However, in this book, we are more concerned with solvent-detergent combinations, in which the detergent acts both as an emulsifying agent for the solvent, and as a detergent in its own right.

In detergent-solvent combinations, the job of the solvent is to dissolve grease and similar oily dirt. The function of the detergent is to act as a penetrating and wetting agent and as an emulsifying agent for carrying off solvent after it has dissolved the oil or grease from the material to be cleaned, but it also keeps solid dirt particles in suspension. In many cleaning operations the detergent, by its surface activity alone, is simply not powerful enough to loosen dirt which is kept strongly attached to the surface by oily or resinous matter. This is very often encountered in metal-cleaning, or during the laundering of oily overalls. On the other hand, it is often impossible for a solvent alone to develop its full dissolving power where the oily matter is covered by solid crusts of insoluble matter. This is the case in the decarbonizing operations carried out on the working parts of internal combustion engines. It is the combination of surface activity

* This same pre-wash is being suggested for household use with the use of enzymes.

plus solvent power which makes solvent-detergents so useful in widely different fields of application.[23]

The most easily produced type of solvent-detergent is a combination of non-ionic detergent with solvents. Very often a simple mixing of solvents with detergents is sufficient to obtain a clear, stable product, which generally forms milky emulsions in water. However, not all non-ionics are soluble in any proportion in any solvent. Very often they are only slightly soluble in non-polar solvents of the aliphatic type. Here it is necessary to use so-called 'co-solvents', together with the non-polar aliphatic solvent, to give the desired results.

The subject of solvency is of the greatest importance in working out effective products. By giving concrete examples, it will be made clear how important this type of solvent is in formulating high-grade products.

Generally speaking, the non-ionic detergents are more easily soluble in solvents than most anionic detergents. It is, nevertheless, quite incorrect to assume that they are soluble in all kinds of solvents. Thus, for example, a condensation product of nonyl phenol with 9–10 moles ethylene oxide is readily soluble in chlorinated solvents, xylene, benzene, and in most polar solvents; it is, however, only slightly soluble in kerosene and white spirit and even less so in dearomatized (deodorized) kerosene. To increase the degreasing power of trichlorethylene, it is possible to add a certain percentage of non-ionic, eg, 3–5 per cent, to the solvent. Furthermore, a stable solution of trichlorethylene may be produced by dissolving 10–15 per cent of non-ionic in trichlorethylene. On dilution of the clear solution with water, a milky-white emulsion will be obtained. In order to prevent corrosion due to free hydrochloric acid, an addition of about 0·5 per cent monoethanolamine to the composition is advisable.

Chlorinated solvents should be used with caution (p 108) preferably they should be used in closed systems, where the solvent is distilled, condensed and recycled, such as in dry cleaning and closed metal degreasing systems. The low heat of vaporization of these chlorinated solvents (210 J/g for perchlorethylene) and non-inflammability are distinct advantages.

As already pointed out, most non-ionic detergents are soluble in aromatic solvents and by dissolving 10 per cent in xylene and diluting the clear xylene solution with water, stable milky white emulsions may be obtained which are very useful for metal-degreasing compounds. Aromatic solvents may serve as co-solvents for dissolving non-ionic in aliphatic non-polar solvents such as dearomatized kerosene and white spirit. To obtain clear solution, a proportion of about 30–40 per cent of aromatic solvents is necessary, eg, 10 parts non-ionic; 30 parts aromatic solvent; 60 parts kerosene.

Formula 46

        Detergent-solvent Combination

Nonyl phenol 9 EO	30 parts
Isopropanol	20 parts
Xylene	50 parts

HARD-SURFACE CLEANERS

Formula 47
Detergent-solvent Combination

Nonyl phenol 9 EO	30 parts
Methylethylketone	35 parts
Deodorized kerosene	50 parts

Mix the materials in the order given. A clear solution is obtained which becomes blue-white opalescent when diluted with water (hard or soft). Nonylphenol-9-ethox. behaves similarly to octyl phenol ethoxylates.

For the internal degreasing of motors, etc, such combinations with non-ionic detergents and solvents should prove very useful, possibly in combination with flushing oils. Even in the unlikely event of solvent-detergent remaining in the engine, practically no danger of subsequent corrosion should exist, because of the complete volatility of the products of combustion, which are hardly any more corrosive than the combustion products of motor fuels. It would even be possible to formulate an 'internal' decarbonizer on the basis of anionic detergents (AB sulphonic acid neutralized with alkanolamine or alkylamine), which does not leave any residues in the engine. Here again, an entire field for further research is open.

Another non-ionic detergent useful for formulating solvent-detergents is alkylolamide. Experiments with this compound (which is a condensation product of ethanolamine with fatty acids) have yielded very efficient solvent-detergents. However, the wetting power of alkylolamide is somewhat lower than that of water-soluble alkyl phenol or fatty alcohol ethoxylates.

A special solvent-detergent combination of interest is one which gives a clear solution of kerosene in water according to Formula 48.

Formula 48
Kerosene Water Solution

Coconut fatty acid diethanolamide	20
Kerosene	20
Water	20

On dilution with soft or hard water, all these products give very stable solvent emulsions of good detergent power.

The alkyl benzene sulphonic acids can also be used as the basis of solvent-detergent combinations.

It is essential to start operations with the unneutralized sulphonic acid. SO_3 sulphonated sulphonic acids, because they contain minimal amounts of inorganic acids, can quite easily be incorporated into these combinations, but oleum sulphonated dodecyl benzene, if treated as described below, can also be turned into an acceptable solvent-detergent.[24]

Formula 49
Solvent-detergent Combination

Mix together:

SO_3 produced alkyl benzene sulphonic acid (ABS)	50 parts

Kerosene	50 parts
Aromatic solvent	25 parts
Then slowly add:	
Caustic soda solution (38°Bé)	17–18 parts
After cooling to about 50°C add:	
Isopropanol	10 parts
Pine oil	2 parts

A clear liquid is obtained, which at low temperatures becomes gel-like.

A small amount of water from both the water of solution of the caustic soda and the water produced by neutralization, will be present. If a completely anhydrous material is required, the lower amines, isopropyl or butyl, can be used for neutralization. Care should be exercised in their use as they are volatile, toxic and inflammable.

The product is completely soluble in soft water, somewhat turbid in hard water, and is completely soluble in organic solvents. It is almost completely soluble in non-polar paraffinic solvents, such as kerosene, petrol, etc. By diluting the solvent-detergent with solvents, eg, 1:10 with white spirit or kerosene, a solvent-emulsion concentrate is formed which gives very stable emulsions on dilution with water. This is important for metal-degreasing operations.

It is interesting to note that the wetting power of the ABS detergents is practically unaffected if only aromatic and polar solvents are used, and only slightly affected when non-polar paraffin solvents are present. In all cases, sinking time as measured by the Draves test is not greater than 25 s at a concentration of 0·1 per cent active detergent matter in distilled water.

A completely anhydrous solvent-detergent can be manufactured from SO_3 sulphonated LAS according to Formula 50.

Formula 50

Solvent-detergent based on 100 per cent ABS (SO_3 produced)

Kerosene	35
ABS (100% basis)	35
Monoethanolamine	15
Trichlorethylene	13
Pine oil	2

All the ingredients, except for the trichlorethylene, are merely mixed together until uniform. When the liquid has cooled to below 50°C, the trichlorethylene is added and the pH adjusted to 8 by using monoethanolamine. This pH was found to be advisable as the trichlorethylene might release traces of hydrochloric acid and the surplus monoethanolamine will absorb this. Trichlorethylene may be replaced by higher boiling perchlorethylene. Formula 50 is an excellent general-purpose combination and can be used for adding to the pre-wash in laundering, as an additive for dry-cleaning fluids, and for degreasing of engine parts, etc.

One of the most important uses of solvent-detergents is as a dry-cleaning aid. The purpose of the dry-cleaning detergent is a multiple one, to increase the effectiveness of the solvent in dissolving solvent-soluble spots and stains, to aid in the removal of water-soluble stains by dispersing or solubilizing a small percentage of water in the solvent, and to increase the dispersion of these water-soluble stains in this water. To achieve this effect generally a combination of non-ionic and anionic detergents is used together with a coupling agent. For example,

Formula 51

Dry-cleaning Detergent

Nonyl phenol-9 ethylene oxide	40
Propylene glycol	17
Butyl cellosolve	3
Monoethanolamine	7
Alkyl benzene sulphonic acid	33
(SO_3 sulphonated to minimize inorganic salts)	

All the ingredients are mixed together and the pH (tested at a dilution of 1:10 in water) is adjusted to between 7 and 8 with small additions of either monoethanolamine or sulphonic acid.

This is added to the dry-cleaning solvent at the rate of $1-1\frac{1}{2}$ per cent with or without the addition of small quantities of water.

If a readily dispersible product is required, the above formula can be diluted 1:1 with a dry-cleaning solvent (for example, perchlorethylene) and is then used at the rate of 2–3 per cent based on the volume of the solvent in the machine.

Although dodecyl benzene sulphonic acid gives excellent results in the above formulations, even better results are obtainable by replacing the ABS with heavy alkylate-sulphonic acid, which has a higher molecular weight and renders the sulphonate more oil soluble. Of course, with a higher molecular weight, the acid value will obviously be lower. The amount of alkali in the above formulations should, therefore, be correspondingly reduced. (This is best determined by the acid value of the heavy alkylate sulphonic acid.)

Carpet and Upholstery Cleaners

Fabric cleaners of this type differ in their operation from other detergent materials in that it is usually very difficult to rinse the material being cleaned. To overcome this, methods of cleaning have been developed where a solution of the detergent is applied to the carpet by 'shampooing' to form copious foam, or a foam is formed first and the carpet sponged with this foam. In either instance the combination of the detergency of the cleaner and the mechanical energy applied lifts and holds the dirt in the foam.

The foam, having very thin walls and enormous surface area dries relatively quickly into brittle particles of dust and this dust is either vacuum cleaned or brushed away.

Initial formulations for these carpet shampoos were normal light-duty detergents with the addition of tetrasodium pyrophosphate, its function being both to increase detergency and to make the dried residue more brittle. A suitable detergent material which is also in itself fairly brittle when dehydrated is the sodium or magnesium salt of one of the fatty alcohol sulphates. Some formulae also called for the incorporation of a solvent but with the newer fabrics and rubberized bases being used, the solvent should be chosen with care or left out.

These formulations were not, however, the complete answer to the problem. A portion of the active matter became absorbed into the fibre and when dried this left a deposit which tended to attract dirt. Carpets cleaned in this manner became soiled very quickly.

This problem has been overcome by using more crystalline detergents as the base. Such detergents are the half ester of sodium sulphosuccinates (p 20) used alone or in admixture with fatty alcohol sulphates. Witco (France) is manufacturing the lithium salt of fatty alcohol sulphate for this specific purpose. The detergent then absorbed on the fibre is more brittle and when dried will shatter and can easily be brushed away. Another line of attack is to add colloidal silica (p 78) to the formula to produce the same effect.

Textile Dressing

Detergents of the anionic type have to a large extent replaced soap in the scouring of wool and cotton. The main factor involved is that anionic detergents do not precipitate their lime and magnesia salts on to the fibres, and thus give a better 'handle' to the finished goods.

As textile processing is a complex process, the detergent manufacturer can only be called upon to supply a concentrated detergent; the various alkalis, phosphates or solvents are added by the texile processer himself. This concentrated detergent is usually manufactured by agreement between the textile scourer and the detergent manufacturer. Fatty alcohol sulphate pastes, neutralized immediately after sulphation, are as often used as are alkyl benzene sulphonate pastes. Depending on the requirements of the textile mill, the alkyl benzene sulphonate can be a concentrated paste neutralized only with caustic soda, or a semi-liquid paste neutralized with ethanolamine. A typical formula can be:

Formula 52

Textile Scouring Paste

Alkyl benzene sulphonic acid (ABS 100%)	375
Diethanolamine	130
Water	495

The ingredients are mixed together and the final pH adjusted according to the buyer's requirements.

On occasion, especially for degumming, detergents compounded with pine oil are preferred for textile scouring, particularly of wools.

We suggest the following formulation:

Formula 53
Textile Degumming Detergent Paste
Pine oil	250
Alkyl benzene sulphonic acid (ABS 100%)	280
Diethanolamine	97
Water	373

MERCERIZING

Cotton cloth is treated for anti-shrink properties by dipping in a concentrated (of the order of 12–18 per cent) caustic soda solution. To allow the solution to penetrate well and evenly into the fibres a detergent or rather a wetting agent is necessary. The bulk of detergents are either not soluble or not effective (or both) in strong caustic soda solutions. A commonly used wetting agent for this purpose is the sodium salt of sulphated 2-ethyl hexanol. By itself it is not completely effective but if blended with 10 per cent each of the unsulphated 2-ethyl hexanol and butanol, the material becomes soluble in caustic soda solutions, the solution remains stable and the wetting properties are greatly enhanced. Ethoxylated glycosides (moderate to high foaming) and the low-molecular-weight phosphate esters (low foaming) are also used for this purpose. Another possibility, which has not yet been exploited commercially, is SO_3 sulphonated olefins.[25]

Food and Dairy Industries

As with any food industry, dairies require spotless cleanliness to prevent spoilage, and sterilization against bacterial contamination. The tolerance for residual bacteria is, however, lower for dairies than for other food industries, as in this case bacteria can cause the milk to turn sour. A further complication is the presence of high amounts of lime salts in the milk.

Many plants use strong alkalis only for cleaning, and sterilization is achieved by steam. These materials have their limitations, as can be seen from Table 7.4, detailing the properties of various alkalis for dairies.

It will be noticed that no one material supplies all the requirements. Chlorinated trisodium phosphate adds a sterilizing action. For this reason it is often desirable to combine several of these alkalis to produce a blended material which will provide all the requirements. The addition of a detergent greatly enhances the performance of these alkalis, but in certain processes foaming will interfere with the operation. Even low-foaming nonionics foam slightly, and the use of detergents is thus sometimes excluded. Formulae 54–56 give some typical formulations for food- and dairy-cleaning alkaline detergents.

TABLE 7.4
Alkalis for Dairy Cleaners

Agent	Wetting power	Solution of heat-deposited milk solids	Emulsification of fats	Buffering power	Rinseability	Water-softening properties	Corrosive action
Caustic soda	Poor	Very good	Poor	Poor	Poor	Good	Strong
Carbonates	Poor	Good	Fair	Poor	Fair	Good	Weak
Trisodium phosphate	Good	Poor	Fair	Good	Good	Good	Weak
Orthosilicate	Good	Good	Very good	Very good	Fair	Fair	Strong
Metasilicate	Good	Good	Good	Fair	Fair	Fair	Weak
Sesquisilicate	Good	Good	Fair	Good	Fair	Fair	Medium
Chlorinated trisodium phosphate	Good	Fair	Fair	Good	Good	Good	Medium

Formulae 54–56
Food and Dairy Alkaline Detergent Cleaner

	Non-foaming	Medium foam	High foam
	54	55	56
Trisodium phosphate	15	—	10
Sodium carbonate	10	39	35
Sodium metasilicate pentahydrate	40	20	20
Tetrasodium pyrophosphate	—	40	—
Sodium tripolyphosphate	35	—	30
Non-ionic detergent	—	1	—
Concentrated anionic detergent powder	—	—	5

To produce a powder that simultaneously sterilizes and cleans, the metasilicates must be of the anhydrous variety, and soda ash and non-ionic detergents should not be used. Subject to the above conditions, sufficient organic-chlorine-bearing materials can be added to the above formulation to give, say, 3 per cent active chlorine in the finished powder.

For bottle-washing in the food industry the pH required is very high. Sodium gluconate should be added for its sequestering action. (See p 92.)

Formula 57
Bottle-washing Compound

Caustic soda flakes	55
Sodium metasilicate	20
Soda ash	17
Sodium gluconate	5
60% anionic detergent concentrate	3

In addition to the above alkaline cleaning materials, in the dairy industry it is necessary as a routine to give the pipelines, etc, an acid wash to dissolve milkstone deposited on the walls.

Very often concentrated hydrochloric acid is used. This is very corrosive and although the dairy plant is invariably stainless steel it is advisable to add an inhibitor (see p 301). Nitric acid is also used on occasion. Although nitric acid is both a strong acid and an oxidizing agent, sight must not be lost of the fact that the original stainless steels were developed to withstand this acid, so no inhibitor is necessary. Both these acids are unpleasant to handle and environmentally unsound, therefore, to overcome these difficulties less corrosive materials have been developed.

Combinations of lactic acid with acid-resistant detergents have found application. So, also, have combinations of phosphoric, citric, and tartaric acid with acid-resistant detergents. The most recent development in this important field is the application of sulphamic acid. This has the structural formula:

$$H_2N-\underset{\underset{O}{\|}}{\overset{\overset{O}{\|}}{S}}-OH$$

It is the half amide of sulphuric acid.

Molecular weight 97·09
Specific gravity 2·03

Figure 7.1 shows the pH of sulphamic acid in comparison with some other acids. The following advantages of sulphamic acid are enumerated.[26]

(1) Its crystalline, non-volatile nature, which makes it easy, safe and economical to handle, and eliminates the evolution of objectionable fumes when solutions are needed.
(2) Its strong acid character gives the 'bite' necessary for the removal of deposits.
(3) All salts of sulphamic acid (eg, calcium, magnesium, iron) are readily soluble in water. Hence, adequate dissolving and thorough rinsing of scale is assured. Further, it is expected that less sulphamic acid is needed than other commonly used acids.
(4) In spite of its strong acid character, dilute sulphamic acid solutions are not unduly corrosive to dairy equipment.
(5) It is simpler and less hazardous in handling.

The following five general steps are recommended for cleaning heat-transfer equipment and vats, using sulphamic acid:

1. Thoroughly rinse or flush the equipment with warm water until the water is no longer milky.

2. Prepare an adequate quantity of sulphamic acid solution of 0·2–2 per cent concentration (1·6–16 kg/m³), heat the solution to 65–70°C, and circulate for 10–30 min. Because of the many variables in equipment, composition of deposits and preferred cleaning practices, precise recommendations suitable for all cleaning applications cannot be given. The above ranges of concentration, temperature and time have developed from experience and should cover most situations.

3. Rinse thoroughly with warm water.

4. Flush the equipment with a cleaning and neutralizing solution consisting of 15–85 g of trisodium phosphate or other alkaline-type detergent compound for each gallon of water. The alkaline treatment should consist of approximately the same concentration, temperature and time as used with the sulphamic acid cleaning solution.

5. Rinse the equipment again with warm water.

In the case of some stubborn or fat-containing deposits which tend to be water-repellent, it is sometimes advantageous to use a small amount of a wetting agent along with the sulphamic acid, so as to enhance the rate at which it penetrates and attacks the material to be moved. Alkyl benzene sulphonic acid is suitable for this purpose.

Since uninhibited sulphamic acid is rather corrosive to ordinary steel and cast iron, it is wise to provide a separate pump for circulating solutions if existing pumps have ferrous metal parts in contact with the fluid.

The authors have conducted experiments with 100 per cent dodecyl

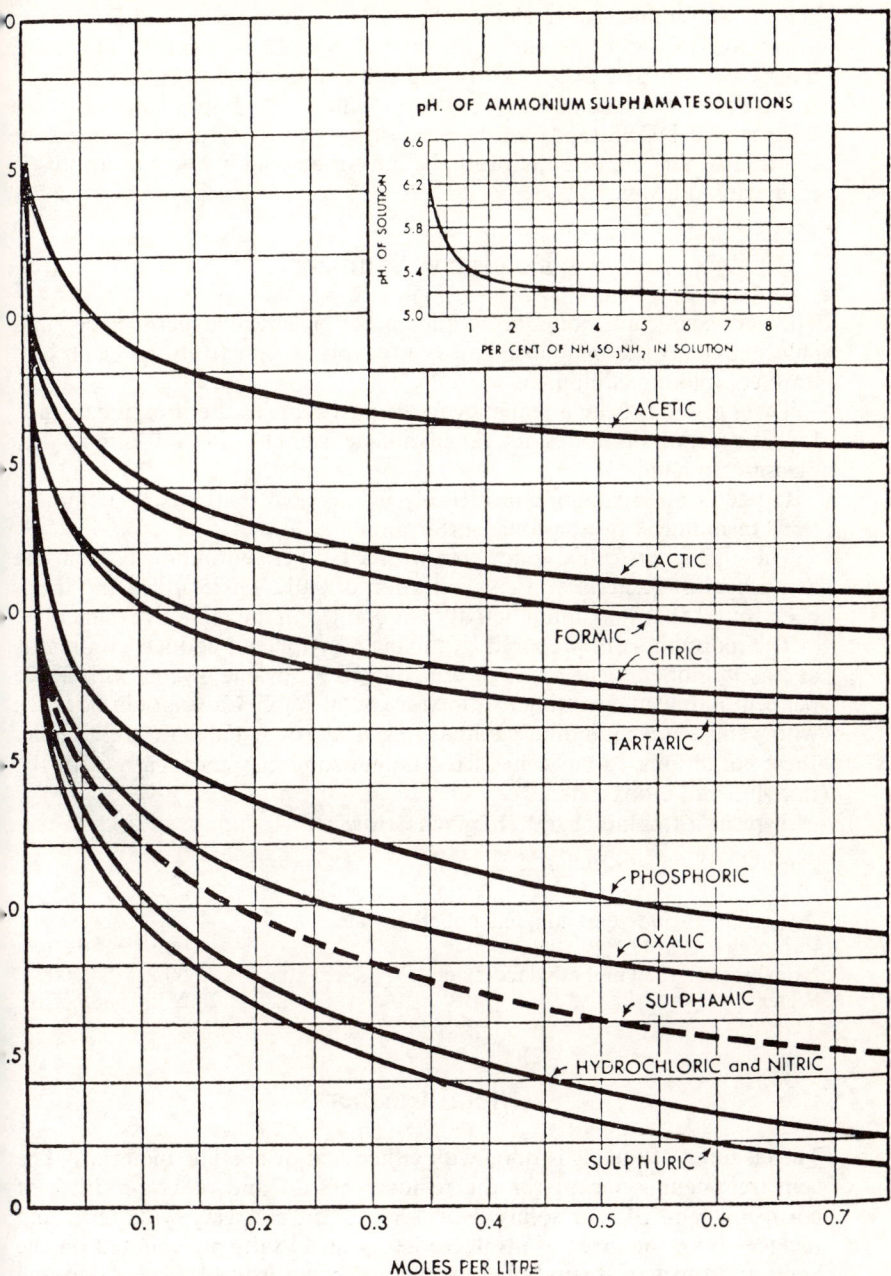

Fig 7.1   pH concentration curves comparing sulphamic acid with other acids

benzene sulphonic acid to clean milkstone. The calcium salts of DDBS are dispersible in water and therefore the acid should be suitable. However, the 100 per cent acid dissolves with difficulty in water in low concentrations. It was found that the addition of 10 per cent xylene sulphonic acid to the 100 per cent DDBS rendered the mixture very easily soluble. A 5 per cent solution of this mixture dissolves the milkstone readily and the surplus is easily rinsed away.

Detergent Sanitizers

The halogens have long been known to be effective germicides, and chlorinated trisodium phosphate is used pretty effectively to clean and disinfect milking equipment.

There also has been a tendency in the past years to use bromine as such for disinfection but it has no real advantage over chlorine and is much less pleasant to handle.

Iodine is now gaining considerably in this field, particularly combinations of iodine with non-ionic surfactants.

These products called iodophors with a 1–3 per cent iodine content are active against bacteria at a concentration of 0·012–0·025 g/litre and it has been found that maximum activity is obtained in an acid environment.

The iodophors are prepared by mixing nonylphenol adducts containing at least 8 mol ethylene oxide or ethoxylated propylene glycols with 20–25 per cent iodine and heating this mixture to 50–60°C. Care should be taken with ventilation as iodine sublimes, even at room temperatures. Under these conditions 75 per cent of the iodine combines chemically with the nonylphenol ethoxylate.

Typical formulae 58, 59 are given below.[27]

Formula 58 and Formula 59

	%	%
20% available iodine in nonylphenol ethoxylate	8·75	8·75
Phosphoric acid (75%)	8·00	0·60
Nonylphenol + 30 mol ethylene oxide	5·00	5·00
Water	78·25	85·65

Metal Cleaners

The cleaning of metals is done with either acid or alkaline materials. The acid treatment is meant for the removal of rust and other products of corrosion, and for the solution of 'scale'. This is a very wide term and includes both the layer of insoluble heavy metal salts precipitated on the walls and tubes of steam boilers, and also the layer of oxides (as opposed to rust) formed on steel surfaces under certain conditions of heating. The alkaline treatment is for the removal of grease, oil, paint and foreign matter, and is also necessary as a preliminary treatment prior to pickling (acid treatment) to remove the film of oil which at best will hinder the pickling acid in its work and at worst will only allow this acid to work in patches, producing an uneven finish.

Scale removal by pickling calls for cleaning compounds (which are added to help the acid to penetrate and give a uniform finish) that will resist the strong acids used in pickling processes. Generally, detergents without any special additions of builders are used in these cases. However, it is of great advantage to use detergents in conjunction with 'inhibitors' which prevent (or retard) the attack of the pickling acid on the metal itself. The detergent has not much inhibiting effect, but its wetting action is apt to prevent pitting to a certain extent—a very important reason for using detergents in the pickling bath. The most suitable detergents are the alkyl benzene sulphonates, but certain petroleum sulphonates or heavy alkylate sulphonates may also be used with advantage. In these instances it is not necessary to neutralize the sulphonic acids, merely to add the material to the bath. If, however, a ready-neutralized material is all that is available, this can be used. These detergents need to be highly acid-resistant and to stand up to severe conditions, such as a 5 per cent sulphuric acid solution at 70°C. The amount of detergent to be added to the bath is on the average between 0·1 and 0·2 per cent, calculated as active detergent matter.

As inhibitors, phenylthiourea and thiourea are very effective. They are solid substances which can be compounded with powdered detergents to increase the rapidity of the pickling effect (in this case a neutralized concentrated detergent powder is used), and to prevent excessive corrosion of the metal proper. The amount of corrosion inhibitor required is 0·01 to 0·05 per cent, based on the sulphuric acid content of the pickling bath. If the concentration of the active detergent in the bath is of the order of 0·1 per cent, it follows that the compounding of an effective pickling bath is not very difficult. An example would be: a concentrated detergent powder containing 40 per cent active dodecyl benzene sulphonate is mixed with 10 per cent of its weight of phenylthiourea or thiourea. This mixture is added to the pickling bath in the proportion of 0·25–0·5 per cent. An alkyl aryl sulphonate of the alkyl naphthalene type was effective in concentrations of 0·1 per cent, both as an inhibitor and a wetting agent.

The following inhibitors were also found to be effective in sulphuric acid pickling baths:

Butyl sulphide, *o*-tolylthiourea, *p*-tolylthiourea, butyl disulphide, amyl mercaptan, ethyl selenide, propyl sulphide, butyl-methyl-sulphide, butyl mercaptan, *p*-thiocresol, iso-butyl-mercaptan, trianyl amine, *m*-thiocresol, trihexylamine, ethyl sulphide, phenyl morpholine, ethyl mercaptan, formaldehyde, methyl sulphide, 2-thionaphthol, crotonaldehyde. (This is only a selection from a list of more than a hundred compounds listed.)[28]

In the past types of inhibitors especially suitable for acid cleaners, such as cleaners for automobile cooling systems, etc, have been developed. The Sharples Chemical Corp brought its alkyl-substituted thioureas, diethyl- and dibutylthiourea, on to the market. These are suitable for use with 10 per cent hydrochloric acid solutions in concentrations as low as

0·05 per cent. Other types of acid inhibitors, marketed by the Hercules Powder Co, are based on rosin amine derivatives, namely polyoxyethylated dehydroabeitylamines. They are sold under the trade names Polyrades. These inhibitors were found to be effective in the range of 0·05–0·2 per cent in hydrochloric acid.[29]

The effect of inhibitors in conjunction with surface-active agents is of great importance when removing water scale from steel pipes with hydrochloric acid. Again it was found that thiourea and phenylthiourea are very effective inhibitors whose effect is strongly enhanced by the use of detergents.[30] A further line of research is the use of surface active agents in conjunction with corrosion inhibitors, such as sodium chromate, in brine solutions, calcium chloride refrigeration solutions and in automobile radiators.

For automobile cooling systems, acid mixtures are often used to remove hard and adherent scale. The effect of these mixtures is often greatly increased by the addition of detergents, which help in the penetration and removal of grease and oil which may have entered the cooling system and which prevent the acid cleaners from exerting their full effect on the scale. Mixtures of the solid acids, such as oxalic acid or sulphamic acid (see p 297), and acid salts such as sodium bisulphate, may be improved by the addition of DDB sulphonate powders. The following formula is a suggestion for such a product:

Formula 60
    Acid Cleaner for Water-cooling Systems

Oxalic acid	80
Sodium bisulphate	10
Concentrated detergent 40 per cent active	10

The concentrated detergent can be either dodecyl benzene sulphonate or an acid-resistant petroleum sulphonate.

Aluminium is often cleaned and brightened by acid solutions containing hydrofluoric acid. One of the few detergents found to be compatible with this highly corrosive acid is an amphoteric of the dicarboxyethylated derivative of cocoimidazoline. A formulation suggested by the Miranol Chemical Company, which we have found to be effective is:

Miranol C2M-SF Conc.	5
Glycol ether	6
Phosphoric acid (85%)	38
Hydrofluoric acid (70%)	8
EDTA	1
Water	42

Although this is called a cleaner, it actually removes the oxides on the surface thus giving a brightening effect. This solution should be used after an alkaline wash to remove the surface film of dirt and oil. Aluminium is attacked by alkali, but silicates effectively inhibit this attack (see p 57).

Besides the acid-type metal cleaners described above, metals are cleaned in strongly alkaline solutions to remove grease, oil and foreign matter prior to plating, painting, enamelling or other protective treatments used nowadays.[31]

The method of removing this foreign matter determines the composition of the alkaline cleaner. It is obvious that if a spray method is to be used the cleaning solution cannot foam. The principal methods of alkaline metal treatment are soaking, spraying and electrolytic, or combinations of two of these processes. In the soaking process, the cleaning effect is obtained by the high concentration of the detergent, and also a possible circulation of the solution by a pump. The spray process adds mechanical energy from the jet to aid in the removal of the dirt. The electrolytic process produces the cleanest surface because of the scrubbing action of the gases evolved and the attraction of the electrically charged dirt particles by the electrodes.

In addition to the types of cleaning process used, the type of metal to be cleaned influences the choice of formulation of the detergent. No detergent can do the universal job of cleaning every surface, and various formulations for different types of metals and also the different cleaning processes are given in Formulae 61–74. These formulae are reprinted by kind permission of the American Society for Metals from *Metals Handbook, Volume 2*.

In these formulae, mention is made of sodium resinate, which is saponified tall oil, obtained as a by-product in the paper industry. This soap is relatively cheap and in addition has good emulsification properties for oil. As its detergency is poor, it is invariably used in conjunction with a detergent. We suggest that the same emulsification can be achieved by the use of a sulphonated heavy alkylate (see p 169). The formulae also call for alkyl aryl sodium sulphonate. This is meant to be a concentrated powder containing 40 per cent sodium dodecyl benzene sulphonate. If another concentration is available the proportion should be varied accordingly.

Miscellaneous Cleaners[32]

LAVATORY CLEANER

The universal ingredient of lavatory cleaners is sodium bisulphate ($NaHSO_4$, technically called nitre cake). The most popular lavatory cleaner is based on this compound. Caking of sodium bisulphate can easily be overcome by adding to the ground salt some 0·5–1 per cent pine oil or a mixture of pine oil and kerosene. This also largely prevents corrosion of the metal containers in which these products may be marketed. (Recently, high density polythene containers have been adopted in many European countries. These are completely unaffected by lavatory-cleaning compounds.)

The cleaning effect can be improved by adding to the sodium bisulphate not only the 0·5–1·0 per cent pine oil, but also approximately 1 per cent alkyl benzene sulphonic acid.

In the last few years liquid toilet bowl cleaners have appeared on the European market, particularly in Germany. These are packed in specially designed plastic bottles so that they can be dispensed on to the inner, invisible, rim of the toilet pan. These liquids are acid, either mineral or organic acids being the acidic component, contain relatively large amounts of surfactants, are dyed and perfumed and have medium viscosity. A typical representative formulation is given by Dragoco:[33]

Non-ionic detergent	11·0
Hydrochloric acid (33%)	10·0
Phosphoric acid (85%)	10·0
Perfume	0·2
Water	68·8

Due to the acidity of the solution, the non-ionics need to be chosen with care and are usually a blend of alkyl phenol ethoxylates of both short and long chains. The perfume needs to be matched to the formulation to render it stable under acid conditions.

A viscous lavatory cleaner with both sanitizing and deodorizing effects may be produced by using LAS (SO_3 sulphonated) and pine oil as its two components:

LAS (acid form)	100
Pine oil	50–70
Water	150

The LAS and the pine oil are mixed and the water added immediately to prevent reaction of the sulphonic acid with the pine oil.

The water and/or the pine oil are varied to produce the desired viscosity.

An oily liquid with lower viscosity but faster dispersibility in water is obtained by adding 10 to 20 parts isopropanol.

Hand Cleansers

The manufacture of hand cleanser is similar to that of other abrasive cleaners. Thus a hand soap in paste form may be produced according to Formula 75:

Formula 75

Hand Cleanser

40% Sodium ABS paste	130 parts
Bentonite	150 parts
Fine sand	200 parts
Sodium carbonate	10 parts

Or it can be manufactured by using as the base a detergent paste according to Formula 76.

Formula 76

Detergent Hand Cleanser

40–50% Sodium alkyl benzene sulphonate paste	100 parts

Formula Nos 61–74

Alkaline Metal Cleaning Compounds

	Aluminium		Copper			Copper-plate	Iron and steel			Magnesium		Zinc		
	Soak	Spray	Soak	Spray	Electro-lytic	Electro-lytic	Soak	Spray	Electro-lytic	Soak	Spray	Soak	Spray	Electro-lytic
Builders				Composition of cleaners, % by weight										
Sodium hydroxide, ground	—	—	20	15	15	55	20	15	15	20	20	—	15	15
Sodium carbonate, dense	—	—	18	—	—	8	18	29	8½	18	29	—	—	—
Sodium bicarbonate	21	24	—	34	34	—	—	—	—	—	—	—	35	34
Sodium tripolyphosphate	30	30	—	10	10	10	20	20	10	20	20	90	10	10
Tetrasodium pyrophosphate	—	—	20	—	—	—	—	—	—	—	—	—	—	—
Sodium metasilicate, anhydrous	45	45	30	40	40	25	30	30	25	30	30	—	40	40
Surface-active (wetting) agents														
Sodium resinate	—	—	5	—	—	—	5	—	—	1	—	5	—	—
Alkyl aryl sodium sulphonate	3	—	5	—	—	—	5	—	1	5	—	5	—	—
Alkyl aryl polyether alcohol	—	—	2	—	—	1	2	—	—	2	—	—	—	—
Nonionic high in ethylene oxide	1	1	—	1	1	1	—	1	½	—	1	—	—	—
				Other conditions										
Operating temperature of solution °F	70	70	80	75	70	80	95	75	80	95	75	80	75	80
Concentration of cleaner, kg/m³ H_2O	25	6	50	6	50	50	50	6	50	50	6	25	6	37

APPLICATION AND FORMULATION OF DETERGENTS

Bentonite	30 parts
Abrasive powder	200 parts
Sodium carbonate	10 parts

The solid materials are incorporated into the soft soap or paste. Some water may need to be added to regulate consistency.

A hand cleanser in powder form may be produced with Formula 77.

Formula 77

Hand Cleanser in Powder Form

Pure powdered soap or concentrated detergent powder	26 parts
Abrasive material	70 parts
Borax	4 parts

Some of the abrasive material may be replaced by vegetable abrasives, sawdust, etc.

A special hand-cleansing compound containing approximately 75 per cent borax and 25 per cent of dry soap was found by the laboratories of Borax Consolidated Ltd to possess desirable fungicidal properties, and yet to be so mild in its action on the skin as to reduce any tendency to dermatitis. The borax is in finely granulated form, so that, when mixed with dry, powdered soap, gumming or caking is avoided. The product is described as a good cleanser, and will effectively, and without the use of waste, remove dirt from hands. The borax is not merely a diluent for the soap. Its hardness of two makes it even softer than chalk; it is readily soluble, and its abrasive action is only temporary, as the sharp edges of the grains become blunted almost immediately. It has detergent and water-softening properties of its own and it is a mild alkaline salt possessing the characteristic properties of imparting to a soap solution a pH value lower than that of soap alone.

WATERLESS HAND CLEANSERS

This type of hand cleanser is specially suitable for motorists, for removing oil, grease and grime after changing tyres or doing repairs while on the road. We confine ourselves to giving some formulae which we have worked out and found suitable for the purpose. (For a review article containing formulae, etc, of many different types of waterless hand cleansers, the reader is referred to an article by Milton A. Lesser.[30]) We found Formula 78 especially useful to motorists for removing oil, paint, tar, etc, from the hands.

Formula 78

Waterless Hand Cleanser

Stearic acid and/or 100% alkyl benzene sulphonic acid	25
Lanolin and/or lecithin are dissolved in:	15
Deodorized kerosene	350

The mixture is kept at a temperature of about 70°C.

To this solution the following mixture, having the same temperature of 70°C, is added

Triethanolamine	25
Water	95

Stir constantly until the mixture is cold

3·5 parts of pine oil is recommended as a cheap perfume and disinfectant.

Formula 79

Waterless Hand Cleanser

White mineral oil	40·5
Oleic acid	10·5
Non-ionic detergent	6·0
Propylene glycol	5·0
Triethanolamine	2·6
Morpholine	1·0
Water	34·4

Formula 80

Waterless Hand Cleanser

Deodorized kerosene	42·8
Lanoline	0·9
Oleic acid	5·9
Cetyl alcohol	0·4
Triethanolamine	2·9
Propylene glycol	2·7
Sodium lauryl alcohol sulphate	1·4
Water	43·0

The present authors are not in favour of adding abrasives to this type of waterless hand cleanser, because they cannot be completely wiped off the hands after use and are likely to choke the pores and thereby irritate the skin. If, for very heavy-duty waterless cleansers, an abrasive cannot be dispensed with, only the softest kinds, such as whiting, bole, kaolin, etc, should be used, and in the instructions for use it should be recommended to wash the hands with soap and water as soon after the application of the cleaner as possible. Neither halogenated solvents nor aromatic solvents should be used.

References

1. *US Patent* 4,192,761, P. Peltre and A. Lefleur (11 Mar 1980).
2. Werdelmann, B. W., *Soap*, 36 (1974).
3. Hatch, L. F., *Hydrocarbon Processing*, **82** (1975); Davis, R. C., **92**, *ibid*.
4. *US Patent* 4,298,491, A. C. Coxon, D. J. Edge and M. L. L. Lapper.
5. Van Wazer, J. R., *Phosphorus and its Compounds*, Vol 1, p 454, Interscience, New York.
6. Schmidlin, *Fachorgan, Textilveredlung*, **8**, 122–6, 183–8 (1953).
7. Batdorf, J., *Soap and Chemical Specialities*, **37**, 1, 61 (1962).
8. *British Patent* 870,081 (1961), Unilever.
9. See 2 above.

10. *British Patent* 877,155 (1961), Thomas Hedley.
11. *Technical Bulletin*, GAF Corp.
12. *British Patent* 1,128,411.
13. Gohlke, F. J. and Bergerhausen, H., *Soap*, **42,** 10, 47 (1967).
14. Cox, M. F. and Matson, T. P., *JAOCS*, **61,** 1273 (1984).
15. Kravetz, L., The Changing Role of Surfactants in Liquid Detergents, Paper presented at CSMA Seminar (June 1985).
16. Rosen, M. J., Current Developments in Surfactants, Paper delivered at CSMA Seminar (June 1985).
17. *US Patent* 4,269,722, D. Joshi and R. Klingaman (26 May 1981).
18. Benson, H. L., Cox, K. R. and Zweig, J. E., *HAPPI*, **22,** 3, 50 (1985).
19. *US Patent* 4,193,886, Schoenholz, D., Petersen, A. W. and Northrop, M. A. (18 Mar 1980).
20. *Technical Bulletin*, Food Machinery Corp.
21. *US Patents* 3,255,177 and 3,350,318, FMC.
22. Davidsohn, A., *SPC*, **25,** 725 (1952).
23. Davidsohn, A., *Seife-Oele-Fette*, **82,** 461–87 (1956) and **83,** 15, 43 (1957), 3rd International Congress on Surface Activity, Cologne, 1960, Vol IV, p 165.
24. Sulfamic Acid Product Information, Grasselli Chemicals Dept, E. I. Du Pont de Nemours & Co, USA.
25. Holtzman, S. and Milwidsky, B. M., *Soap*, **42,** 11, 64 (1967).
26. Mann Trans Electrochem Soc, **69,** 115 (1936), Laue and Hultin, *Ind and Eng Eng Chem*, **28,** 159 (1936). See also Uhlig, *Handbook of Corrosion*, 905–16, 'Inhibitors and Passivators', Eldridge & Warner (1948).
27. Schick, N. J. (ed), *Non-ionic Surfactants*, p 415, Marcel Dekker, New York (1967).
28. Stauffer, Winn and Wagner, *Soap*, **30,** 161, 163, 165, 181 (1954).
29. Gardwell and Eiler—paper read in 1947 at a local meeting of the American Society, reported Jan 1948 in *Forecast*.
30. Lesser, Milton A., *Drug and Cosmetic Industry*, **72,** 326–7, 508–414.
31. Innes, W. P., *Metal Cleaning, Metal Finishing Guidebook*, 118–40 (1986).
32. Chalmers, L., *Domestic and Industrial Chemical Specialities*, Leonard Hill, London (1966).
33. Ohrmann, R., Dragoco Report, 1/1984, p 20.

Selected Bibliography

Theoretical Aspects of Detergency

Surface Activity, J. L. Moillet, B. Collie and W. Black, 2nd ed: Spon Ltd, London (1961).
Fundamentals of Detergency, W. W. Niven: Reinhold Publishing Corp, New York (1950).
Surface Chemistry, L. I. Osipov: Reinhold Publishing Corp, New York (1962).

General Detergents

Surface Active Agents, A. M. Schwartz and J. W. Perry: Interscience Publishers Inc, New York, 2 vols (1948) and (1958).
Synthetic Detergents, J. W. McCutcheon: MacNair-Dorland Company, New York (1950).
Encyclopaedia of Surface Active Agents, J. P. Sisley and P. J. Wood, 2nd ed: Chemical Publishing Co Inc, New York (1972).
Industrial Detergency, Ed W. W. Niven: Reinhold Publishing Corp, New York (1954).
Tenside-Textilhilfsmittel Waschrohstoffe, Lindner, Wissenschaftliche Verlagsgesellschaft mbH. Stuttgart, Vols I & II (1964), Vol III (1971).
Surface Active Ethylene Oxide Adducts, N. Schönfeldt: Pergamon Press, London (1969).
Grenzflächenaktive Substanzen, H. Bueren and H. Grossman: Verlag Chemie (1971).
Cationic Surfactants, E. Jungerman: Marcel Dekker (1970).
Non-ionic Surfactants, N. J. Schick, Ed: Dekker Publ, New York (1967).
Tenside, G. Gawalek: Berlin DDR (1975).
Anionic Surfactants (in two parts), W. M. Linfield, Ed: Marcel Dekker, New York (1976).
Amphoteric Surfactants, B. R. Bluestein and C. L. Hilton, Eds: Marcel Dekker, New York (1982).
Tensid-Taschenbuch, H. Stache, Ed: 2nd Auflage, Carl Hanser Verlag (1981).
The Manufacture of Soaps, Other Detergents and Glycerine, E. Woollatt: Halsted-John Wiley & Sons, New York (1985).
Alkansulfonate, H. G. Hauthal, Ed: Leipzig DDR (1985).

Analytical Aspects

Systematic Analysis of Surface Active Agents, Rosen, Goldsmith: Interscience Publishers, New York (1960).

SELECTED BIBLIOGRAPHY

Methods of Analysis of Non-Soapy Detergent Products, Longman and Hilton: Society for Analytical Chemistry, Monograph.
'Analysis of Synthetic Detergents', W. P. Smith: *Analyst*, **84**, 77–89 (1959).
'Analyse der Tenside', *Infrarotspektroskopie und Chemische Methoden*, D. Hummel: C. Hanser Verlag, München (1962).
Practical Detergent Analyses, B. M. Milwidsky: MacNair-Dorland, New York (1970).
The Analysis of Detergents and Detrgent Products, G. F. Longman: John Wiley, New York (1975).
Detergent Analysis, B. M. Milwidsky and D. M. Gabriel: George Godwin, London (1982).

Index

AB sulphonates *see* Alkyl aryl sulphonates
Abrasive cleaner, 251–3, 286–8
Abrasive materials, 252
Absorption and neutralisation process *see* Powders
Acid cleaners, 297, 300
Acrysol, 84
Acyl sarcosides, 26
Alcohol ether sulphates *see* Sulphated ethers
Alcohol ethoxylates *see* Nonionic detergents
Alcohol ethylene oxide condensates, 29, 185
Alcohol sulphates, 17, 170
Alcohols, fatty, 17, 112–17
Alcohols, lower, 108
Alcohols, Ziegler, 17, 116
Alkaline metal cleaners, 295
Alkane sulphonates, 23, 175
Alkanolamines *see* Alkylolamines
Alkene sulphonates *see* Olefin sulphonates
Alkyl aryl sulphonates, 14–16
Alkyl benzene, 124–9
  neutralisation with soda ash, 203
  sulphonic acid, 2, 14, 149
Alkyl isethionates, 26–7
Alkyl naphthalene sulphonates, 1, 15
Alkyl phenol, 30, 181
  ether sulphates, 20
  ethoxylates, 181
Alkyl phosphates *see* Phosphate esters
Alkyl pyridinium bromide (chloride), 196
Alkyl sulphates and sulphonates *see* Alkane sulphonates
Alkyl taurides, 27
Alkyl trimethylammonium chloride, 195
Alkylamines, 104
Alkylate heavy *see* Heavy alkylate
Alkylation processes, 125
Alkylolamides, 37–9, 191–5
Alkylolamines, 103
Aluminium silicate *see* Zeolites
Amides, ethylene oxide condensates, 33
Amido sulphonic acid *see* Sulphamic acid
Amine acetates, 195
Amine oxides, 39
Amines, 103
Amino carboxylic acids, 88

Ammonia, 73
Amphiphyllic compounds, 1
Amphoteric detergents, 41
Anhydrides, 152
Anionic detergents, 12, 14–28
Anti-chlor, 205
Anti-redeposition agents, 82–4
Anti-static agents *see* Fabric softeners
ADS *see* Olefin sulphonates

Bacteriostats, 104
Bentonite, 69, 84
Betaines, 42
Biodegradation, 5, 45, 118
  of nonionics, 30
Block polymers, 34–6, 189–91
Borax, 67, 201
Bouveault-Blanc reduction, 113
Buffering action, 76
Builders, 4, 47–68

Calcium silicate, colloidal, 78
Calcium sulphonate, 244
Carbomer, 84
Carbonates, 61–4
Carbopol, 84
Carboxy methylcellulose, 4, 82
  degree of substitution, 82
  in liquid detergents, 264
Carpet cleaners, 293
Cationic detergents, 28, 195–8
Caustic alkalis, 69
Caustic potash, 69
Caustic soda, 69
Cetyl pyridinium chloride, 196
Cetyl trimethylammonium chloride, 195
Chelating agents, 88–97
  guide to, 94
Chlorinate trisodium phosphate, 49
Chlorine, active, 80
  in the presence of detergents, 268
Chlorine-releasing compounds, 80, 284
Chlorocyanurics, 81
Chlorosulphonic acid, 147
  sulphonation by means of, 170
Citric acid, 93

INDEX

Cloud point, 97, 277–8
  of nonionics, 19, 30–1, 238
CMC see Carboxymethyl cellulose
Cold-water washing, 279–81
Colloidal silica, 78
Colloidal silicates see Silicates
Complex phosphates see Condensed phosphates
Condensed phosphates, 49–54
Converter gas see SO_3 sulphonation
Co-sulphonation, 161

Dairy cleaners, 295
DDB see Alkyl benzene
DDBS see Alkyl benzene sulphonic acid
Definitions, 1
Deflocculation, 51
DEG, dihydroxyethyl glycine, 91
Degreasing, 289
Descaling, 300
Detergent bars, 245–8
Detergent paste, 241–4
Detergent sanitizers, 300
Detergent–solvent combinations, 289–93
Dialkyl dimethylammonium chloride, 196
Diethanolamine, 37, 103, 191–3
Diethyl ethanolamine, 103
Dishwashing machine detergents, 284–6
Dodecane, 10
Dodecyl benzene, 14
Dodecyl benzene sulphonate, 15–16
  neutralised, properties, 237
Dodecyl beta alanine, 11
Dry-cleaning detergents, 170, 293
Drying
  drum, 235
  spray, 209–27
DTPA, diethyline triamine pentacetic acid, 89
Dyeing of powders, 206

EDTA, ethylene diamine tetra-acetic acid, 88
Enzymes, 8, 98–102
Ester interchange, 130
Ethanolamines, 103
Ether linkage, 30
Ethoxylated alcohols, 29–32
Ethoxylation, 185
  continuous, 187
  laboratory, 184
2-Ethyl hexanol sulphate, 295
Ethylene glycol, 31
Ethylene oxide, 29–36, 178–80
Eutrophication, 6–8, 52

Fabric softeners, 29, 249–51
Fatty acids, 182
  ethylene condensates, 32

Fatty alcohols, 17, 112–17
  sulphates, 17–18, 150, 170
Fatty amine oxides, 39
Fine wash detergents see Light-duty detergents
Floor cleaners, 202, 282
Fluorosurfactants, 28
Foam, 79, 254
  boosting, 38–40
  control, 258
Food industry, detergents for, 295
Friedel-Crafts reaction, 125–9

Gantrez, 94, 267
General-purpose detergents, 274–6
Gluconic acid, 92
Glycols and glycol ethers, 108–11
Guar gum, 85

Hand cleaners
  mechanics', 304
  waterless, 306
Hard-surface cleaners, 281–4
Heavy alkylates, 127
  sulphonation of, 169
Heavy-duty laundering, 256–68
Heavy-duty liquid bleach and detergent, 268
Heavy-duty liquid detergents, 262–7
Hectorites, 84
HEDTA, hydroxyethylene diamine tetra acetic acid, 91
Hydrochloric acid
  absorption, 170–1
  inhibitors for, 301
Hydrofluoric acid, 126–7
Hydrogen chloride see Hydrochloric acid
Hydrogenation, high pressure, 113–15
Hydrotropes, 97
Hydroxyaminocarboxylic acids, 91
Hydroxycarboxylic acids, 92

Igepon, 2
Industrial cleaning materials, 294–303
Inhibitors, acid, 301
Interox H84, 88
Invert solubility, 19, 30–1, 239
Iron salts
  contamination of detergents, 238
  removal of phosphates by, 54
Isopropanolamides, 38, 194–6
Isopropanolamines, 103

Jergens process, 153

Laponite, 84
LAS, linear alkyl benzene sulphonates, 5, 14
  2-phenyl isomer, 5, 16, 121
Laundering, commercial, 288

312

INDEX

Laundry powders, heavy-duty, 256–8
Lauryl alcohol see Fatty alcohols
Lavatory cleaners, 303
Leblanc process, 64
Light alkylate, 126
Light-duty detergents, 269–274
Linear alkyl benzene, 5
Liquid detergents, 235–9
Lithium hypochlorite, 80
Locust Bean gum, 85

Magnesium dodecyl benzene sulphonate, 68
Magnesium sulphate, 68–9
Magnesium monoperoxyphthallate, 88
Maleic esters, 20, 173–5
Mechanics' hand-cleaners, 304
Mercerising, 295
Mersolate, 2, 23
Metal cleaners, 300
Methyl esters, 130, 162
  sulphonates, 168
Miranol, 43
Mixers, 200, 203, 229, 231–4, 242
Molecular sieves, 58, 123
Monoethanolamine, 37, 103, 194
Morpholine, 103
Myristic alcohols see Fatty alcohols

Naphthalene sulphonates, 14–15
Narrow-range ethoxylates see Peaked ethoxylates
Nekal, 2
Neutralisation, 171, 241
  dry, 202–5
  heavy alkylates, 170
  systems, 160
Nonionic detergents, 29–36, 178–191, 276–8
  modified low-foaming, 190
Nonyl phenol, 30, 181
Nonyl phenol ethoxylates, 30, 185
  sulphates, 20
NRE see Peaked ethoxylates
NTA, nitrilo triacetic acid, 6, 7, 89

Olefins, 119–124
  alkylation with, 125
  SHOP process, 121
  sulphonates, 18, 166
  Ziegler, 116, 120
Oleum, 135
Opacifying agents, 238
Optical brighteners, 85–7
  in fabric softeners, 251
Oven cleaners, 283–4
OXO alcohols, 17, 115
  sulphation with sulphamic acid, 173
Oxygen-releasing compounds, 64–7, 88
  activators for, 66

Paraffins, normal, 24, 124
  sulphonates see Alkane sulphonates
Paste detergents, 241–4
Peaked ethoxylates, 31
Peptization, 51
Perborate, 64
Perfuming of powders, 215
pH
  of acids, 299
  of alkaline builders, 75
Phosphates, 48–54
  esters, 25, 177
  in spray drying, 212–13
  replacement, 6, 54
  treatment in effluent, 54
Phosphation, 25, 177
Phosphonates, 95–7
Photoactivated bleaches, 87
Pickling, 301
Pine oil, 105, 206
Pluronics see Block polymers
P_2O_5, 25
Polyalkyl benzene see Heavy alkylates
Polyethylene glycol, 34
  in ethoxylates, 31–2
Polyphosphoric acid, 25, 177
Polypropylene glycols, 34, 190
Polyvinyl pyrrolidine, 83
Potassium carbonate, 63
Potassium phosphates, 51–2
Powder manufacture, 200–35
  fines, 227
  processes
    absorption, 200
    combined absorption and neutralisation, 202
    combined spray drying and absorption, 227–34
    drum drying, 235
    dry mixing, 209
    spray drying, 209–227
Powders
  bleaching, 204
  dyeing, 206
  general-purpose, 275–6
  heavy-duty, 208, 257–60
  light-duty, 273
  perfuming, 215
  spray-dried
    bulk density, 219
    colour, 216
    particle size, 217
    residual moisture, 222
    separation, 225
    stickiness, 224
    uniformity, 225
Propylene oxide, 34–6, 180
PVP see Polyvinyl pyrrolidine

INDEX

Quaternary ammonium salts, 29, 195

Reed reaction, 23, 175

Sanitizers, 268, 300
Scouring liquid, 287
Scouring powder, 287
  containing chlorine, 287
Scrub bars, 247
Sequestration of ferric ions, 91–2, 94
Sequestration of metallic ions see Chelating agents; Condensed phosphates
Sewage problems see Biodegradation; Eutrophication; Phosphate treatment in effluent
Shampoos, 239–41
Silica, colloidal, 78
Silicates, 54–8
  colloidal, 55
Slurry preparation, 210–14
SO_3, 144
Soda ash, 65, 203
Soda, modified see Sodium sesquicarbonate
Sodium bicarbonate, 63
Sodium carbonate see Soda ash
Sodium chloride, 68
Sodium hexametaphosphate, 49
Sodium hypochlorite, 78
  with detergents, 268
Sodium metasilicate, 55
Sodium orthosilicate, 55
Sodium perborate, 64
  activators, 66
  addition to spray-dried powders, 214, 233
  in powders, 205
Sodium percarbonate, 67
Sodium reduction process, 113
Sodium resinate, 303
Sodium sesquicarbonate, 63
Sodium sesquisilicate, 55
Sodium silicate see Silicates
Sodium sulphate, 47
Sodium tetraborate see Borax
Sodium tripolyphosphate, 4, 48–54
  incorporation in spray-dried powders, 213
Solid detergents, 245
Soluble glass see Silicates
Solvay process, 62
Solvent detergents, 289–93
Solvents, 108–11
  alcohols, 108
  chlorinated, 108, 290
  glycols and glycol ethers, 108
  hydrocarbons, 111
  incorporation into powders, 205
Sorbitan esters, 36
Spent acid, 150
Spray drying, 209–227

Spray mixing, 231
Sucrose esters, 36
Sulphamic acid, 20, 145, 297
Sulphated alcohols, 170
Sulphated ethers, 19, 171, 173
Sulphated monoglycerides, 19
Sulphated nonionics see Sulphated ethers
Sulphation see Sulphonation
Sulphobetaine, 42
Sulphochlorination, 23, 175
Sulphonated methyl esters see Methyl esters, sulphonates
Sulphonated olefins, 18, 166
Sulphonation, 132–73
  conditions, 150
  falling film units, 162
  processes, 157–73
  with acid, 149
  with chlorsulphonic acid, 170
  with SO_3, 151
  with sulphamic acid, 172
Sulphones, 152
Sulphosuccinamates, 22, 173
Sulphosuccinates, 20, 173
Sulphoxidation, 175
Sulphur, 145
  effect of impurities, 146
Sulphur-burning plant, 155
Sulphur trioxide
  from oleum stripping, 145
  liquid, 144
Sulphuric acid, 135
  handling and storage of, 140
Sultones, 18, 166
Super amides, 193
Surface active agents, 1
Surfactants, 1
Synergism, 50, 58

TAED, tetra-acetyl ethylene diamine, 66
Tartaric acid, 93
Teepol, 2
Tenside, 1
Tetrapotassium pyrophosphate, 51
Tetrasodium pyrophosphate, 4, 49
Textile dressing, 294
Thickening agents, 84
Toilet bars, 201
Toilet products, 239
Toluene sulphonates, 97–8, 161
1,1,1, trichlorethane, 108, 293
Trichlorethylene, 108, 290, 292
Tridecylbenzene sulphonate, 15
Triethanolamine, 92, 105, 265
Trisodium phosphate, chlorinated, 49
Turbo drymex, 230

314

INDEX

Urea, 97, 122
  in liquid detergents, 264

Water-glass *see* Silicate, colloidal
Waterless hand cleaner, 306
Window cleaner, 288

Xanthan gums, 85
Xylene sulphonates, 97, 161

Zeolites, 7, 58–61, 84
Ziegler alcohols, 17, 116
Ziegler olefins, 116, 120
Zwitterionics, 41